Frontiers in Clinical Drug Research - CNS and Neurological Disorders

(Volume 12)

Edited by

Zareen Amtul
The University of Windsor
Department of Chemistry and Biochemistry
Windsor, ON
Canada

Frontiers in Clinical Drug Research - CNS and Neurological Disorders

(Volume 12)

Editor: Zareen Amtul

ISSN (Online): 2214-7527

ISSN (Print): 2451-8883

ISBN (Online): 978-981-5179-84-2

ISBN (Print): 978-981-5179-85-9

ISBN (Paperback): 978-981-5179-86-6

First published in 2024.

Bentham Science Publishers Pte. Ltd.
80 Robinson Road #02-00
Singapore 068898
Singapore
Email: subscriptions@benthamscience.net

CONTENTS

PREFACE

Brain disorders are a major public health and economic concern. The causes of these disorders are heterogeneous, and so as the treatment options. State-of-the-art advances in scientific techniques, evolving clinical observations, and diagnostic accuracies to study the neurobiology of different diseases are very illuminating, and assist in not only deciphering the physiology of brain disorders but also the mechanisms that make people more vulnerable to these disorders.

As our understanding is evolving, our knowledge is expanding, and so is our book series *Frontiers in Clinical Drug Research - CNS and Neurological Disorders*. The organization and content of volume 12 of our book series like its predecessors sets the physiology of the brain by outlining the pathology of neurodegenerative disorders, the role of neuroinflammation in neurological disorders, aging in brain tumors, comorbidities in epilepsy as well as placebo effect based on available empirical evidence by actively formulating, synthesizing, and analyzing these pieces of evidence. The relationships between these medical illnesses are complex, multifactorial, and bidirectional.

Since the aim of the series is to include current advancements in the field, rather covering the materials exhaustively, this volume also precisely educates readers about the salient examples of the latest advancement happening in the translational research of various brain aliments, encompassing neurodegenerative disorders and neurological disorders. The volume also contextualizes the engaging and accessible introduction to drug repurposing for neurological disorders. The last chapter of the volume also discusses the organophosphorus poisoning or intoxication emergency, a very scarcely discussed topic, with reference to neuroactive drugs that can be used to treat it.

Briefly, **Chapter 1** highlights the current approaches used for tracking the new drugs by reviewing recent drugs tested in clinical trials for the treatment of Alzheimer's and Parkinson's diseases. **Chapter 2** summarizes an intellectual and conceptual framework for understanding the neurobiology of placebo by interpreting its evolutionary origin, meanings, mechanisms, monitoring, and implications from therapeutics' perspective. **Chapter 3** discusses the role of gut microbiota in neuroinflammation and neurological disorders. **Chapter 4** glances the role of age in pediatric tumors of the central nervous system. **Chapter 5** comprehensively and critically assesses the current knowledge about epilepsy when it is comorbid with the full range of medical disorders with reference to drug repurposing in clinical trials. **Chapter 6** reviews the progress on the development of oxime derivatives as a potential antidote for organophosphorus poisoning.

Therefore, each chapter in the book very ostensibly introduces beginners in the basic science to the primary as well as advanced knowledge in the field. At the same time, these chapters are equally valuable not just for other mental health professionals, and psychiatrists but also for a wide range of medical specialists as well. Besides, it offers researchers, postdoctoral fellows, and students in diverse fields of neurobiology, neurology and neuroscience the tools they need to obtain a fundamental background in the major neurodegenerative, and neurological disorders. In conclusion, the volume provides a glimpse into the future that we are moving toward by exploring a greater understanding of the common pathways that mediate neuropathological illnesses, symptoms, and the relevant pathophysiological and neurophysiological phenomenon.

The timely assistance provided by the editorial staff; Mr. Mahmood Alam (Director Publications), and Ms. Asma Ahmed (Senior Manager Publications) at Bentham Science Publishers is highly appreciated.

Zareen Amtul
The University of Windsor
Department of Chemistry and Biochemistry
Windsor, ON
Canada

List of Contributors

Ankit Chaudhary	Department of Pharmacology, ISF College of Pharmacy, Moga, Punjab-142001, India
Anupam Awasthi	Department of Pharmacology, ISF College of Pharmacy, Moga, Punjab-142001, India
Antonio Carlos Pinheiro de Oliveira	Departamento de Farmacologia, Instituto de Ciências Biológicas, Universidade Federal de Minas Gerais, Brazil
Akash Marathakam	Department of Pharmaceutical Chemistry, National College of Pharmacy Kozhikode, Kerala 673602, India
Alice Martins	MARE, Marine and Environmental Sciences Centre, ESTM, Polytechnic of Leiria, Peniche, Portugal
Bruno Dahmer	Graduate Program in Biotechnology, University of Vale do Taquari (Univates), Lajeado, Brazil
B.P. Nandeshwarappa	Department of Studies in Chemistry, Davangere University, Shivagangothri, Tholhunase - 577 007, Karnataka, India
Celso Alves	MARE, Marine and Environmental Sciences Centre, ESTM, Polytechnic of Leiria, Peniche, Portugal
Dylan V. Scarton	George Mason University, Interdisciplinary Program in Neuroscience, Fairfax, Virginia, USA
Fernanda Majolo	Graduate Program in Biotechnology, University of Vale do Taquari (Univates), Lajeado, Brazil
Guilherme Liberato da Silva	Graduate Program in Biotechnology, University of Vale do Taquari (Univates), Lajeado, Brazil
G.D. Gupta	Department of Pharmaceutics, ISF College of Pharmacy, Moga, Punjab-142001, India
Girish Bolakatti	Department of Pharmaceutical Chemistry, GM Institute of Pharmaceutical Sciences and Research, Davangere - 577 006, Karnataka, India
Gabriela Machado Parreira	Departamento de Bioquímica e Imunologia, Instituto de Ciências Biológicas, Universidade Federal de Minas Gerais, Brazil
Joana Silva	MARE, Marine and Environmental Sciences Centre, ESTM, Polytechnic of Leiria, Peniche, Portugal
Jennifer Fernandes	Department of Pharmaceutical Chemistry, NGSM Institute of Pharmaceutical Sciences, Mangalore-574 160, Karnataka, India
Khadga Raj	Department of Pharmacology, ISF College of Pharmacy, Moga, Punjab-142001, India
Leonardo de Oliveira Guarnieri	Departamento de Fisiologia e Biofísica, Instituto de Ciências Biológicas, Universidade Federal de Minas Gerais, Brazil
Lavynia Ferreira Hoffmann	Graduate Program in Biotechnology, University of Vale do Taquari (Univates), Lajeado, Brazil

Luís Fernando Saraiva Macedo Timmers	Graduate Program in Biotechnology, University of Vale do Taquari (Univates), Lajeado, Brazil Graduate Program in Medical Sciences, University of Vale do Taquari (Univates), Lajeado, Brazil
Mayank Patel	Department of Pharmacology, ISF College of Pharmacy, Moga, Punjab-142001, India
M.L Sujatha	Department of Studies in Chemistry, Davangere University, Shivagangothri, Tholhunase - 577 007, Karnataka, India
Márcia Inês Goettert	Graduate Program in Biotechnology, University of Vale do Taquari (Univates), Lajeado, Brazil Graduate Program in Medical Sciences, University of Vale do Taquari (Univates), Lajeado, Brazil
Manjunatha S. Katagi	Department of Pharmaceutical Chemistry, Bapuji Pharmacy College, Davangere - 577 004, Karnataka, India
MK Unnikrishnan	Department of Pharmacy Practice, NGSM Institute of Pharmaceutical Sciences, Nitte (Deemed to be University), Deralakatte, 575018 Mangaluru, Karnataka, India
Nesibe S. Kutahyalioglu	Karabuk University, Faculty of Health Science, Karabuk, Turkey
Navneet Arora	Department of Pharmacy Practice, ISF College of Pharmacy, Moga, Punjab-142001, India
Rui Pedrosa	MARE, Marine and Environmental Sciences Centre, ESTM, Polytechnic of Leiria, Peniche, Portugal
Rafael Pinto Vieira	Departamento de Bioquímica e Imunologia, Instituto de Ciências Biológicas, Universidade Federal de Minas Gerais, Brazil
Rohit	Department of Pharmacy Practice, ISF College of Pharmacy, Moga, Punjab-142001, India
Shamsher Singh	Department of Pharmacology, ISF College of Pharmacy, Moga, Punjab-142001, India
S.N. Mamledesai	Department of Pharmaceutical Chemistry, PES's Rajaram & Tarabai Bandekar College of Pharmacy, Farmagudi-Ponda - 403 401, Goa, India
Vimal Mathew	Department of Pharmaceutics, National College of Pharmacy Kozhikode, Kerala 673602, India
Wilian Luan Pilatti Sant'Ana	Graduate Program in Medical Sciences, University of Vale do Taquari (Univates), Lajeado, Brazil

CHAPTER 1

Recent Drugs Tested in Clinical Trials for Alzheimer´s and Parkinson´s Diseases Treatment: Current Approaches in Tracking New Drugs

Fernanda Majolo[1], Lavynia Ferreira Hoffmann[1], Wilian Luan Pilatti Sant'Ana[2], Celso Alves[3], Joana Silva[3], Alice Martins[3], Rui Pedrosa[3], Bruno Dahmer[1], Guilherme Liberato da Silva[1], Luís Fernando Saraiva Macedo Timmers[1,2] and **Márcia Inês Goettert[1,2,*]**

[1] *Graduate Program in Biotechnology, University of Vale do Taquari (Univates), Lajeado, Brazil*

[2] *Graduate Program in Medical Sciences, University of Vale do Taquari (Univates), Lajeado, Brazil*

[3] *MARE, Marine and Environmental Sciences Centre, ESTM, Polytechnic of Leiria, Peniche, Portugal*

Abstract: Affecting more than 50 million people worldwide and with high global costs annually, neurological disorders such as Alzheimer's disease (AD) and Parkinson's disease (PD) are a growing challenge all over the world. Globally, only in 2018, AD costs reached an astonishing $ 1 trillion and, since the annual costs of AD are rapidly increasing, the projections estimate that these numbers will double by 2030. Considering the industrial perspective, the costs related to the development of new drugs are extremely high when compared to the expected financial return. One of the aggravating factors is the exorbitant values for the synthesis of chemical compounds, hindering the process of searching for new drug candidates. In the last 10-year period, an average of 20 to 40 new drugs were approved per year, representing a success rate of less than 6%. However, the number of referrals for new drug orders and/or applications remained at approximately 700 each year, reinforcing the difficulty in the process of identifying and developing novel drugs. Regarding neurodegenerative diseases, the FDA (USA) approved 53 new therapies in 2019, including 48 new molecules and, from these, three are medicines and two are vaccines. The main drugs recommended for the treatment of these disorders are included in the following classes: Dopamine supplement (Levodopa), Monoamine oxidase (MAO) inhibitor (Selegiline, Rasagiline), Dopamine agonist (Apomorphine, Pramipexole), and Acetylcholinesterase inhibitor (Donepezil, Rivastigmine, Galantamine). Additionally, the current pharmacological treatments are not able to cure these patients and considering the etiological complexity and the prevalence of neurological disorders, scientists have a

* **Corresponding author Márcia Inês Goettert:** Graduate Program in Biotechnology, University of Vale do Taquari (Univates), Lajeado, Brazil and Graduate Program in Medical Sciences, University of Vale do Taquari (Univates), Lajeado, Brazi; E-mails: marciagoettert@gmail.com and m.goettert@uni-tuebingen.de

Zareen Amtul (Ed.)

great challenge in exploring new therapies and new molecules to find an adequate and viable treatment for these diseases. Clinical trials are essential in this process and thus, this chapter describes the most important drugs that were targets of phase III and IV clinical studies in the last five years, associated with the most common neurological disorders worldwide, AD and PD. Information about mechanisms of action, experimental studies in other diseases that support their use, and chemical structure of the drugs are included in this chapter. Additionally, nature as a source of valuable chemical entities for PD and AD therapeutics was also revised, as well as future advances in the field regarding tracking new drugs to get successful results and critical opinions in the research and clinical investigation.

Keywords: Acetylcholinesterase inhibitor, Clinical trials, FDA, Neurological disorders.

INTRODUCTION

Neurological diseases (ND) are a heavy burden carried by patients, their families, communities, and governments [1]. As the world population grows old, especially in developed nations, the combined annual costs of ND are rapidly rising [2]. In the United States, the social burden of ND is up to $ 800 billion, and disorders like Parkinson's disease (PD), Alzheimer's disease (AD), and other dementias represent more than one-third of these ND [2]. For elderly, ND can dramatically increase health care costs due to other associated comorbidities, such as Idiopathic Parkinson's Syndrome (IPS) disease and fall-related fractures [3]. It is estimated that about 61% of the IPS patients will have at least one fall during the course of the disease, and 39% will suffer multiple falls, generating high disease-specific costs [3]. In 2015, German data showed that IPS patients' treatment cost was more than € 3.2 billion, which amounted to about 1% of Germany's total annual medical expenses [3]. Projections estimate that by 2050 only AD dementia will have a devastating impact, affecting 131 million people worldwide [4]. In 2018, AD costs were nearly $ 1 trillion and, since the annual costs of AD are rapidly increasing, the projections estimate that these numbers will double by 2030 [4].

To reduce this social burden related to ND as a whole, not only for AD and PD, a global effort towards discovering new drug therapies that may reduce these costs is welcome. That is a concern mainly because many ND still have poorly defined or even undefined etiopathogenesis [5]. In addition, many ND present subjective and context-dependent clinical manifestations, making the sample selection for treatment trials, using clinical criteria, inevitably heterogeneous [5]. Due to this heterogeneity, the inclusion criteria for the studies are often more rigorous, adding to the time, cost, and risk to the drug development process [6]. All these aspects, when combined, reflect the success rates of new drugs for ND, which are the lowest for any therapeutic area [6]. For example, in the early 2010 decade, less than 10% of the potential drugs that started clinical testing reached the market, and from the compounds that eventually moved on to phase III testing, less than 50% got approval [6]. This scenario explains why clinical research programs for ND tend to be longer and more complex than those for other diseases [6].

In 2008, Pharmaceutical Research and Manufacturers of America (PhRMA) presented a report that contained more than five hundred drugs for neurological disorders, which were still in the development stage [6]. When analyzed in detail, the reports data demonstrated that the research and development (R&D) pipeline contained previously known drugs, undergoing repurposing processes, *i.e.*, tested for new indications [6].

Regarding the pipeline of drugs and biologics in clinical trials for the treatment of AD, a recent study that has utilized a survey of annual pipeline reports of the past five years provided a longitudinal insight into clinical trials and drug development for AD [4]. According to the Common Alzheimer's and Related Dementias Research Ontology (CADRO) for classifying treatment targets and mechanisms of action, the results revealed that, in 2020, there were 121 agents in clinical trials to treat AD [4]. Among them, there were 29 agents in phase 3, 65 in phase 2, and 27 in phase I trials. Also, the data showed testing of twelve agents in trials targeting cognitive enhancement, twelve intended to treat neuropsychiatric and behavioral symptoms, and 97 agents in disease modification trials [4]. For example, compared to the 2019 pipeline, these data showed a growth in the number of disease-modifying agents targeting pathways other than the amyloid or the tau pathways [4]. Finally, the clinical trials' data from the last five years showed a progressive emphasis on non-amyloid targets. Those candidate treatments aim at targets involving mechanisms like inflammation, synapse, neuronal protection, vascular factors, neurogenesis, and interventions on epigenetics [4]. Also, data revealed significant growth in the repurposed agents' pipeline as well [4].

In recent years, several drugs with the potential to modify the disease and with neuroprotective effects are being evaluated in preclinical and clinical studies. The United States stands out for conducting the largest number of phase III and phase IV clinical studies, both for AD and PD Fig (**1**). Thus, this chapter summarizes the most important drugs that were targets of clinical studies from 2015 to 2020, associated with the most common neurological disorders worldwide: AD and PD. The approached clinical studies are related to phase III and IV studies registered on *clinicaltrials.gov*. Data like mechanisms of action, experimental studies in other diseases that support their use, and chemical structure of the drugs are included in this chapter. Also, a revision focused on nature as a source of valuable chemical entities for PD and AD therapeutics is also reported. Finally, future advances in the field regarding tracking new drugs to get successful results in the research and clinical investigation are highlighted.

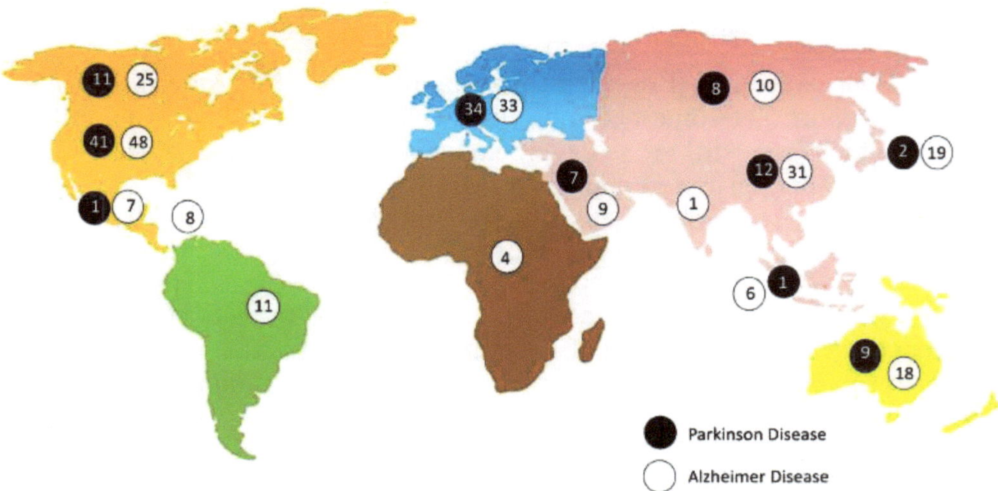

Fig. (1). Illustrative map indicating the number of clinical studies (phases III and IV) carried out from 2015 to 2020 related to Alzheimer's and Parkinson's diseases.

ALZHEIMER'S DISEASE

Alzheimer's disease (AD) is a neurodegenerative dysfunction which causes and pathogenesis are not fully understood. It is the most frequent cause of dementia [8], representing 60% to 80% of cases [8], affecting essentially older adults. Several primary studies, including a moderate-quality systematic review, reported that age significantly predicts AD incidence [7]. *Dementia* is a general term that describes neurocognitive symptoms like difficulties with memory and language, diminished problem-solving capabilities, and other thinking-related skills that directly impact the ability to perform daily activities [8].

Etiopathogenesis and Physiopathology

Alzheimer's is a gradually progressive brain disease that starts years before its symptoms appearance [8]. AD dementia has multiple etiopathogeneses, such as autosomal dominant forms, primarily attributable to mutations in proteases that release the amyloid-beta (Aβ) peptide or familial AD amyloid precursor protein (APP) mutations, further implicating amyloid metabolism [9]. Other invoked mechanisms to explain the stereotypical AD spread within the brain are the prion-like seeding of amyloid fibrils and neurofibrillary tangles [9]. AD's pathological hallmark feature is the accumulation of the beta-amyloid protein fragments (plaques) outside brain neurons and tau protein twisted strands (tangles) inside neurons [8]. These changes are associated with neuron death and brain tissue damage [8]. In addition, an up-to-date meta-analysis result, which included several recent studies that aimed at the association of levels of homocysteine and folic acid and the development of AD, has shown that homocysteine and folic acid are potential predictors for both occurrence and development of AD [8, 9].

Diagnosis

Alzheimer's *dementia* is a clinical syndrome that results from the AD pathophysiological process [10]. This process encompasses the *antemortem* biological changes manifested in the *postmortem* neuropathological diagnosis of AD [10]. For obvious reasons, the diagnosis of AD dementia must first meet all-cause dementia criteria, which fits to comprise the spectrum of severity, ranging from the mildest to the most severe stages of dementia [10]. According to these criteria, the diagnosis of all-cause dementia requires some degree of interference in the individual's ability to function at work or usual activities, representing a decline from a previous level of functioning and performing, which not explains itself by delirium or major psychiatric disorder [10].

The differentiation between dementia and mild cognitive impairment rests on determining whether there is a significant disturbance in functioning at work or in habitual daily activities [10]. Inherently, this is a clinical judgment made by a trained clinician based on the patient's circumstances and the description of day-to-day affairs of the patient [10]. Patients and knowledgeable informants are necessary to obtain the clinical information [10]. A combination of history-taking from the patient and a knowledgeable informant, and an objective cognitive assessment, either a bedside mental status examination or neuropsychological testing, are the basis of detection and diagnosis of cognitive impairment [10].

Two of the following domains, at least, must be present to characterize cognitive or behavioral impairment: acquiring and remembering new information ability impairment, impaired reasoning and handling of complex tasks; poor judgment, impaired visuospatial abilities, impaired language functions, changes in personality, behavior, or comportment [10]. A series of criteria established by the National Institute on Aging and the Alzheimer's Association (NIA-AA), most recently updated in 2011, is the currently proposed method to diagnose and classify AD dementia [10]. This classification encompasses three possibilities: (1) Probable AD dementia, (2) Possible AD dementia, and (3) Probable or possible AD dementia with evidence of the AD pathophysiological process [10]. The first two are fitted for clinical settings usage, while the third one currently fits the research field [10].

Treatment – Approved Drugs

Until today, there is no available pharmacological treatment capable to delay or arrest the injury and loss of neurons that cause AD manifestations and make the condition fatal [8]. Nevertheless, the U.S. Food and Drug Administration (FDA) already approved some drugs for AD treatment — galantamine, rivastigmine, memantine, donepezil, and memantine combined with donepezil [8]. Except for memantine, these drugs temporarily improve the cognitive symptoms by increasing neurotransmitters in the brain [8]. Memantine acts by blocking excess stimulation in specific receptors in the brain that can injure nerve cells [8]. The effectiveness of these drugs varies from person to person and is limited in duration [8]. Also, with US approval, in 2021, aducanumab can be prescribed for the treatment of AD in those with mild cognitive impairment or mild dementia stage, the population in which treatment was initiated in clinical trials [11].

The FDA approves explicitly no drugs to treat behavioral and psychiatric symptoms that may develop in the moderate and severe stages of AD dementia [8]. If non-pharmacologic treatment is not successful, and there is the possibility that these symptoms may harm the individual or others, physicians are authorized to prescribe drugs approved for similar symptoms in people with other diseases [8].

Prevention

The primary prevention of AD refers to prevent resulting dementia in cognitively normal subjects and is the final purpose for AD management [11]. AD has numerous well-instituted risk factors [12]. Some are not modifiable, like age, sex, and genotype, but other are potentially modifiable, such as vascular risk factors and traumatic brain injury [8, 12]. In contrast, there are suggested protective factors involving pharmacological mechanisms, like the use of antihypertensives, non-steroidal anti-inflammatories, statins, and hormone replacement therapy [12]. There are also environmental and behavioral factors, like diet, physical activity, high education, and engagement in social and intellectual activities [12]. However, how modifying these factors will reduce the risk of dementia is not yet known [12]. The secondary prevention of AD refers to preventing AD development in non-demented subjects with some evidence of cognitive impairment [12]. In this regard, the most often studied groups are those with Mild Cognitive Impairment (MCI), but there are no treatments that have demonstrated efficacy for preventing or delaying AD in MCI subjects until now [12]. At the same time, evidence shows that cholinesterase inhibitors, Vitamin E, *Ginkgo biloba*, and anti-inflammatories are not substantively helpful [12].

Clinical Trials

Below, we address individually the drugs used in clinical trials (phases III and IV) in the last 5 years related to studies that have ended normally, and participants are no longer being examined or treated (that is, the last participant's last visit has occurred) Table (**1**). The chemical structures of these drugs can be found in (Fig. **2**). Additionally, Table (**2**) shows all the ongoing clinical trials in the last 5 years, from phases III and IV to treat this disease.

Table 1. Summary of clinical trials performed in the last 5 years related to studies that have ended normally from phases III and IV concerning drugs for Alzheimer's and Parkinson's disease treatments registered in *ClinicalTrials.gov*. These studies have ended normally, and participants are no longer being examined or treated.

Drug	Dosage	Number of Participants	Ages Eligible for Study	Study	ID*** Study Start
ALZHEIMER`S DISEASE					
PHASE 3					
Pimavanserin	34 mg / 20 mg	392	50 - 90	To evaluate the efficacy of pimavanserin by comparing pimavanserin with placebo in preventing a relapse of psychotic symptoms in subjects with dementia-related psychosis and who were stabilized after 12 weeks of open label pimavanserin treatment.	NCT03325556 2017
AXS-05	AXS-05 tablets taken by mouth for 5 weeks.	366	65 - 90	To assess the efficacy and safety of AXS-05 in the treatment of agitation in patients with AD.	NCT03226522 2017
Umibecestat (CNP520)	50 mg / 15 mg	1145	60 - 75	To determine the effects of CNP520 on cognition, global clinical status, and underlying AD pathology, as well as the safety of CNP520, in people at risk for the onset of clinical symptoms of AD based on their age, APOE genotype and elevated amyloid.	NCT03131453 2017

(Table 1) cont.....

Drug	Dosage	Number of Participants	Ages Eligible for Study	Study	ID*** Study Start
Zolpidem *Zoplicone*	10 mg 7.5 mg	62	55 - older	To determine whether Zolpidem and Zoplicone are efective in the treatment of sleep disorders in AD.	NCT03075241 2016
Suvorexant 1	0 mg to 20 mg	285	50 - 90	To examine the safety and efficacy of Suvorexant (MK-4305) to improve sleep in individuals with AD. The primary hypothesis for the study is that Suvorexant is superior to placebo in improving insomnia as measured by change from baseline in polysomnography (PSG)-derived total sleep time (TST) at week 4.	NCT02750306 2016
Methylphenidate	5 mg	200	All	To examine the efficacy and safety of methylphenidate as treatment for clinically significant apathy in AD participants.	NCT02346201 2016
PHASE 4					
Donepezil	10 mg	241	50 - 85	To evaluate the safety and efficacy of donepezil in AD patients in China and investigate the relationship between Apo-E gene type with adverse events of donepezil.	NCT02787746 2016

(Table 1) cont.....

Drug	Dosage	Number of Participants	Ages Eligible for Study	Study	ID*** Study Start
Rivastigmine	18 mg	118	50 - 85	To evaluate the efficacy of rivastigmine patch with 1-step titration on cognitive function measured as change from baseline to week 24 in the total score of Mini-Mental State Examination (MMSE) in mild to moderate AD patients who failed to benefit from other cholinesterase inhibitors (ChEIs).	NCT02703636 2016
PARKINSON`S DISEASE					
PHASE 3					
Inosine	500 mg of inosine / placebo comparator	298	30 - older	To determine whether oral inosine dosed to moderately elevate serum urate (from ≤5.7 mg/dL to 7.1-8.0 mg/dL) over 2 years slows clinical decline in early PD.	NCT02642393 2015
LY03003 (Rotigotine)	Microspheres / placebo comparator	294	18 - older	To evaluate the efficacy and safety of rotigotine in the treatment of patients with early stage of primary Parkinson's disease	NCT04455555 2020
Pimavanserin 34 mg and *20 mg*	34 mg and 20 mg / placebo comparator	392	50 - 90	To evaluate the efficacy of pimavanserin by comparing pimavanserin with placebo in preventing a relapse of psychotic symptoms in subjects with dementia-related psychosis.	NCT03325556 2017
PHASE 4					

(Table 1) cont.....

Drug	Dosage	Number of Participants	Ages Eligible for Study	Study	ID*** Study Start
Droxidopa	600 mg / 200 mg	15	18 - older	To find out whether droxidopa, a medication that increases norepinephrine levels, may be effective in improving some aspects of cognition and movement in PD.	NCT03115827 2017
Pramipexole SR Pramipexole IR	-	98	30 - older	To explore, evaluate and confirm the efficacy between Pramipexole Sustained Release (SR) *versus* Pramipexole Immediate Release (IR) on nocturnal symptoms in L-dopa+ treated patients with advanced PD.	NCT03521635 2018
Opicapone (BIA 9-1067)	50 mg opicapone and levodopa/dopa decarboxylase inhibitor (L-dopa/DDCI)	518	30 - older	To evaluate the change in subject's condition accor-ding to the Investigator's Global Assessment of Change after three months of treatment with 50 mg opicapone.	NCT02847442 2016
Safinamide	50 mg	20	30 - 80	To explore the safety and tolerability of the immediate switch from rasagiline (irreversible MAO-B inhibitor) to safinamide, with the expectation that there will be no adverse events or increased risk of hypertensive crisis for patients with PD or signs of serotonin syndrome.	NCT03843944 2019
DA-9701(Motilitone)	30 mg / placebo comparator	147	50 - 80	To determine whether DA-9701 is effective and safe for the treatment on health-related quality of life in PD patients with gastrointestinal symptoms.	NCT02775591 2016

*ID: Clinical trial identification.
Abbreviations: AD, Alzheimer´s Disease; PD, Parkinson´s Disease.

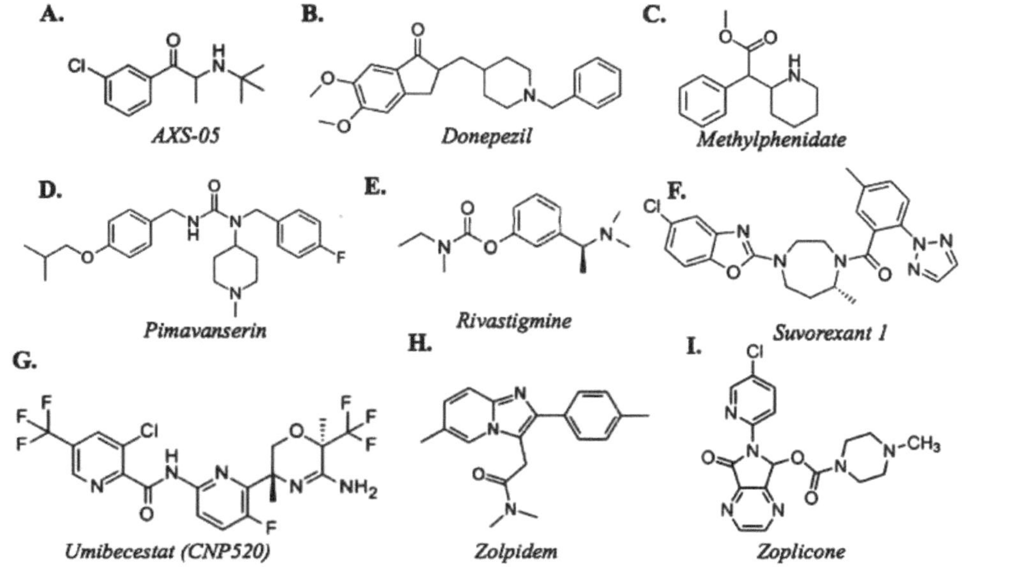

A. *AXS-05* B. *Donepezil* C. *Methylphenidate* D. *Pimavanserin* E. *Rivastigmine* F. *Suvorexant 1* G. *Umibecestat (CNP520)* H. *Zolpidem* I. *Zoplicone*

Fig. (2). Chemical structures of target drugs submitted to clinical studies from 2015 to 2020 (phases III and IV) related to Alzheimer's disease that have ended normally, and participants are no longer being examined or treated.

Table 2. Summary of ongoing clinical trials along the last 5 years (phases III and IV) concerning drugs for Alzheimer's and Parkinson's diseases treatment registered in *ClinicalTrials.gov*.

Drug	Origin	ALZHEIMER		PARKINSON	
		Phase III	Phase IV	Phase III	Phase IV
ABBV-951	Mimetic	-	-	NCT03781167	-
-	Mimetic	-	-	NCT04379050	-
-	Mimetic	-	-	NCT04380142	-
Aducanumab	Mimetic	NCT04241068	-	-	-
AGB101	Mimetic	NCT03486938	-	-	-
Agomelatine	Mimetic	-	-	-	NCT03977441
Alfoatirin®	Mimetic	-	NCT03441516	-	-
Ampreloxetine	Mimetic	-	-	NCT03750552	-
-	Mimetic	-	-	NCT03829657	-
-	Mimetic	-	-	NCT04095793	-
ANAVEX2-73	Mimetic	NCT04314934	-	-	-
-	Mimetic	NCT03790709	-	-	-

(Table 2) cont.....

Drug	Origin	ALZHEIMER		PARKINSON	
		Phase III	Phase IV	Phase III	Phase IV
Angiotensin II receptor blocker	Mimetic	NCT02913664	-	-	-
APL-130277	Mimetic	-	-	NCT03391882	-
Apomorphine	Mimetic	-	-	NCT02864004	-
Aricept	Mimetic	NCT03197740	-	-	-
-	Mimetic	NCT03283059	-	-	-
Aripezil®	Mimetic	-	NCT03441516	-	-
AVP-786	Mimetic	NCT04464564	-	-	-
-	Mimetic	NCT04408755	-	-	-
-	Mimetic	NCT03393520	-	-	-
Botulinum toxin type A	Mimetic	-	-	NCT04277247	-
BPDO-1603	Mimetic	NCT04229927	-	-	-
Brexpiprazole	Mimetic	NCT03620981	-	-	-
-	Mimetic	NCT03594123	-	-	-
-	Mimetic	NCT03724942	-	-	-
-	Mimetic	NCT03548584	-	-	-
Bydureon	Mimetic	-	-	NCT04232969	-
Caffeine	Mimetic	NCT04570085	-	-	-
Carbidopa	Mimetic	-	-	NCT04380142	-
-	Mimetic	-	-	NCT04006210	-
COR388	Mimetic	NCT03823404	-	-	-
Donepezil	Mimetic	NCT03197740	-	-	-
-	Mimetic	NCT04661280	-	-	-
-	Mimetic	-	-	-	NCT04117178
-	Mimetic	-	NCT04117178	-	-
-	Mimetic	-	NCT03454646	-	-
Droxidopa	Mimetic	-	-	-	NCT02586623
-	Mimetic	-	-	-	NCT03229174
-	Mimetic	-	-	-	NCT04510922
Ebicomb	Mimetic	-	NCT03954613	-	-
Elenbecestat	Mimetic	NCT02956486	-	-	-
Enacarbil	Mimetic	-	NCT03082755	-	-
Escitalopram	Mimetic	NCT03108846	-	-	-

(Table 2) cont.....

Drug	Origin	ALZHEIMER		PARKINSON	
		Phase III	Phase IV	Phase III	Phase IV
Escitalopram	Mimetic	-	-	NCT03652870	-
Gabapentin	Mimetic	-	NCT03082755	-	-
Galantamine	Mimetic	-	NCT03454646	-	-
Gantenerumab	Mimetic	NCT04374253	-	-	-
-	Mimetic	NCT03444870	-	-	-
-	Mimetic	NCT03443973	-	-	-
-	Mimetic	NCT04339413	-	-	-
Ginkgo biloba	Natural origin	NCT03090516	-	-	-
GOCOVRI	Mimetic	-	-	-	NCT04387773
Guanfacine	Mimetic	NCT03116126	-	-	-
GV-971	Mimetic	NCT04520412	-	-	-
Huperzine A	Natural-derived	-	NCT02931136	-	-
Hydrogen	Mimetic	-	-	NCT03971617	-
Icosapent ethyl (IPE)	Mimetic	NCT02719327	-	-	-
Idebenone	Mimetic	-	-	NCT04152655	-
-	Mimetic	-	-	-	NCT03727295
IPX203	Mimetic	-	-	NCT03877510	-
IPX203 ER CD-LD	Mimetic	-	-	NCT03670953	-
IR CD-LD	Mimetic	-	-	NCT03670953	-
Lecanemab	Mimetic	NCT03887455	-	-	-
-	Mimetic	NCT04468659	-	-	-
Levodopa	Mimetic	-	-	NCT02601586	-
-	Mimetic	-	-	NCT04380142	-
-	Mimetic	-	-	NCT04006210	-
Lithium Carbonate	Mimetic	-	NCT03185208	-	-
LY03003	Mimetic	-	-	NCT04571164	-
Memantine	Mimetic	-	NCT04117178	-	-
-	Mimetic	-	NCT03703856	-	-
-	Mimetic	-	-	NCT03858270	-
-	Mimetic	-	-	-	NCT04117178
Metformin	Natural-derived	NCT04098666	-	-	-
Methanesulfonate	Mimetic	-	-	-	NCT03841604
Methylphenidate	Mimetic	-	NCT03811847	-	-

(Table 2) cont.....

Drug	Origin	ALZHEIMER		PARKINSON	
		Phase III	Phase IV	Phase III	Phase IV
Mirtazapine	Mimetic	NCT03031184	-	-	-
Nabilone	Mimetic	NCT04516057	-	-	-
-	Mimetic	-	-	NCT03773796	-
NE3107	Mimetic	NCT04669028	-	-	-
Nortriptyline	Mimetic	-	-	NCT03652870	-
Octohydroaminoacridine Succinate	Mimetic	NCT03283059	-	-	-
Omega-3	Mimetic	NCT03691519	-	-	-
Oxycodone	Mimetic	-	-	NCT02601586	-
P2B001	Mimetic	-	-	NCT03329508	-
Pimavanserin	Mimetic	-	-	-	NCT04292223
-	Mimetic	-	-	-	NCT04373317
Pramipexole	Mimetic	-	-	NCT03329508	-
-	Mimetic	-	-	NCT03329508	-
Rasagiline	Mimetic	-	-	NCT03329508	-
Rivastigmine	Mimetic	-	NCT02989402	-	-
-	Mimetic	-	NCT03454646	-	-
-	Mimetic	-	-	NCT04226248	-
-	Mimetic	-	-	-	NCT03840837
Safinamide	Mimetic	-	-	NCT03881371	-
-	Mimetic	-	-	-	NCT03841604
-	Mimetic	-	-	-	NCT03968744
Sinemet CR	Mimetic	-	-	-	NCT03111485
Solifenacin	Mimetic	-	-	NCT03149809	-
Spironolactone	Mimetic	-	NCT04522739	-	-
Suvorexant	Mimetic	-	-	-	NCT02729714
Tavapadon	Mimetic	-	-	NCT04201093	-
-	Mimetic	-	-	NCT04223193	-
-	Mimetic	-	-	NCT04542499	-
Tricaprilin	Mimetic	NCT04187547	-	-	-
Troriluzole	Mimetic	NCT03605667	-	-	-
TRx0237	Mimetic	NCT03446001	-	-	-
Vortioxetine	Mimetic	-	-	-	NCT04301492

(Table 2) cont.....

Drug	Origin	ALZHEIMER		PARKINSON	
		Phase III	Phase IV	Phase III	Phase IV
Zoledronic	Mimetic	-	-	-	NCT03924414

Donepezil

Mechanisms of Action: Currently, the piperidine-based drug Donepezil is the most used pharmacological agent for the treatment of AD, being a reversible inhibitor of acetylcholinesterase, centrally acting in a rapid way This pharmacological compound exerts its neuroprotective property through the upregulation of the nicotinic receptors in the cortical neurons, inhibiting voltage-activated sodium currents reversibly and delaying rectifier potassium currents and fast transient potassium currents. The cholinergic transmission is enhanced when Donepezil causes the hydrolysis of acetylcholine, thus increasing the availability of acetylcholine at the synapses [13 - 15].

The use of Donepezil is approved by FDA in mild, moderate, and severe AD, when initially it was tested for patients with mild-to-moderate AD [16 - 18]. This type of drug (cholinesterase inhibitors) was the first category approved by the FDA for this indication [19]. Until now, no evidence proves that Donepezil can alter the progression of AD but can improve the symptoms such as cognition and behavior [13 - 20]. Donepezil is a benefit to patients that show the two ends of the AD spectrum: moderate-to-severe impairment [21] and those very mild, early-stage disease [22] and, in this case, patients can be included in nursing homes [23].

Experimental Studies in other Diseases that Support their Use: The cholinesterase inhibitor Donepezil is suggested to be effective in other diseases, such as vascular dementia [24, 25], dementia associated with PD [26, 27] and other conditions [28]. Although not still approved by FDA, the use of Donepezil includes dementia associated with PD, vascular dementia, and Lewy body dementia, where some studies show the improvement of cognition and executive function. Additionally, the use for traumatic brain injury was tested and it was observed an improvement in memory dysfunction in patients.

The clinical trial performed by Jia *et al.* (2020) concluded that donepezil (10 mg/day) can be tolerated and is effective in patients with mild-to-moderate AD (ClinicalTrials.gov NCT02787746) [29].

Methylphenidate

Mechanisms of Action: The mechanism of action of Methylphenidate is still unclear to researchers but it is important to many therapeutic effects [30]. It is known that this drug causes the increasing of extracellular dopamine levels by the binding to the dopamine transporter in the presynaptic cell membrane, thus blocking the uptake of dopamine [31 - 34]. Also, Methylphenidate binds to the specific transporter of serotonin and norepinephrine, blocking the uptake too [35]. It's worth mentioning that the effects are weaker on serotonin that dopamine [36].

For more than 50 years the psychostimulant Methylphenidate has been considered the first-line pharmacological treatment for the Attention-Deficit/Hyperactivity Disorder (ADHD) patients [37, 38]. The drug presents favorable effects on reducing the core symptoms of excessive hyperactivity, impulsivity, and inattention in children and adolescents with ADHD [39].

Experimental Studies in other Diseases that Support their Use: Some evidence gives support to the use of Methylphenidate not only to ADHD but on physical and psychological symptoms in cancer patients. The prognostic is unclear, but some findings suggest that Methylphenidate improve cognitive symptoms and reduction of fatigue in cancer patients. It's worth mentioning that the results could vary according to the profile of population, outcomes measures, and study design. Also, it is approached in Depression treatment, but evidences are insufficient [40].

Patients with brain tumors that present cognitive abnormalities are frequent, and some studies suggested that the use of Methylphenidate improves those conditions. Meyers *et al.* (1998) [41] reported that half of the patients with malignant gliomas evidenced an improvement in cognitive function receiving this psychostimulant during nonrandomized studies. A study on children with brain tumors or acute lymphocytic leukemia that displayed association with neurologic injury, the use of methylphenidate presented an improvement in language skills, attention, memory, and academic performance in 10 of 12 pediatric patients [42].

The improvement of cognitive conditions using Methylphenidate were observed in patients with advanced cancer and hypoactive delirium [43 - 46]. Another symptom frequently observed by cancer patients is fatigue [47]. Studies reporting the effect of Methylphenidate on fatigue in cancer patients concluded that this drug improved significantly several skills following the Brief Fatigue Inventory (BFI), like anxiety, appetite, pain, nausea, depression, and drowsiness [48].

The clinical trial performed by Mintzer *et al.* (2021) suggested that Methylphenidate is an efficacious medication and can be considered safe to use in the treatment of apathy in AD (ClinicalTrials.gov Identifier: NCT02346201) [49].

Rivastigmine

Mechanisms of Action: Rivastigmine acts in postsynaptic clefts, inhibiting the metabolism of acetylcholine, and it has been considered a selective cholinesterase inhibitor, thus enhancing cholinergic neurotransmission. Rivastigmine is different from other cholinesterase inhibitors because it inhibits both enzymes acetylcholinesterase (AChE) and butyrylcholinesterase (BuChE). This fact results in an increase of acetylcholine levels. AChE is mainly present in areas of the cerebral cortex that have high activity and at synaptic nerve junctions. However, the BuChE helps mediate cholinergic activity being found in the glial cells of the brain. In neurological disorders like AD and PD, it is observed the upregulation of these cholinesterase enzymes, being higher than in normal conditions [50].

This drug was approved by FDA to treat mild to moderate dementia of Alzheimer's type. Also, it can be used for the treatment of mild to moderate dementia in PD [51].

Experimental Studies in other Diseases that Support their Use: Rivastigmine influences gain stability, reducing the risk of falls in patients with PD [52] and other studies proved that this drug, in older patients, decreases the incidence of postoperative delirium [53]. In multiple sclerosis patients it was tested the effectiveness of rivastigmine in order to check if the cognitive dysfunction was improved, but the studies showed no significant benefits for cognitive function when compared with placebo [54]. However, studies using Rivastigmine demonstrated cognition improvements in patients with Lewy bodies when compared to the placebo group. Such patients that received rivastigmine showed an improvement of 30% or more, as well as overall fewer episodes of hallucinations and anxiety [55].

The clinical trial (Identifier: NCT02703636) has no results published until this moment, where this work aims to evaluate the efficacy of rivastigmine on cognitive function in the total score of MMSE in mild to moderate AD patients who failed to benefit from other cholinesterase inhibitors (ChEIs).

Suvorexant

Mechanisms of Action: Suvorexant is a basic diazepane structure considered dual orexin receptor antagonist and has a function for blocks of both OX1R and OX2R. Through this block occurs the inhibition of orexin A and B that promote wakefulness. The orexins are neuropeptides and their receptors are blocked using suvorexant, inducing sleep. The drug is hepatically metabolized mainly by CYP3A from two hours after ingestion during the half-life of 12 h [56]. These orexin neurons (more than 70,000), located in the perifornical lateral hypothalamus in the human brain, send signals to the spinal cord and spread to the brain [57, 58].

Experimental Studies in other Diseases that Support their Use: Studies demonstrated the benefits, but side effects were observed, for example fatigue, next day somnolence, xerostomia and peripheral oedema being rarely reported cases of hallucinations, sleep paralysis and somnambulism. Suvorexant is expensive and has not been trailed against other hypnotics. It's worth mentioning that it is very useful to insomnia but in patients with psychiatric disorders the real action is still unclear. The use of suvorexant should be considered where the established treatments demonstrated to be inappropriate [56].

Outcomes from clinical trials using suvorexant in the prevention of delirium during the acute hospitalization have been positive [59]. The use of suvorexant can improve subjective sleep quality without inducing delirium in acute stroke patients during the ramelteon therapy [60].

The clinical trial performed by Herring *et al.* (2020) [61] reported that suvorexant improved total sleep time (TST) in patients with clinical diagnostic criteria suggesting AD dementia and insomnia (ClinicalTrials.gov NCT02750306) [61].

Zolpidem

Mechanisms of Action: Zolpidem is FDA approved for short-term treatment of insomnia mainly to patients with difficulty to start sleep and is a non-benzodiazepine receptor modulator [62 - 64]. The medication is an imidazo-pyridine compound that acts like a GABAA receptor chloride channel modulator/agonist that increases GABA inhibitory effects by interaction with Omega-1 receptor subtype, upregulating these receptors leading to sedative effects and deep sleep [65, 66]. It also has other properties as anticonvulsant, anxiolytic, and minor myorelaxant, consequently increasing sleep duration in patients with insomnia [67, 68]. The use of Zolpidem does not cause any residual cognitive impairment in the next morning due to its pharmacokinetic profile [65].

However, due to the high potential for abuse, this drug is not recommended for the general population as first-line treatment. Other drugs like controlled release melatonin and doxepin could be used as first line and still are recommended for sleep hygiene and cognitive behavioral therapy [69, 70].

The clinical efficacy of this drug has been characterized by a rapid action on sleep-maintenance parameters, increasing stage 2 and slow-wave sleep, decreasing sleep-onset latency, thus, significantly improves sleep quality and duration, and reduces night-time awakenings [71]. Therefore, to patients that present troubles to achieving sustained sleep, zolpidem can be considered an appropriate treatment [65].

Experimental Studies in other Diseases that Support their Use: Zolpidem has frequently been related to treatment of a variety of neurologic disorders such as movement disorders and disorders of consciousness [72]. Other improvements related to effects of Zolpidem are on JFK Coma Recovery Scale-Revised, the Unified Parkinson´s Disease Rating Scale, and the Burke-Fahn-Marsden Dystonia Rating Scale [72].

Zolpidem did not show positive effects in patients with traumatic brain injury or post-anoxic encephalopathy, however, on brain functions in patients with non-brain stem injuries it was observed an improvement. Also, it was not detected a consistent positive effect of zolpidem in patients with disorders of consciousness [73].

Zopiclone

Mechanisms of Action: Zopiclone is considered a cyclopyrrolone agent indicated for use in short-term insomnia that improves the sleep time, awakenings, sleep quality and latency [74]. The drug involves allosteric modulation of the GABAA receptor, displacing the binding of [(3)H]-flunitrazepam with an affinity of 28 nM, and enhances the binding of the channel blocker [(35)S]-TBPS. This medication is metabolized by the cytochrome P450 enzymes CYP3A4 and CYP2C8 generating several metabolites (SPC). The consequence of this interaction with the GABAA receptor is to potentiate responses to GABA, as can be demonstrated by electrophysiological methods [75, 76].

Experimental Studies in other Diseases that Support their Use: Zopiclone improves sleep quality in patients with chronic obstructive pulmonary disease [76]. In addition, it was proved that Zopiclone demonstrated positive aspects on early morning performance and residual sedative activity [77 - 79].

The clinical trial performed by Louzada *et al.* (2022) [80] evaluated the use of zolpidem and zopiclone by AD patients considered insomniacs, suggesting that the short-term use appears to be clinically helpful, but it is worth mentioned that safety and tolerance remain subjects to be discussed and personalized in healthcare settings and in future investigations (ClinicalTrials.gov Identifier: NCT03075241) [80].

AXS-05

Mechanisms of Action: According to Axsome (2017) [81], the compound AXS-05 is composed of dextromethorphan and bupropion. The bioavailability of dextromethorphan following the inhibition of CYP2D6 is affected by addition of bupropion. AXS-05 is also in phase III studies being tested in patients with treatment resistant depression (TRD). A Fast Track Designation has been addressed this compound to TRD and AD.

AXS-05 has been granted U.S. Food and Drug Administration Breakthrough Therapy designation for major depressive disorder, Fast Track Designation for treatment resistant depression, and Breakthrough Therapy and Fast Track designations for AD agitation. AXS-05 is not approved by the FDA.

Experimental Studies in other Diseases that Support their Use: There are no experimental studies in other diseases.

The clinical trial (Identifier: NCT03226522) has no results posted until this moment aiming to evaluate the effectiveness and safety of AXS-05 use in the treatment of Agitation in AD patients with Dementia.

Umibecestat (CNP520)

Mechanisms of Action: Umibecestat (named CNP520) is an orally active inhibitor of beta-site amyloid precursor protein (APP) cleaving enzyme (BACE) that is involved in APP processing and has approximately 3-fold selectivity for BACE-1 over BACE-2 and no relevant off-target binding or activity [82].

Experimental Studies in other Diseases that Support their Use: Following single and multiple dose administration of CNP520 in animals and humans, Aβ concentrations in cerebrospinal fluid were reduced by up to 95% compared with baseline, and results of toxicology studies have not raised major safety concerns [83].

The clinical trial (Identifier: NCT03131453) was terminated due to safety issues where the main objective was to evaluate the effects of CNP520 on cognition, global clinical status, and underlying AD pathology.

PARKINSON'S DISEASE

Multiple anatomical structures degeneration within the brain is the hallmark of a neurodegenerative disorder that manifests as Parkinsonism [84]. Human *postmortem* studies show that individuals have a neuronal loss in the *substantia nigra pars compacta*, *locus coeruleus*, and other neuronal populations [85]. It is a clinical syndrome based on three cardinal motor manifestations, which present themselves with any combination of bradykinesia, rest tremor, and rigidity [86]. These features must be observable and not attributable to confounding factors [86]. Postural instability and other motor and non-motor symptoms may also be present [85]. The most prevalent form of parkinsonism is PD, with an incidence rate of 14.2 per 100,000 person-years [86]. The incidence and the duration of illness are the determinants of its prevalence—the incidence of PD links to the risk factors and protective ones [87].

Etiopathogenesis and Physiopathology

There is a debate about the pathogenesis of PD and the relative contribution of genetics, environmental, and lifestyle factors [85]. PD's most critical risk factor is age, with a median onset of 60 years old [85]. The frequency is higher in men than in women (ratio ranges from 1.3 to 2.0), but differences in the prevalence of cigarette smoking behavior, postmenopausal hormones, and caffeine intake may influence the incidence [85]. As in other neurodegenerative disorders, age-related biological dysfunction may support and clear the way for neuronal demise [85]. These factors include ubiquitin-proteasome and autophagy-lysosomal system disturbances, mitochondrial defects, telomere dysfunction, genomic instability, and epigenetic changes [85]. In addition, several gene mutations are genetic risk factors for developing PD, like SNCA, PINK1, PARK2, PARK7, PLA2G6, FBXO7, ATP13A2, and LRRK2 [88].

The association of industrial chemicals and pollutants, such as pesticides and solvents, exposure to heavy metals, living in rural areas, agricultural occupation, consumption of dairy products, history of melanoma, traumatic head injury, type 2 diabetes mellitus, among many other factors, have already been identified as risk factors [85, 87]. However, despite this long list, PD's most critical risk factor is age [87].

The increasing longevity of societies also translates into longer disease duration, but the factors that interfere in the disease course are less well-known [87]. As stunning as it may sound, some data suggest that smoking is associated with a decreased risk of PD, but whether this association is causal is debatable [87]. Therefore, as societies grow in aging, industrialization increases globally, and smoking decreases in some regions, the prevalence of PD seems inclined to increase [87].

Diagnosis

Clinical features are the basis for PD diagnosis, and its diagnosis remains a clinical one [87]. However, during life, total diagnostic certainty is virtually impossible [86]. Between 75% and 95% of patients diagnosed with PD by experts have their diagnosis confirmed on autopsy [86]. Although the basis of PD diagnosis during life is its distinctive clinical features discerned by the history and neurologic examination, there is an inherent conflict between sensitivity and specificity of the clinical diagnosis [86].

When considering only the initial diagnosis, the misdiagnosis is even higher since the diagnostic accuracy improves over time, during follow-up visits, with a continuous diagnostic re-evaluation process [87]. Unlike the common opinion that detecting PD is an easy task, seminal clinicopathological studies have shown that up to one-fourth of patients diagnosed with PD during life have an alternative diagnosis at *postmortem* [86].

The Movement Disorder Society (MDS) recently proposed updated diagnostic criteria, intending to use them as the official "The International Parkinson and MDS Clinical Diagnostic Criteria for Parkinson's disease" (MDS-PD Criteria) [86]. MDS-PD Criteria encompass the presence of parkinsonism and its cardinal manifestations, like bradykinesia, rigidity, and rest tremor [86].

Treatment – Approved Drugs

It is a central clinical issue to approach the relative effectiveness of first-line treatments for PD motor symptoms, *e.g*., amantadine, dopamine agonists (DA), levodopa, and monoamine oxidase B (MAO-B) inhibitors, considering that their comparative effectiveness is unclear [89]. No known theoretical reasons show one class of drugs to be more effective than another [89]. Levodopa approaches the dopamine loss of PD when converted into dopamine by the human metabolism [89]. Therefore, it helps to enhance dopamine availability [88]. DA stimulate neuronal cells similarly to dopamine [89]. MAO-Bs, in turn, break down the enzyme responsible for dopamine metabolism within the brain increasing dopamine availability [89]. Lastly, amantadine both enhances dopamine release and hinders dopamine reuptake [89]. Levodopa is the most prescribed drug to treat motor symptoms of PD [89]. However, its effectiveness declines over time, and crucial adverse motor complications may arise [89]. Therefore, clinicians often aim to keep the dose of levodopa as low as possible to maintain good function and reduce motor complications [89]. In extension to the drugs above described (dopamine agonists, MAO-Bs, and amantadine), catechol-*O*-methyltransferase (COMT) inhibitors and anticholinergics have also been used in the treatment pathway [89]. COMT inhibitors prolong levodopa effects by blocking an enzyme that breaks down, thereby allowing lower levodopa doses [89]. Anticholinergics intend to improve motor symptoms, most commonly in the earlier stages of PD [89].

For monotherapy of early PD, non-ergot dopamine agonists, oral levodopa preparations, selegiline, and rasagiline are clinically helpful [89]. In early/stable PD, non-ergot dopamine agonists, rasagiline, and zonisamide are clinically helpful as adjunct therapy [90]. Selegiline and rasagiline are used as monotherapy in the early stages of Parkinson's disease. In PD patients, these monoamine oxidase B inhibitors exhibit antiprotective properties and patients targeting dopamine metabolising, thus increasing the time available for dopamine neurotransmission, thus increasing the time that dopamine is available for neurotransmission [91]. Zonisamide is a reversible MAO-B inhibitor. It is used as a complementary treatment to overcome the deficiencies of general therapies used to solve motor and non-motor problems of PD [92]. Rivastigmine is possibly helpful as an adjunct therapy in optimized PD for general or specific motor symptoms, including gait, and physiotherapy is clinically useful; exercise-based movement strategy training and formalized patterned exercises are possibly helpful [90]. Most non-ergot dopamine agonists, including pergolide, levodopa ER, levodopa intestinal infusion, entacapone, opicapone, rasagiline, zonisamide, safinamide, and bilateral subthalamic nucleus (STN) and GPi deep brain stimulation (DBS) are clinically useful to treat motor fluctuations [90].

As for the non-motor symptoms, there are clinically practical or possibly useful interventions to treat depression, psychosis, apathy, dementia, insomnia, daytime sleepiness, impulse control, fatigue, orthostatic hypotension, pain, erection and urinary dysfunctions [93].

Prevention

There are no clinically helpful interventions to prevent or delay PD progression nor differences in the outcomes for the prevention/delay of motor complications [90].

Clinical Trials

Below, we address individually the drugs used in clinical trials in the last 5 years related to studies that have ended normally, and participants are no longer being examined or treated (that is, the last participant's last visit has occurred) from phases III and IV (Table **1**). The chemical structure of these drugs can be found in (Fig. **3**). In addition, Table **2** shows all the ongoing clinical trials of the last 5 years, from phases III and IV.

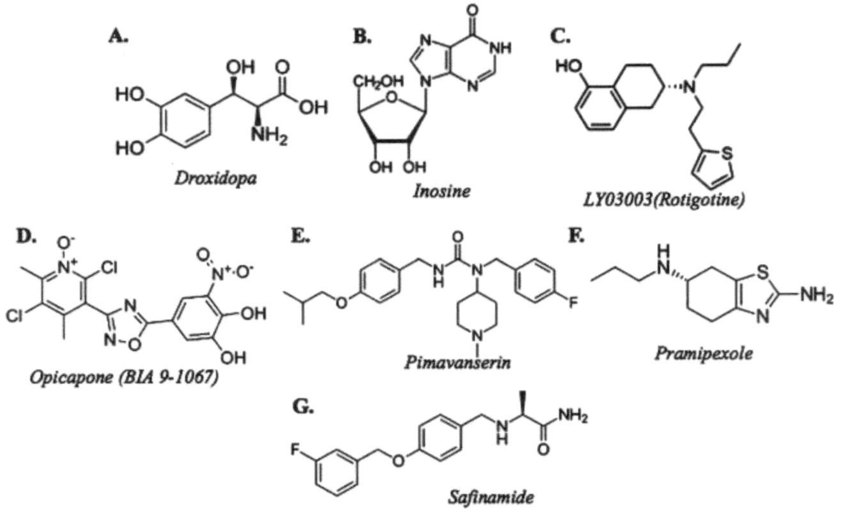

Fig. (3). Chemical structures of target drugs submitted to clinical studies from 2015 to 2020 (phases III and IV) related to Parkinson's disease that have ended normally, and participants are no longer being examined or treated.

Inosine

Mechanisms of Action: Inosine is a purine naturally formed from the degradation of adenosine by adenosine deaminase [94]. It can bind to adenosine receptors and initiates intracellular signaling effects. Furthermore, it can also affect cell function *via* receptor-independent pathways, increases uric acid production, inhibits poly (ADP ribose) polymerase, and upregulates GAP-43 in neurons [95]. Inosine is a urate precursor and dietary supplement at the onset of PD. Oxidative damage is thought to play a role in dopaminergic neuron degeneration of PD, and urate protects dopaminergic neurons in PD cellular and animal models [96].

Experimental Studies in other Diseases that Support their Use: Inosine has been analyzed for the treatment of AD and Amyotrophic Lateral Sclerosis (ALS). In the case of AD, Inosine showed protection against memory impairment in an experimental model induced by streptozotocin in rats [97]. In turn, the Inosine pilot test to increase urate levels in Amyotrophic Lateral Sclerosis has been shown to be safe and effective in increasing urate levels in ALS patients, which provides justification for clinical trials [98].

At the end of the clinical trial, in addition to the various adverse effects presented by patients treated with Inosine such as, upper abdominal pain, diverticulitis, increased salt levels in the blood, incope, testicular mass and knee arthroplasty, diarrhea, fatigue, sinusitis, blood creatinine, and increased spasms, among others.

LY03003

Mechanisms of Action: Rotigotine is a dopaminergic agonist that has a high affinity *in vitro* for all dopamine subtypes, showing preference for D3>D2>D1, acting through the activation of these receptors [99]. The extended-release rotigotine microspheres (RTGT-MS, LY033003) exert analgesic effects on inflammatory pain, and this mechanism may be associated with the regulation of the AC-cAMP-PKA pathway through the peripheral D2 receptor [100]. It is commonly used in combination with levodopa and has been reported to be effective in patients with PD under 70 years old, as well as in advanced stage PD [101].

Experimental Studies in other Diseases that Support their Use: The studies developed using Rotigotine extended-release microspheres are all focused on PD and possible manifestations that affected patients may be present due to the disease. Li *et al.* (2020) [100], for example, analyzed and proved the analgesic effect on inflammatory pain of RTGT-MS in PD patients.

No results about the study have been published to date.

Pimavanserin

Mechanisms of Action: Unlike other atypical antipsychotics, Pimavanserin has no affinity for dopaminergic receptors and serotonin transporters (in relation to selective serotonin reuptake inhibitors). It represents a new class of drugs that selectively target specific serotonin receptor subtypes. It is responsible for specifically modulating serotonergic neurotransmission as an agonist / antagonist inverse of 5-hydroxytryptamine2A (5-HT2A) and 5-HT2C in the case of higher doses [102]. It is the first US Food and Drug Administration (FDA)-approved treatment for Parkinson's disease psychosis [103].

Experimental Studies in other Diseases that Support their Use: Pimavanserin has been analyzed for the treatment of psychosis in cases of AD, where it has been shown to be effective in primary outcome (week 6 of the study), presenting an acceptable tolerability profile and without negative effects on cognition [104].

No results about the study have been published to date.

Droxidopa

Mechanisms of Action: Droxidopa is an orally active synthetic amino acid that is converted by the aromatic L-amino acid decarboxylase to norepinephrine [105]. It acts by several mechanisms to increase the levels of norepinephrine in post-ganglionic sympathetic neurons, leading to increased stimulation of adrenergic receptors. Through the ability to cross the blood-brain barrier, it increases the production of norepinephrine in the CNS, which leads to an increase in the activation of pre-ganglionic neurons in the spinal cord. Finally, norepinephrine synthesized from Droxidopa acts as a circulating hormone, exerting a pressure effect [106]. Norepinephrine is a prodrug available for the treatment of systomatic neutrogenic orthostatic hypotension (nOH) caused by primary autonomic failure, including PD [106].

Experimental Studies in other Diseases that Support their Use: Droxidopa has been analyzed for the treatment of Postural Ostatic Tachycardia Syndrome (POTS), where the drug shows a modest improvement in symptoms, as in the case of dizziness, syncope, and fatigue, but only 27% of the disorders being treated improve the functional state, which open further studies [107].

Throughout the study, some adverse events were recorded, which may or may not be related to drug administration, such as eye disorders, gastrointestinal problems, brief confusional episode, and nausea, among others, none exceeding 13.33% of patients. No deaths were recorded.

Pramipexole

Mechanisms of Action: Pramipexole is a non-ergoline, oral, active dopaminergic agonist with selective activity for dopamine receptors that belong to the D2 receptor subfamily (subtypes D2, D4 with preferential affinity for D3) [108]. Through the connection to dopamine auto receptors, Pramipexole exerts negative feedback in the synthesis of endogenous dopamine and this process leads to a reduction in oxidative stress, attenuating the damage in the nigrostriatal pathways [109]. It is effective in the treatment of motor and non-motor symptoms in patients with PD [110].

Experimental Studies in other Diseases that Support their Use: Pramipexole is studied as a drug / treatment for several specific diseases and conditions, such as in cases of bipolar and unipolar depression. In a study by Tundo *et al.* (2019) [111], it has been shown that the use of Pramipexole as an antidepressant is safe, and the overall short-term response is similar to the response rates of traditional antidepressants.

Patients treated with sustained-release (SR) of Pramipexole had 2.04% of serious adverse events such as gastrointestinal problems and hepatobiliary diseases, while subjects treated with immediate-release (IR) of Pramipexole had 0%. Some common adverse events were also presented as dizziness and drowsiness, in a maximum of 14.29% of patients. No deaths were recorded. The transition to stage II was halted due to study results, so only stage I results were reported.

Opicapone (BIA 9-1067)

Mechanisms of Action: Opicapone is a peripheral, selective, and reversible inhibitor of catechol-*O*-methyltransferase (COMT). It has a high lack of binding affinity (sub-picomolar Kd), which results in a more constant and long action *in vivo* and a long residence time of the reversible COMT-OPC complex and translates into a constant slow dissociation rate of the complex [112, 113]. This enzyme catalyzes the *O*-methylation of catecholamines, such as dopamine and adrenaline [112]. Catechol-*O*-methyl transferase inhibitors are currently used as first-line adjunct therapy to levodopa for the treatment of end-of-dose motor fluctuations in patients with Parkinson's disease, as they increase the bioavailability of levodopa [114]. No results about the study have been published to date.

Safinamide

Mechanisms of Action: Safinamide is an α-aminoamide that has dopaminergic and non-dopaminergic properties. It combines potent, selective, and reversible MAO-B inhibition with voltage-dependent Na^+ blocking and Ca^{2+} channels, and inhibition of glutamate release [115 - 117]. PD is a multi-directed disease in which drugs with multimodal action may be of particular value. Thus, safinamide was recently proposed as a promising new multimodal PD drug, because catechol-*O*-methyl transferase (COMT) inhibitors are currently used as adjunctive therapy to L-dopa to improve end-of-dose motor fluctuations as they inhibit the peripheral metabolism of L-dopa and increase its release to the brain [118].

Experimental Studies in other Diseases that Support their Use: In addition to studies already registered for the treatment of PD, the drug has been studied as a drug intervention and treatment of other diseases, for example the study by Xu *et al.* (2020) [119] which demonstrates that Safinamide exhibits neuroprotection in animals with acute induced stroke, showing *in vitro* that Safinamide benefits the survival of endothelial cells and improves stress-induced hypermeability having a preventive effect on the occurrence of stroke.

No results about the study have been published to date.

DA-9701

Mechanisms of Action: Gastrointestinal (GI) dysfunction is a common non-motor feature of PD. DA-9701 has been shown to have an agonist effect on 5-HT 1A, 5-HT 4 and α-2 receptors and antagonistic effect on the D2 receptor in the GI tract. In *in vivo* studies, DA-9701 increases gastric motility and accommodation, also reducing visceral hypersensitivity [120].

Experimental Studies in other Diseases that Support their Use: In addition to studies registered for the treatment of PD, a study by Kwon & Son (2013) [121] was developed to evaluate its action in patients with functional dyspepsia (FD). DA-9701 proved to be beneficial in gastric accommodation, visceral hypersensitivity, gastrointestinal motility and, possibly, also in *in vivo* stress-induced disorder. These effects neutralize the main physiological causes of FD.

No results about the study have been published to date.

NATURE AS A VALUABLE SOURCE OF CHEMICAL ENTITIES FOR PD AND AD THERAPEUTICS

Despite the increasing research efforts and significant advances achieved in recent years, the current therapeutic strategies for NDs are mostly symptomatic alleviating disease symptoms, but without capacity to control, retard, or prevent the NDs progression. Moreover, several of these drugs induce significant side-effects following over long periods treatment [122]. Therefore, there is an urgent need to discover and to develop new therapeutic agents with ability to act on diverse biochemical targets, with new mechanisms of action and low toxicity that can result in more effective therapies.

Over the decades, nature has revealed to be an enormous and rich source of new chemical entities with great pharmacological activities inspiring the development of numerous therapeutic agents currently in the market [123]. To date, quite a wide range of natural products (NPs) have been isolated from different terrestrial and marine organisms, including plants, sponges, seaweeds, and microorganisms, among others [123 - 125]. Distinct chemical classes have been identified such as, alcohols, alkaloids, amino acid derivatives, aromatic compounds, fatty acids, lactones, peptides, polyacetylenes, polyketides, quinones, quinolones, sphingolipids, sterols, and terpenes [126]. Opposite to chemically synthetic compounds, NPs have, presumably, evolved over billions of years in close association with the biological systems, conferring adaptive advantages to their producers in the surrounding environment where they are inserted. Their high specificity and affinity to interact with biological target structures (*e.g.*, receptors, DNA, membranes, proteins, *etc.*) make them excellent candidates to inspire the development of new drugs [127]. Thus, a large focus has been placed on the potential of the NPs as neuroprotective agents, essentially due to their enormous scaffold diversity, structural complexity, ability to act in multi-key intracellular signaling pathways (mitochondrial dysfunction, apoptosis, excitotoxicity, neuroinflammation, oxidative stress, protein misfolding, among others), distinct mechanisms of action, and minimal side-effects. Therefore, this topic is focused on the therapeutic potential of some NPs isolated in the last five years (2017 –

2021) and evaluated on different *in vitro* and *in vivo* PD and AD models Tables (**3** and **4**), respectively, reporting the neuroprotective effects and the activated intracellular signaling pathways. The bibliographic research was carried out in the ISI Web Knowledge database using the following keywords: "Neuroprotective", "Alzheimer's", "Parkinson's", "Natural products" and "Marine natural products". Due to the adopted bibliographic search strategy, some studies may be not described. Tables 3 and 4 summarize several reports attesting the potential of NPs isolated from plants, sponges, sea cucumbers, bacteria and macroalgae to act on several hallmarks linked to PD and AD development.

Table 3. Neuroprotective activities of natural products and their mode of action in Parkinson's disease evaluated on *in vitro* and *in vivo* models. *The bibliographic search was performed using the ISI Web Knowledge database with the keywords: "Neuroprotective", "Parkinson's", "Natural products" and "Marine natural products" (2017-2021).

Organism	Compound	Chemical Class	*In vitro*/ *In vivo* Model	Effects/ Mechanisms of Action	Ref.
PLANTS					
Not defined	Boswellic acids	Pentacyclic terpenes	Adult male albino rats	Increases motor functions. Increases percent of viable neurons in the *substantia nigra pars compacta*, reduces inflammatory markers (TNF-α, IL-6, COX-2), decreases NF-κB levels, and increases striatal dopamine level and nigral tyrosine hydroxylase (TH) immunostaining.	[151]
Chrysanthemi indici, Calamintha, Linaria spp.	Acacetin	Flavonoid	SH-SY5Y cells	Protective effects against 6-OHDA-induced neurotoxicity, inhibiting mitochondrial dysfunction, apoptosis (decreases Caspase-3 (Casp-3) and Caspase - 9 activity), and intracellular ROS production. Reduces JNK, p38-MAPK, PI3K/Akt and GSK-3β proteins expression.	[152]

(Table 3) cont.....

Organism	Compound	Chemical Class	*In vitro*/ *In vivo* Model	Effects/ Mechanisms of Action	Ref.
PLANTS					
Aegle marmelos	Aegelin	Alkaloid	Yeast SNARE protein Sec22p model	Inhibits the ROS generation and decreases mitochondrial dysfunction and nuclear DNA fragmentation.	[153]
Centella asiatica	Asiatic acid	Pentacyclic terpene	C57BL/6 mice	Inhibits the MPTP-induced phosphorylation of MAPK/P38 related proteins JNK and ERK. Increases the phosphorylation of PI3K, Akt, GSK-3β and mTOR proteins and activates PI3K/Akt/mTOR signaling pathway.	[154]
Ampelopsis grossedentada	Dihydromyricetin	Flavonoid	MES23.5 cells, C57BL/6 mice	*In vitro*: Protective effects against MPTP -induced neurotoxicity. Reduces intracellular ROS production. Inhibits GSH -3β activation and decreases tyrosine hydroxylase (TH) and vesicular monoamine transporter 2 (VAMT) expression. *In vivo*: Attenuates MPTP-induced mouse behavioral impairments, dopaminergic neuron loss and deficit in movement balance. Attenuates the decrease of TH and VAMT 2 expression in the striatum and *substantia nigra pars compacta.*	[155]

(Table 3) cont.....

Organism	Compound	Chemical Class	*In vitro/ In vivo* Model	Effects/ Mechanisms of Action	Ref.
PLANTS					
Cannabis	β-Caryophyllene	Bicyclic sesquiterpene	Male Wistar rats	Prevents dopaminergic neuronal loss in the *substantia nigra.* Reduces Iba-1 and GFAP protein expressions. Decreases the number of activated astrocytes and microglia. Attenuates proinflammatory cytokines (IL-1β, IL-6, and TNF-α) in the midbrain tissues and inflammatory mediators (COX-2 and iNOS expressions). Restores antioxidant enzymes, glutathione depletion and inhibits lipid peroxidation	[156]
Tobacco leaves	4*R*-Cembranoid	Cyclic diterpene	Neuro-2a cells and Sprague Dawley rat	In vivo: Exhibits significant neuroprotective activity in the rat unilateral 6-OHD--induced PD model. Reduces forelimb asymmetry scores and corner test scores 4 weeks after injection of 6-OHDA. Decreases depletion of TH in the striatum and *substantia nigra* on the side injected with 6-OHDA. In vitro: Protects differentiated neuro-2a cells from 6-OHD--induced cytotoxicity. Activates p-AKT, HAX-1 expression, and inhibits Casp-3 expression and endothelial inflammation.	[157]

(Table 3) cont.....

Organism	Compound	Chemical Class	*In vitro*/ *In vivo* Model	Effects/ Mechanisms of Action	Ref.
PLANTS					
SPONGE					
Haliclona (Rhizoniera) sarai	Sarain A	Alkaloid	SH-SY5Y cells	Protective effects against H_2O_2 - induced neurotoxicity. Blocks mPTP and activates Nrf2 pathway.	[158]
Hymeniacidon sp.	Hymenin	Alkaloids	Primary cortical neurons	Protective effects against H_2O_2 - induced neurotoxicity. Inhibits mitochondrial dysfunction, reduces the intracellular reactive oxygen species (ROS) levels, and increases the glutathione levels, decreases Casp-3 expression, and activates the Nrf2/ARE pathway.	[159]
	Hymenialdisine				
Psammaplysilla sp.	Psammaplysene A	Alkaloid	HEK293 cells and *Drosophila in vivo*	Neuroprotective effects in the HEK293 cells and *Drosophila* by modifying processes dependent on heterogeneous nuclear ribonucleoprotein (HNRNPK) that controls biological aspects of RNA.	[160]
Jaspis splendens	Jaspamycin	Nucleoside analogue	hONS cells	Increases lysosomal staining and decreases EEA-1 number of that are phenotypic responses on a hONS cellular model of PD.	[161]

(Table 3) cont.....

Organism	Compound	Chemical Class	*In vitro/ In vivo* Model	Effects/ Mechanisms of Action	Ref.
PLANTS					
Cladiella australis	1-Tosylpentan-3-one	Tosylated compound	SH-SY5Y cells	Neuroprotective effect on SH-SY5Y cells against 6-OHDA – induced neurotoxicity. Activates p38 mitogen-activated protein kinase (MAPK) and decreases Caspase-3 and nuclear factor erythroid 2-related factor 2 (Nrf2) protein expression.	[162]
SEA CUCUMBER					
Cucumaria frondosa	Glucocerebrosides -1 Glucocerebrosides- 2 Glucocerebrosides- 3	Cerebrosides	PC12 cells	Neuritogenic effects in a dose-dependent and structure-selective manner. Increases the ratio of neurite-bearing cells and expression of axonal (GAP-43) and synaptic (synaptophysin) proteins.	[163]
BACTERIA					
Streptomyces sp.	Piloquinone A	Quinone	Monoamine oxidase A and B activity	Inhibits monoamine oxidase (A and B) activity	[164]
MACROALGAE					
Porphyra haitanensis	Porphyran	Sulfated polysaccharide	MES 23.5 cells	Neuroprotective effects against 6-OHD- -induced neurotoxicity.	[165]
Hypnea musciformis	κ-Carrageenan	Sulfated polysaccharide	SH-SY5Y cells	Neuroprotective effect against 6-OHDA – induced neurotoxicity. Restores mitochondrial membrane potential and decreases Caspase -3 activity.	[166]

(Table 3) cont.....

Organism	Compound	Chemical Class	*In vitro*/ *In vivo* Model	Effects/ Mechanisms of Action	Ref.
PLANTS					
Gracilaria cornea	Agaran	Sulfated polysaccharide	Adult male Wistar rats	Neuroprotective effect against 6-OHDA – induced *in vivo* neurotoxicity on rat. Reduces the oxidative/nitro active *stress* and through alteration in the MAO contents induced by 6-OHDA.	[167]
Codium tomentosum	Loliolide	Monoterpene lactone	SH-SY5Ycells; RAW264.7 cells	Protective effects against 6-OHDA - induced neurotoxicity. Reduces intracellular ROS and H_2O_2 levels, inhibits mitochondrial dysfunction, decreases Casp-3 activity, increases catalase activity, and inhibits NF-κB activation. Protective effects against LPS - induced inflammation on RAW264.7 cells,	[168]
Bifurcaria bifurcata	Eleganolone	Diterpene	SH-SY5Y cells; RAW264.7 cells	Protective effects against 6-OHDA - induced neurotoxicity. Reduces intracellular ROS levels, inhibits mitochondrial dysfunction, decreases Casp-3 activity and ATP levels, increases catalase activity, and inhibits NF-κB activation. Protective effects against LPS - induced inflammation on RAW264.7 cells, reducing TNF-α and IL-6 cytokines levels.	[169]

(Table 3) cont.....

Organism	Compound	Chemical Class	*In vitro*/ *In vivo* Model	Effects/ Mechanisms of Action	Ref.
			PLANTS		
Not defined	Agaropentose / Agaro-oligosaccharide	Polysaccharides	SH-SY5Y cells	Protective effect against 6-OHDA - induced neurotoxicity. Reduces intracellular ROS levels and inhibits mitochondrial membrane potential loss and NF-κB activation. Downregulates p38MAPK signaling pathway. Enhances superoxide dismutase, glutathione reductase and glutathione peroxidase activities. Reduces malondialdehyde levels and apoptotic cells number. Suppresses Casp-3 expression and decreases Bax/Bcl-2 ratio.	[170]

Table 4. Neuroprotective activities of natural products and their mode of action in Alzheimer's disease on *in vitro* and *in vivo* models. *The bibliographic search was performed using the ISI Web Knowledge database with the keywords: "Neuroprotective", "Alzheimer´s", "Natural products" and "Marine natural products" (2017-2021).

Organism	Compound	Chemical Class	*In vitro*/ *In vivo* Model	Effects/ Mechanisms of Action	Ref.
			PLANTS		
Stephania tetrandra S.Moore	Fangchinoline	Alkaloid	SH-SY5Y cells	Protective effects against neurotoxicity induced by sodium nitroprusside, sodium dithionate and potassium chloride. *In vitro* acetylcholine esterase (AChE) inhibitory activity	[171]

(Table 4) cont.....

Organism	Compound	Chemical Class	*In vitro*/ *In vivo* Model	Effects/ Mechanisms of Action	Ref.
Menispermum dauricum DC	Dauricine	Alkaloid	N2a cells and SH-SY5Y cells transfected with Swedish mutant β-amyloid precursor protein (APP); GMC101 worms	*In vitro*: Inhibits amyloid precursor protein processing, decreases Aβ accumulation and reduces the tau hyperphosphorylation *via* PP2A, p35/25, and CDK5 pathways. Decreases the $A\beta_{1-42}$ release levels and oxidative stress induced by Cu^{2+} treatment on SH-SY5Y/APP cells, recovering mitochondrial membrane potential (MMP) and superoxide dismutase (SOD) activity. Inhibits apoptosis suppressing Caspase-3 activation (Casp3) and increasing Bcl-2 expression and regulates the Nrf2, and Kelch-like ECH-associated protein 1 (Keap1) levels, essential for Nrf2 activation. *In vivo*: Protective effects against oxidative stress condition mediated by Aβ or paraquat (PQ) on GMC101 worms.	[172]
Scutellaria baicalensis Georgi	Wogonin	Flavonoid	Triple transgenic AD mice (h-APPswe, h-Tau P301L, and h-PS1 M146V); Tet-On Aβ-42-GFP SH-SY5Y cells	*In vivo*: Improvement of triple transgenic AD mice performance validated by the means of different behavioural tests linked to learning and memory (Morris water maze, Y-maze, novel object recognition) *In vitro*: Promotes neurite outgrowth increasing their length and complexity and mitigates the amyloidogenic pathway decreasing β-secretase, APP β-C-terminal fragment, Aβ-aggregation, and phosphorylated Tau levels. Increases MMP and blocks programmed cell death attenuating Bax expression and cleaved poly (ADP-ribose) polymerase (PARP).	[173]
Camellia japonica	17β,29-Dihydroxy-3,16-dioxo-28-norolean-12-ene;1β,11α,17β-trihydroxy-3,16-dioxo-28-norolean-12-ene;1β, 17β-Dihydroxy-11α-methoxy- 3,16-dioxo-28-norolean-12-ene;17β-hydroxy-11α-methoxy3, 16-dioxo-28-noroleana-1,12-diene	Triterpenoids	HT22-transfected cells; BV2 microglial cells	Protective effects against Aβ- and glutamate-induced neurotoxicity. Reduce intracellular glutamate levels, increase cell viability, inhibit BACE1 activity, and inhibit nitric oxide (NO) production.	[174]
Corydalis tomentella	(13*R*,14*R*)-13-Hydroxy-13-methyl-8-oxosinactine; (13*S*,14*S*)-tomentelline E	Isoquinoline alkaloids	BV2 microglial cells	Decrease LPS- induced cytotoxicity and the release of inflammatory cytokines, including tumour necrosis factor (TNF)-α, interleukin (IL)-6 and IL-1β.	[175]
Morus alba	Mulberrofuran C, Mulberrofuran K, Mulberrofuran G, and Isomulberrofuran G	Benzofurans	BV2 microglial cells; HT22 cells	Inhibition of NO production on LPS-induced BV2 cells; Protective effects against glutamate-induced neuronal cell death on HT22 cells. Mulberrofuran K displayed the best performance to cross the blood-brain barrier (BBB), as well as neuroprotective activity against glutamate-induced toxicity on HT22 cells preventing apoptosis and mediating an antioxidative mechanism, decreasing the ROS production and increasing the glutathione (GSH) levels.	[176]

(Table 4) cont.....

Organism	Compound	Chemical Class	*In vitro*/ *In vivo* Model	Effects/ Mechanisms of Action	Ref.
Codonopsis pilosula	*Codonopsis pilosula* polysaccharide (CPPs)	Polysaccharides	HEK293/tau cells; Adeno-associated virus serotype 2 (AAV2) infected C57/BL6 mice	*In vitro*: Increases the cell viability and PP2A activity and decreases tau phosphorylation on HEK 293/tau cells. *In vivo*: Decreases tau phosphorylation (Ser199, Ser202/Thr205 (AT8), Thr231) on hippocampus region. Recovery of cognitive impairments induced by hTau overexpression, attenuating the fEPSP slope decrease, and increases synaptic proteins levels (synaptotagmin and synaptophysin).	[177]
Lawsonia inermis	1,2,4-Trihydroxynaphthalene -2-*O*-β-D-glucopyranoside	Phenol glycoside	SH-SY5Y cells	Reduces $A\beta_{42}$ - induced cytotoxicity decreasing ROS production and Ca^{2+} intracellular levels.	[178]
Crocus sativus L.	Crocin	Carotenoids	HT22 cells; BALB/c mice	*In vitro*: Neuroprotective effect against glutamate-induced toxicity, accompanied by a decrease of ROS and intracellular Ca^{2+} levels, mitochondrial disfunction, and inhibition of the apoptosis by the means of a down-regulation of Bax, Bad and cleaved Casp3 proteins expression, and up-regulation of the Bcl-xL, Akt and mTOR proteins expression. *In vivo*: Improves the cognitive and memory capabilities. At the biochemical level, increases antioxidant enzymes (glutathione peroxidase and superoxide dismutase), acetylcholine, and choline acetyltransferase levels. Reduces $A\beta_{1-42}$, ROS and acetylcholinesterase levels in the serum, cerebral cortex, and hypothalamus brain regions. Decrease of $A\beta_{1-42}$ deposition on the hippocampus region was also observed.	[179]
MACROALGAE					
Not defined	Fucosterol	Phytosterol	SH-SY5Y cells	Neuroprotective effects against Aβ-induced cytotoxicity, enhancing neuroglobin mRNA expression and decreasing APP mRNA expression, intracellular Aβ levels and apoptosis.	[180]
Ecklonia maxima	Eckmaxol	Phlorotannin	SH-SY5Y cells	Induces neuroprotective effects inhibiting glycogen synthase kinase 3β and extracellular signal-regulated kinase pathways.	[181]
Ecklonia stolonifera	Fucosterol	Phytosterol	Primary hippocampal neurons; aging rats	*In vitro*: Decreases $A\beta_{1-42}$-induced cytotoxicity reducing the expression of glucose-regulated protein 78 (GRP78) and activating tyrosine receptor kinase B-mediated ERK1/2 signaling pathway. *In vivo*: Protective effects against $A\beta_{1-42}$-induced cognitive injury in aging rats *via* the upregulation of BDNF-TrkB-ERK1/2 signaling pathway in the dentate gyrus.	[182]
Sargassum horneri	Fucoxanthin	Carotenoid	SH-SY5Y cells	Protective effects against Aβ oligomer-induced neurotoxicity reducing ROS levels, apoptosis and regulating positively and negatively the PI3K/Akt and the ERK pathways, respectively.	[183]

(Table 4) cont.....

Organism	Compound	Chemical Class	*In vitro/ In vivo* Model	Effects/ Mechanisms of Action	Ref.
Sargassum horneri	Fucoxanthin	Carotenoid	SH-SY5Y cells; Male ICR (Institute of Cancer Research)mice	In vitro: Ability to change the shape of $A\beta_{1-42}$ assemblies, converting to less toxic forms. In vivo: Improves spatial learning, memory, and recognition performances. At biochemical level, increases SOD, catalase (CAT) and glutathione activities and decreases malondialdehyde (MDA) levels. Up-regulate the proteins expression of brain-derived neurotrophic factor and choline acetyltransferase.	[184]
Not defined	Fucoidan	Sulfated polysaccharide	PC12 cells; Male ICR mice	In vitro: Inhibits neurotoxicity induced by the combined treatment of $A\beta_{25-35}$ and D-galactose on PC-12 cells and cell death by apoptosis preventing cytochrome c release, caspases activation, and increasing the expression of anti-apoptotic proteins (livin, X-linked IAP). Improves antioxidant defenses, increasing SOD activity and GSH levels. In vivo: Improves learning and memory performance. At biochemical level, increases acetylcholine (ACh), choline acetyl transferase (ChAT), and SOD activities as well as GSH levels, and decrease AChE activity.	[185]
Not defined	κ-Carrageenan	Sulfated polysaccharide	SH-SY5Y cells	Protective effects against $A\beta_{25-35}$-induced neurotoxicity. Inhibition of apoptosis decreasing Casp-3 expression. Inhibition of JNK signaling pathway activation.	[186]
Not defined	Fucoidan	Sulfated polysaccharide	*Caenorhabditis elegans*	Reduces the paralyzed phenotype induced by Aβ and decreases its accumulation on model tissues, possibly due to the increase of proteasomes activity. Decrease ROS generation.	[187]
Ecklonia cava	Phloroglucinol	Benzenetriol	6-Month-old 5X familial AD (5XFAD) mice	Improves cognitive function performance. Reduces the number of amyloid plaques, and BACE1, GFAP and Iba-1 protein levels, as well as the TNF-α and IL-6 mRNA levels. Promotes the increase of dendritic spine density and the number of mature spines on the hippocampus brain region.	[188]
Padina gymnospora	α-Bisabolol	Monocyclic sesquiterpene alcohol	Neuro2a cells; *Caenorhabditis elegans*	In vitro: Mitigates cholinesterase and β-secretase activity, ROS and reactive nitrogen species (RNS) generation. Decreases the levels of pro-apoptotic proteins (Bax, Casp3) In vivo: Protective effects inhibiting the genes expression (ace-1, hsp-4 and Aβ) linked to Aβ synthesis	[189]
Sargassum fusiforme	24(S)-Saringosterol	Sterol	APPswePS1ΔE9 mice	Improves cognitive performance without affecting the number of Aβ plaques. Decreases Iba1 inflammatory marker levels.	[190]
Pyropia haitanensis	Porphyran	Sulfated polysaccharide	90 Kunming male mice	Improves cognitive functions. At biochemical level, increases ChAT activity and decreases AChE activity in the cortical and hippocampal brain regions.	[191]

(Table 4) cont.....

Organism	Compound	Chemical Class	*In vitro/ In vivo* Model	Effects/ Mechanisms of Action	Ref.
SEA CUCUMBER					
Acaudina molpadioides	Cerebroside – L; Cerebroside - H	Cerebrosides	Male SD rats	Amelioration on the impaired cognitive function. Further findings indicate that Cer ameliorates Aβ 1–42 -induced neuronal damage and suppresses the induced apoptosis by decreasing the level of Bax/Bcl-2. Additionally, Cer enhances the expressions of PSD-95 and synaptophysin by activating BDNF/TrkB/CREB signaling pathway, thereby ameliorating Aβ 1–42 -induced synaptic dysfunction. Furthermore, Cer attenuates Aβ 1–42 -induced tau hyperphosphorylation by activating the PI3K/Akt/GSK3β signaling pathway amelioration on the impaired cognitive function. Further findings indicate that Cer ameliorates Aβ 1–42 -induced neuronal damage and suppresses the induced apoptosis by decreasing the level of Bax/Bcl-2. Additionally, Cer enhances the expressions of PSD-95 and synaptophysin by activating BDNF/TrkB/CREB signaling pathway, thereby ameliorating Aβ 1–42 -induced synaptic dysfunction. Furthermore, Cer attenuates Aβ 1–42 -induced tau hyperphosphorylation by activating the PI3K/Akt/GSK3β signaling pathway. Improves cognitive function impairment and decreases neuronal damage mediated by $A\beta_{1-42}$ treatment. Suppress mitochondria-dependent apoptosis increasing and decreasing of Bcl-2 and Bax protein levels, respectively. Enhances synaptic function and protective effects by acting BDNF/TrkB/CREB pathway and upregulating PSD-95 and synaptophysin protein expression levels. Activates PI3K/Akt/GSK3β signaling pathway mitigating tau hyperphosphorylation.	[192]
SEA URCHIN					
Strongylocentrotus nudus	GM4(1S), GD4(1S), GD4(2G), GD4(2A)	Gangliosides	PC12 cells Male SAMP8 mice	*In vitro:* Protective effects against $A\beta_{25-35}$-induced neurotoxicity and neurite length improvement. Upregulation of the synaptophysin and GAP-43 mRNA expression levels. Inhibition of mitochondrial apoptosis pathway downregulating the Bax, Casp9, and Casp3 mRNA and protein expression and upregulating the Bcl-2 mRNA and protein expression. *In vivo*: Improves memory functions. Decreases the $A\beta_{1-42}$ levels and increases the synaptophysin and GAP-43 mRNA expression levels on the hippocampus brain region. mRNA levels of apoptosis-related proteins expression exhibited a similar profile to the observed on *in vitro* assays.	[193]
MICROORGANISMS					-

(Table 4) cont.....

Organism	Compound	Chemical Class	*In vitro*/ *In vivo* Model	Effects/ Mechanisms of Action	Ref.
Streptomyces caniferus	Caniferolide A	Macrolide	BV2 cells; SH-SY5Y cells; SH-SY5Y tau441 cells	Attenuates neuroinflammatory markers on LPS-activated microglial cells blocking NFκB-p65 translocation and activating Nrf2 pathway. Mitigates IL-1β, IL-6 and TNF-α production, ROS, and NO release, and iNOS, JNK and p38 activities. Inhibits tau phosphorylation by targeting p38 and JNK MAPK kinases. Increases GSH content.	[194]
Streptomyces sp.	Streptocycline A and B	Polyketides	BV2 cells; SH-SY5Y cells; SH-SY5Y tau441 cells	Actives Nrf2 factor. Reduces pro-inflammatory factors (iNOS, IL-1β, TNF-α; ROS, NO) and increase anti-inflammatory factor (IL-10) inhibiting NFκB and MAPKs, and promotes Nrf2 translocation. Mitigates tau hyperphosphorylation inhibiting MAPK kinases.	[195]

The neurodegeneration process observed in PD and AD is triggered by a set of distinct biological mechanisms (*e.g.*, protein aberrant aggregation, mitochondrial deficits, oxidative stress, defective protein quality-control and degradation pathways, stress granules, maladaptive immune response, *etc.*) leading to malfunctions and, consequently, to cellular death [128]. In Tables **1** and **2** it is possible to observe that compounds belonging to distinct chemical classes, including terpenes, alkaloids, sterols, polysaccharides, carotenoids, and others, exhibit neuroprotective capabilities on different *in vitro* and *in vivo* models. Those compounds displayed capacity to improve cognitive functions like memory and learning on *in vivo* PD and AD models, and mechanisms of action underlying their activities are linked to the mitigation of mitochondrial dysfunction, oxidative stress, neuroinflammation and apoptosis inhibition. Furthermore, those compounds also demonstrated ability to up- and/or down-regulate the expression of key proteins/factors associated to PD and AD development, for example, ace-1, hsp-4, JNK, p38-MAPK, PI3K/Akt and GSK-3β, and improvement of cognitive functions (*e.g.*, synaptophysin, GAP-43). Despite the promising results of NPs, it is also true that only five drugs are clinically approved for AD treatments, one from natural origin and other nature-derived, while for PD therapy, among eleven drugs, none was obtained from a natural resource [129]. The translation of promising preclinical results to clinical application has proven to be challenging, especially due to some NPs features like low bioavailability and limited water solubility, physicochemical instability, rapid metabolism, and ability to cross the blood-brain barrier, which can limit their efficacy. However, efforts to overcome these challenges should be accomplished due to the great potential of NPs to act in multitargets of PD and AD [130].

FUTURE ADVANCES IN THE FIELD - TRACKING NEW DRUGS

Efforts to bring new AD drugs to the market have failed due to several causes such as an incomplete understanding of the pathogenesis of AD, the multifactorial etiology and complex pathophysiology of the disease, the slowly progressive nature of AD and the high rate of comorbidity in elderly population [131]. Other reasons include methodological difficulties, course of treatment in relation to the development of the disease, appropriate use of validated biological and neuropsychological markers. In fact, many efforts have been directed towards identifying the most appropriate and sensitive biological and neuropsychological markers that can predict the progression of AD: neuroimaging markers (atrophy of the hippocampus and amyloid positron emission tomography), markers CSF (association of high tau with low levels of amyloid β peptide) and neuropsychological markers (episodic memory deficits and executive dysfunction) [132].

In the case of PD, the main cause of the variety of motor symptoms observed in patients is the selective neuronal loss in brain areas involved in fine adjustment of movement. Aimed at keeping motor symptoms aligned, but with limited efficacy, current treatments are unable to stop, reverse, or delay PD. Thus, scientific research is strongly focused on the development of new therapeutic strategies, considering that the development of small molecules for the treatment of these neurodegenerative diseases has a high failure rate [133].

As in cancer, AD has a so-called clinical syndrome, that is, genetic and biological heterogeneity between individuals who share the same clinical characteristics. It is highly frequent in multifactorial polygenic diseases with complex and non-linear pathophysiological dynamics. It is recognized that the adaptive and innate immune systems have enormous individual heterogeneity, interfering with the individual's specific response to vaccines and other immunomodulatory therapies. Therefore, some drugs, administered regularly, may benefit only a specific subset of patients, even having detrimental effects for some defined ethnic groups. Taking this point into account, it is essential to identify the molecular/cellular and environmental factors that indicate the presence and type of reaction of a single patient with AD to a specific therapy [134].

Structural bioinformatics and computational molecular biophysics are promising research fields to gather information regarding molecular aspects of different diseases. These areas comprise techniques such as, comparative modeling, molecular docking, virtual screening, molecular dynamics simulation, and so on. Nowadays, these approaches can be widely used due to the continuously evolving fields of cheminformatics and computational biology. The goal of computer-aided drug design is to reduce costs and time, while speeding up the process and improving the accuracy and efficiency of new drug candidates [135].

The first step in any virtual screening protocol is the selection of a protein target or receptor, however, in the case of AD, due to its polygenic and multifactorial nature, the act of selecting a specific target is a challenging task. From a molecular point of view, AD is considered multifactorial since it is made distinctive by various molecular and cellular processes including protein aggregation, oxidative stress, cell cycle deregulation, and neuroinflammation, the causes of which are still rather unclear [136]. In neurodegenerative diseases, processes like endosomal-lysosomal autophagy, neuroinflammatory responses, mitochondrial homeostasis, proteostasis, and metabolic profiling are usually dysregulated. Specifically, alterations in homeostasis mechanisms and an increase in misfolded protein aggregation are major factors in AD and PD alike. The positive feedback loop between oxidative stress, misfolded proteins, and mitochondrial dysfunction is a crucial target in therapeutic interventions [137].

Besides being characterized by the aggregation of amyloid-beta (Aβ) plaques and neurofibrillary tangles comprised of hyperphosphorylated tau protein, an AD-afflicted brain also shows a significant decrease in the concentration of the neurotransmitter acetylcholine (ACh). According to Carpenter *et al.* (2018) [135], these two factors comprise the main ideas that guide the general treatment of AD: first, the idea that cognitive decline in patients with AD is caused by the aggregation of Aβ plaques; and second, that it is caused by the loss of ACh, and most treatments focus on remedying the latter [135].

In recent years, for instance, several attempts have been made to identify novel acetylcholinesterase (AChE) and butyrylcholinesterase (BChE) inhibitors using computational methods. These enzymes are responsible for the hydrolysis of ACh, and its inhibition would raise the levels of ACh in the brain. However, these approaches only treat the symptoms associated with AD, and are unable to cure or to prevent neurodegeneration [136, 137]. Nevertheless, *in vitro* studies have suggested that AChE also plays a role in Aβ fibril formation *via* a peripheral binding site. This was further supported by the analysis of the enzyme's tridimensional structure, which revealed a peripheral anionic binding site (PAS) besides the classical catalytic anionic binding site (CAS) [138]. Therefore, the development of compounds with dual binding effects, capable of targeting both active sites, to avoid both ACh hydrolysis and Aβ aggregation could display even higher therapeutic efficacy by increasing ACh levels, delaying the formation of Aβ aggregates, and improving cognitive deficit [139].

Techniques based on QSAR (Quantitative Structure-Activity Relationship) models, which are mathematical methods based on the established relationship between molecular structure and chemical activity, have been shown to also be effective in the search for new anti-AD drugs. The correlation of chemical structures and their respective properties *via* mathematical equations allows researchers to use the properties of known compounds to screen for novel compounds with potentially better activities, while the inverse process is also made possible: the *de novo* design of novel structures based on the desired properties, including multipotent activity [131]. One example of this approach is the application of QSAR in the design, synthesis, and analysis of donepezil-pyridyl hybrids in the study performed by Bautista-Aguilera *et al.* (2014) [120]. The authors found that a donepezil-indolyl hybrid molecule was a potent and selective MAO A inhibitor, while also being able to moderately inhibit MAO B, AChE, and BChE, comparable to its parent compound ASS234 [140, 141], which was found to reduce amyloid plaque burden and gliosis in the cortex of mice, as well as decreasing or even reversing scopolamine-induced memory loss, according to the authors [139]. Wang *et al.* (2014) [142] also employed QSAR models based on molecular topology and found several compounds with Aβ-lowering and anti-oligomerization activities, eight of which were subsequently patented.

Furthermore, some studies point out that only the soluble, aqueously diffusible subset of Aβ oligomers is actually bioactive and neurotoxic, meaning that the shortcomings in several anti-oligomer-based therapies can be related to the large pool of inactive oligomers and low specificity towards the soluble ones [143, 144]. Additionally, the pilot study performed by Sideris *et al.* (2021) [144] points out the presence of soluble Aβ aggregates in several regions of the AD afflicted brain, indicative of a global pathology, even in early stages of the disease. The authors highlight that these aggregates induce inflammation in the brain, driving synaptic dysfunction and eventual neuronal cell death, and suggest that targeting the aggregate induced inflammation instead of only specific aggregates could be another strategy in delaying AD progression.

As discussed, besides the decrease in ACh, one of the hypotheses behind the development of novel treatments for AD is that the disease can originate from the hyperphosphorylation of tau protein and the resulting aggregation of these proteins, caused by an imbalance between the catalytic activity of kinases and phosphatases that act on tau protein [145]. As such, numerous diverse enzymes are involved in regulating tau phosphorylation, but the inhibition of specific tau kinases remains one of the most important strategies in reversing tau phosphorylation and treating AD. The highest number of tau phosphorylation sites involved in AD is present in GSK3β and CK1/2, corresponding to 70% and 60%

of pathological phosphorylation sites, respectively, which makes these two enzymes promising targets. Moreover, it is suggested that GSK3β and CK1 co-localize with neurofibrillary tangles, and in AD brains GSK3β level, GSK3β activity, CK1δ mRNA level and CK1 activity are largely increased. It has also been noted that overactivation of either GSK3β or CK1 leads to an increase in Aβ production, whereas Aβ peptides lead to an increase in the enzymatic activity of GSK3β or CK1 in return. Thus, ideally, the combined inhibition of GSK3β and CK1 could be a promising way to search for new compounds [146].

In the case of PD, some recent studies show that fatty acid-binding proteins (FABPs) have been implicated in α-synuclein regulation, such as in the uptake of extracellular α-synuclein monomers and their aggregation and oligomerization into the insoluble fibrils which are characteristic of the neurodegeneration associated with PD. However, it appears that the exact mechanisms involved have not yet been completely elucidated [147 - 149].

The reuse of medications has been calling attention in recent years, both from pharmaceutical companies and government agencies, being a very promising approach to find new, effective, but low-cost and safe drugs to treat highly unmet medical needs. To generate new drugs to meet growing medical needs that are not being met quickly and economically, there are several approaches to identifying new indications for known drugs. Most of these approaches are based on rational selection, and the known or a new mechanism of action of a drug [150].

Unfortunately, from the use of the drugs present in the current market, different side effects and aggressiveness appear during these treatments. According to Troncoso-Escudero (2020) [133], in relation to PD, new formulations have been developed to control the psychomotor complications produced by some drugs, however, it is important to reduce all events that can affect or deteriorate the quality of life of patients, and perhaps the answers are outside the classical pharmacology approach [133]. In fact, individualized therapy needs more attention, as well as targeted treatments through exploratory, impartial, high-performance, integrative analyzes and in cohorts with a larger number of people due to the characteristics of individuals with the disease [134].

CONCLUDING REMARKS

Considering the etiological complexity and the prevalence of neurological disorders, scientists have a great challenge in exploring new therapies and new molecules to find an adequate and viable treatment for these diseases. Because AD is a gradually progressive brain disease that begins many years before its symptoms are revealed, and that the pathogenesis of PD has a contribution of several factors, such as genetic, environmental, and lifestyle, it still has age as the most critical risk factor. Here, we presented and discussed treatment possibilities in development for AD and PD. Data like mechanisms of action, experimental studies in other diseases that support their use and chemical structure of the drugs are included in this chapter.

Currently, there are no effective neuroprotective treatments that can prevent or delay disease progression to AD and PD in individuals, strengthening the idea that more studies, and new therapeutic approaches are needed. The reuse of medications has been calling attention, being a very promising approach to find new, effective, but low-cost and safe drugs to treat highly unmet medical needs.

Finally, we need to better understand which changes in metabolism and etiology cause the diseases, and perhaps it will be possible to design or adapt individual metabolism-based therapies that could work as alternative treatments for these kinds of disorders.

REFERENCES

[1] Le HND, Le LKD, Nguyen PK, *et al.* Health-related quality of life, service utilization and costs of low language: A systematic review. Int J Lang Commun Disord 2020; 55(1): 3-25. [http://dx.doi.org/10.1111/1460-6984.12503] [PMID: 31556211]

[2] Gooch CL, Pracht E, Borenstein AR. The burden of neurological disease in the United States: A summary report and call to action. Ann Neurol 2017; 81(4): 479-84. [http://dx.doi.org/10.1002/ana.24897] [PMID: 28198092]

[3] Verboket RD, Mühlenfeld N, Woschek M, *et al.* Inpatient treatment costs, cost-driving factors and potential reimbursement problems due to fall-related fractures in patients with Parkinson's disease. Chirurg 2020; 91(5): 421-7. [http://dx.doi.org/10.1007/s00104-019-01074-w] [PMID: 31807819]

[4] Cummings JL, Morstorf T, Zhong K. Alzheimer's disease drug-development pipeline: few candidates, frequent failures. Alzheimers Res Ther 2014; 6(4): 37. [http://dx.doi.org/10.1186/alzrt269] [PMID: 25024750]

[5] Gomez-Mancilla B, Marrer E, Kehren J, *et al.* Central nervous system drug development: An integrative biomarker approach toward individualized medicine. NeuroRx 2005; 2(4): 683-95. [http://dx.doi.org/10.1602/neurorx.2.4.683] [PMID: 16489375]

[6] Craven R. The risky business of drug development in neurology. Lancet Neurol 2011; 10(2): 116-7. [http://dx.doi.org/10.1016/S1474-4422(11)70004-7] [PMID: 21256451]

[7] Hersi M, Irvine B, Gupta P, Gomes J, Birkett N, Krewski D. Risk factors associated with the onset and progression of Alzheimer's disease: A systematic review of the evidence. Neurotoxicology 2017; 61: 143-87. [http://dx.doi.org/10.1016/j.neuro.2017.03.006] [PMID: 28363508]

[8] 2020 Alzheimer's disease facts and figures. Alzheimers Dement 2020; 16(3): 391-460. [http://dx.doi.org/10.1002/alz.12068]

[9] Gandy S, DeKosky ST. Toward the treatment and prevention of Alzheimer's disease: rational strategies and recent progress. Annu Rev Med 2013; 64(1): 367-83. [http://dx.doi.org/10.1146/annurev-med-092611-084441] [PMID: 23327526]

[10] McKhann GM, Knopman DS, Chertkow H, *et al.* The diagnosis of dementia due to Alzheimer's disease: Recommendations from the National Institute on Aging-Alzheimer's Association workgroups

on diagnostic guidelines for Alzheimer's disease. Alzheimers Dement 2011; 7(3): 263-9. [http://dx.doi.org/10.1016/j.jalz.2011.03.005] [PMID: 21514250]

[11] Dhillon S. Aducanumab: First Approval. Drugs 2021; 81(12): 1437-43. [http://dx.doi.org/10.1007/s40265-021-01569-z] [PMID: 34324167]

[12] Hort J, O'Brien JT, Gainotti G, *et al.* EFNS guidelines for the diagnosis and management of Alzheimer's disease. Eur J Neurol 2010; 17(10): 1236-48. [http://dx.doi.org/10.1111/j.1468-1331.2010.03040.x] [PMID: 20831773]

[13] Seltzer B. Donepezil: a review. Expert Opin Drug Metab Toxicol 2005; 1(3): 527-36. [http://dx.doi.org/10.1517/17425255.1.3.527] [PMID: 16863459]

[14] Seltzer B. Donepezil: an update. Expert Opin Pharmacother 2007; 8(7): 1011-23. [http://dx.doi.org/10.1517/14656566.8.7.1011] [PMID: 17472546]

[15] https://www.ncbi.nlm.nih.gov/books/NBK513257/

[16] Rogers SL, Friedhoff LT. The efficacy and safety of donepezil in patients with Alzheimer's disease: results of a US Multicentre, Randomized, Double-Blind, Placebo-Controlled Trial. Dementia 1996; 7(6): 293-303. [PMID: 8915035]

[17] Rogers SL, Farlow MR, Doody RS, Mohs R, Friedhoff LT. A 24-week, double-blind, placebo-controlled trial of donepezil in patients with Alzheimer's disease. Neurology 1998; 50(1): 136-45. [http://dx.doi.org/10.1212/WNL.50.1.136] [PMID: 9443470]

[18] Burns A, Rossor M, Hecker J, *et al.* The effects of donepezil in Alzheimer's disease - results from a multinational trial. Dement Geriatr Cogn Disord 1999; 10(3): 237-44. [http://dx.doi.org/10.1159/000017126] [PMID: 10325453]

[19] Takada Y, Yonezawa A, Kume T, *et al.* Nicotinic acetylcholine receptor-mediated neuroprotection by donepezil against glutamate neurotoxicity in rat cortical neurons. J Pharmacol Exp Ther 2003; 306(2): 772-7. [http://dx.doi.org/10.1124/jpet.103.050104] [PMID: 12734391]

[20] Kumar A, Sidhu J, Goyal A, Tsao JW, Svercauski J. Alzheimer Disease (Nursing). StatPearls 2021.

[21] Feldman H, Gauthier S, Hecker J, Vellas B, Subbiah P, Whalen E. A 24-week, randomized, double-blind study of donepezil in moderate to severe Alzheimer's disease. Neurology 2001; 57(4): 613-20. [http://dx.doi.org/10.1212/WNL.57.4.613] [PMID: 11524468]

[22] Seltzer B, Zolnouni P, Nunez M, *et al.* Efficacy of donepezil in early-stage Alzheimer disease: a randomized placebo-controlled trial. Arch Neurol 2004; 61(12): 1852-6. [http://dx.doi.org/10.1001/archneur.61.12.1852] [PMID: 15596605]

[23] Tariot PN. A randomized, double-blind, placebo-controlled study of the efficacy and safety of donepezil in patients with Alzheimer's disease in the nursing home setting. J Am Geriatr Soc 2001; 49(12): 1590-9. https://pubmed.ncbi.nlm.nih.gov/11843990/ [PMID: 11843990]

[24] Black S, Román GC, Geldmacher DS, *et al.* Efficacy and tolerability of donepezil in vascular dementia: positive results of a 24-week, multicenter, international, randomized, placebo-controlled clinical trial. Stroke 2003; 34(10): 2323-30. [http://dx.doi.org/10.1161/01.STR.0000091396.95360.E1] [PMID: 12970516]

[25] Wilkinson D, Doody R, Helme R, *et al.* Donepezil in vascular dementia: A randomized, placebo-controlled study. Neurology 2003; 61(4): 479-86. [http://dx.doi.org/10.1212/01.WNL.0000078943.50032.FC] [PMID: 12939421]

[26] Leroi I, Brandt J, Reich SG, *et al.* Randomized placebo-controlled trial of donepezil in cognitive impairment in Parkinson's disease. Int J Geriatr Psychiatry 2004; 19(1): 1-8. [http://dx.doi.org/10.1002/gps.993] [PMID: 14716693]

[27] Aarsland D, Laake K, Larsen JP, Janvin C. Donepezil for cognitive impairment in Parkinson's disease: a randomised controlled study. J Neurol Neurosurg Psychiatry 2002; 72(6): 708-12. [http://dx.doi.org/10.1136/jnnp.72.6.708] [PMID: 12023410]

[28] Shigeta M, Homma A. Donepezil for Alzheimer's disease: pharmacodynamic, pharmacokinetic, and clinical profiles. CNS Drug Rev 2001; 7(4): 353-68. [http://dx.doi.org/10.1111/j.1527-3458.2001.tb00204.x] [PMID: 11830754]

[29] Jia J, Wei C, Chen W, *et al.* Safety and Efficacy of Donepezil 10 mg/day in Patients with Mild to Moderate Alzheimer's Disease. J Alzheimers Dis 2020; 74(1): 199-211. [http://dx.doi.org/10.3233/JAD-190940] [PMID: 31985467]

[30] Solanto MV. Neuropsychopharmacological mechanisms of stimulant drug action in attention-deficit hyperactivity disorder: a review and integration. Behav Brain Res 1998; 94(1): 127-52. [http://dx.doi.org/10.1016/S0166-4328(97)00175-7] [PMID: 9708845]

[31] Challman TD, Lipsky JJ. Methylphenidate: its pharmacology and uses. Mayo Clin Proc 2000; 75(7): 711-21. [http://dx.doi.org/10.1016/S0025-6196(11)64618-1] [PMID: 10907387]

[32] Hurd YL, Ungerstedt U. *In vivo* neurochemical profile of dopamine uptake inhibitors and releasers in rat caudate-putamen. Eur J Pharmacol 1989; 166(2): 251-60. [http://dx.doi.org/10.1016/0014-2999(89)90066-6] [PMID: 2477259]

[33] Volkow ND, Wang GJ, Fowler JS, *et al.* Dopamine transporter occupancies in the human brain induced by therapeutic doses of oral methylphenidate. Am J Psychiatry 1998; 155(10): 1325-31. [http://dx.doi.org/10.1176/ajp.155.10.1325] [PMID: 9766762]

[34] Volkow ND, Ding YS, Fowler JS, *et al.* Is methylphenidate like cocaine? Studies on their pharmacokinetics and distribution in the human brain. Arch Gen Psychiatry 1995; 52(6): 456-63. [http://dx.doi.org/10.1001/archpsyc.1995.03950180042006] [PMID: 7771915]

[35] Ferris RM, Tang FL. Comparison of the effects of the isomers of amphetamine, methylphenidate and deoxypipradrol on the uptake of l-[3H]norepinephrine and [3H]dopamine by synaptic vesicles from rat whole brain, striatum and hypothalamus. J Pharmacol Exp Ther 1979; 210(3): 422-8. [PMID: 39160]

[36] Gatley SJ, Pan D, Chen R, Chaturvedi G, Ding YS. Affinities of methylphenidate derivatives for dopamine, norepinephrine and serotonin transporters. Life Sci 1996; 58(12): PL231-9. [http://dx.doi.org/10.1016/0024-3205(96)00052-5] [PMID: 8786705]

[37] Kadesjo B. ADHD in Children and Adults. *ADHD hos barn och vuxna Stockholm:*, (2002).

[38] Attention Deficit Hyperactivity Disorder: Diagnosis and Management of ADHD in Children, Young People and Adults - PubMed. Available at: https://pubmed.ncbi.nlm.nih.gov/22420012/. (Accessed: 26th May 2021).

[39] Storebø OJ, Pedersen N, Ramstad E, *et al.* Methylphenidate for attention deficit hyperactivity disorder (ADHD) in children and adolescents - assessment of adverse events in non-randomised studies. Cochrane Libr 2018; 5(5): CD012069. [http://dx.doi.org/10.1002/14651858.CD012069.pub2] [PMID: 29744873]

[40] Andrew BN, Guan NC, Jaafar NRN. The Use of Methylphenidate for Physical and Psychological Symptoms in Cancer Patients: A Review. Curr Drug Targets 2018; 19(8): 877-87. [http://dx.doi.org/10.2174/1389450118666170317162603] [PMID: 28322161]

[41] Meyers CA, Weitzner MA, Valentine AD, Levin VA. Methylphenidate therapy improves cognition, mood, and function of brain tumor patients. J Clin Oncol 1998; 16(7): 2522-7. [http://dx.doi.org/10.1200/JCO.1998.16.7.2522] [PMID: 9667273]

[42] DeLong R, Friedman H, Friedman N, Gustafson K, Oakes J. Methylphenidate in neuropsychological sequelae of radiotherapy and chemotherapy of childhood brain tumors and leukemia. J Child Neurol

1992; 7(4): 462-3.
[http://dx.doi.org/10.1177/088307389200700425] [PMID: 1469256]

[43] Gagnon B, Low G, Schreier G. Methylphenidate hydrochloride improves cognitive function in patients with advanced cancer and hypoactive delirium: a prospective clinical study. J Psychiatry Neurosci 2005; 30(2): 100-7.
[PMID: 15798785]

[44] Fernandez F, Adams F. Methylphenidate treatment of patients with head and neck cancer. Head Neck Surg 1986; 8(4): 296-300.
[http://dx.doi.org/10.1002/hed.2890080410] [PMID: 3744858]

[45] Weitzner MA, Meyers CA, Valentine AD. Methylphenidate in the treatment of neurobehavioral slowing associated with cancer and cancer treatment. J Neuropsychiatry Clin Neurosci 1995; 7(3): 347-50.
[http://dx.doi.org/10.1176/jnp.7.3.347] [PMID: 7580196]

[46] Masand PS, Tesar GE. Use of stimulants in the medically ill. Psychiatr Clin North Am 1996; 19(3): 515-47.
[http://dx.doi.org/10.1016/S0193-953X(05)70304-X] [PMID: 8856815]

[47] Sood A, Moynihan TJ. Cancer-related fatigue: An update. Curr Oncol Rep 2005; 7(4): 277-82.
[http://dx.doi.org/10.1007/s11912-005-0051-8] [PMID: 15946587]

[48] Sood A, Barton DL, Loprinzi CL. Use of methylphenidate in patients with cancer. Am J Hosp Palliat Care 2006; 23(1): 35-40.
[http://dx.doi.org/10.1177/104990910602300106] [PMID: 16450661]

[49] Mintzer J, Lanctôt KL, Scherer RW, *et al.* Effect of Methylphenidate on Apathy in Patients With Alzheimer Disease. JAMA Neurol 2021; 78(11): 1324-32.
[http://dx.doi.org/10.1001/jamaneurol.2021.3356] [PMID: 34570180]

[50] Kandiah N, Pai MC, Senanarong V, *et al.* Rivastigmine: the advantages of dual inhibition of acetylcholinesterase and butyrylcholinesterase and its role in subcortical vascular dementia and Parkinson's disease dementia. Clin Interv Aging 2017; 12: 697-707.
[http://dx.doi.org/10.2147/CIA.S129145] [PMID: 28458525]

[51] https://www.ncbi.nlm.nih.gov/books/NBK557438/

[52] Henderson EJ, Lord SR, Brodie MA, *et al.* Rivastigmine for gait stability in patients with Parkinson's disease (ReSPonD): a randomised, double-blind, placebo-controlled, phase 2 trial. Lancet Neurol 2016; 15(3): 249-58.
[http://dx.doi.org/10.1016/S1474-4422(15)00389-0] [PMID: 26795874]

[53] Youn YC, Shin HW, Choi BS, Kim S, Lee JY, Ha YC. Rivastigmine patch reduces the incidence of postoperative delirium in older patients with cognitive impairment. Int J Geriatr Psychiatry 2017; 32(10): 1079-84.
[http://dx.doi.org/10.1002/gps.4569] [PMID: 27561376]

[54] Cotter J, Muhlert N, Talwar A, Granger K. Examining the effectiveness of acetylcholinesterase inhibitors and stimulant-based medications for cognitive dysfunction in multiple sclerosis: A systematic review and meta-analysis. Neurosci Biobehav Rev 2018; 86: 99-107.
[http://dx.doi.org/10.1016/j.neubiorev.2018.01.006] [PMID: 29406017]

[55] Hershey LA, Coleman-Jackson R. Pharmacological Management of Dementia with Lewy Bodies. Drugs Aging 2019; 36(4): 309-19.
[http://dx.doi.org/10.1007/s40266-018-00636-7] [PMID: 30680679]

[56] Keks NA, Hope J, Keogh S. Suvorexant: scientifically interesting, utility uncertain. Australas Psychiatry 2017; 25(6): 622-4.
[http://dx.doi.org/10.1177/1039856217734677] [PMID: 28994603]

[57] Cox CD, Breslin MJ, Whitman DB, *et al.* Discovery of the Dual Orexin Receptor Antagonist [(7 *R*)--

-(5-Chloro-1,3-benzoxazol-2-yl)-7-methyl-1,4-diazepan-1-yl][5-methyl-2-(2 *H* -1,2,3-triazo--2-yl)phenyl]methanone (MK-4305) for the Treatment of Insomnia. J Med Chem 2010; 53(14): 5320-32.
[http://dx.doi.org/10.1021/jm100541c] [PMID: 20565075]

[58] Mieda M, Sakurai T. Orexin (hypocretin) receptor agonists and antagonists for treatment of sleep disorders. Rationale for development and current status. CNS Drugs 2013; 27(2): 83-90.
[http://dx.doi.org/10.1007/s40263-012-0036-8] [PMID: 23359095]

[59] Adams AD, Pepin MJ, Brown JN. The role of suvorexant in the prevention of delirium during acute hospitalization: A systematic review. J Crit Care 2020; 59: 1-5.
[http://dx.doi.org/10.1016/j.jcrc.2020.05.006] [PMID: 32480359]

[60] Kawada K, Ohta T, Tanaka K, Miyamura M, Tanaka S. Addition of Suvorexant to Ramelteon Therapy for Improved Sleep Quality with Reduced Delirium Risk in Acute Stroke Patients. J Stroke Cerebrovasc Dis 2019; 28(1): 142-8.
[http://dx.doi.org/10.1016/j.jstrokecerebrovasdis.2018.09.024] [PMID: 30322756]

[61] Herring WJ, Ceesay P, Snyder E, *et al.* Polysomnographic assessment of suvorexant in patients with probable Alzheimer's disease dementia and insomnia: a randomized trial. Alzheimers Dement 2020; 16(3): 541-51.
[http://dx.doi.org/10.1002/alz.12035] [PMID: 31944580]

[62] Sharma MK, Kainth S, Kumar S, *et al.* Effects of zolpidem on sleep parameters in patients with cirrhosis and sleep disturbances: A randomized, placebo-controlled trial. Clin Mol Hepatol 2019; 25(2): 199-209.
[http://dx.doi.org/10.3350/cmh.2018.0084] [PMID: 30856689]

[63] Kim H M, *et al.* Predictors of long-term and high-dose use of zolpidem in veterans. J Clin Psychiatry 2019 Feb 5; 80(2): 18m12149.
[http://dx.doi.org/18m12149]

[64] Bjurström MF, Irwin MR. Perioperative Pharmacological Sleep-Promotion and Pain Control: A Systematic Review. Pain Pract 2019; 19(5): 552-69.
[http://dx.doi.org/10.1111/papr.12776] [PMID: 30762974]

[65] Dang A, Garg A, Rataboli PV. Role of zolpidem in the management of insomnia. CNS Neurosci Ther 2011; 17(5): 387-97.
[http://dx.doi.org/10.1111/j.1755-5949.2010.00158.x] [PMID: 20553305]

[66] Neumann E, Rudolph U, Knutson DE, *et al.* Zolpidem activation of alpha 1-containing GABAA receptors selectively inhibits high frequency action potential firing of cortical neurons. Front Pharmacol 2019; 9: 1523.
[http://dx.doi.org/10.3389/fphar.2018.01523] [PMID: 30687091]

[67] Nigam G, Camacho M, Riaz M. The effect of nonbenzodiazepines sedative hypnotics on apnea–hypopnea index: A meta-analysis. Ann Thorac Med 2019; 14(1): 49-55.
[http://dx.doi.org/10.4103/atm.ATM_198_18] [PMID: 30745935]

[68] Greenblatt DJ, Roth T. Zolpidem for insomnia. Expert Opin Pharmacother 2012; 13(6): 879-93.
[http://dx.doi.org/10.1517/14656566.2012.667074] [PMID: 22424586]

[69] Swainston Harrison T, Keating GM. Zolpidem. CNS Drugs 2005; 19(1): 65-89.
[http://dx.doi.org/10.2165/00023210-200519010-00008] [PMID: 15651908]

[70] Zolpidem - StatPearls - NCBI Bookshelf. Available at: https://www.ncbi.nlm.nih.gov/books/NBK442008/. (Accessed: 26th May 2021)

[71] Buscemi N, Vandermeer B, Friesen C, *et al.* Manifestations and management of chronic insomnia in adults. Evid Rep Technol Assess (Summ) 2005; (125): 1-10.
[PMID: 15989374]

[72] Bomalaski MN, Claflin ES, Townsend W, Peterson MD. Zolpidem for the treatment of neurologic

disorders: A systematic review. JAMA Neurol 2017; 74(9): 1130-9.
[http://dx.doi.org/10.1001/jamaneurol.2017.1133] [PMID: 28655027]

[73] Noormandi A, Shahrokhi M, Khalili H. Potential benefits of zolpidem in disorders of consciousness. Expert Rev Clin Pharmacol 2017; 10(9): 983-92.
[http://dx.doi.org/10.1080/17512433.2017.1347502] [PMID: 28649875]

[74] Wadworth AN, McTavish D. Zopiclone. Drugs Aging 1993; 3(5): 441-59.
[http://dx.doi.org/10.2165/00002512-199303050-00006] [PMID: 8241608]

[75] Döble A, Canton T, Malgouris C, *et al.* The mechanism of action of zopiclone. Eur Psychiatry 1995; 10(S3) (Suppl. 3): 117s-28s.
[http://dx.doi.org/10.1016/0924-9338(96)80093-9] [PMID: 19698408]

[76] Hesse LM, von Moltke LL, Greenblatt DJ. Clinically important drug interactions with zopiclone, zolpidem and zaleplon. CNS Drugs 2003; 17(7): 513-32.
[http://dx.doi.org/10.2165/00023210-200317070-00004] [PMID: 12751920]

[77] Holmedahl NH, Øverland B, Fondenes O, Ellingsen I, Hardie JA. Zopiclone effects on breathing at sleep in stable chronic obstructive pulmonary disease. Sleep Breath 2015; 19(3): 921-30.
[http://dx.doi.org/10.1007/s11325-014-1084-8] [PMID: 25501294]

[78] Pinto LR Jr, Bittencourt LRA, Treptow EC, Braga LR, Tufik S. Eszopiclone *versus* zopiclone in the treatment of insomnia. Clinics (São Paulo) 2016; 71(1): 5-9.
[http://dx.doi.org/10.6061/clinics/2016(01)02] [PMID: 26872077]

[79] Ponciano E, Freitas F, Camara J, Faria M, Barreto M, Hindmarch I. A comparison of the efficacy, tolerance and residual effects of zopiclone, flurazepam and placebo in insomniac outpatients. Int Clin Psychopharmacol 1990; 5 (Suppl. 2): 69-77.
[PMID: 2201731]

[80] Louzada LL, Machado FV, Quintas JL, *et al.* The efficacy and safety of zolpidem and zopiclone to treat insomnia in Alzheimer's disease: a randomized, triple-blind, placebo-controlled trial. Neuropsychopharmacology 2022; 47(2): 570-9.
[http://dx.doi.org/10.1038/s41386-021-01191-3] [PMID: 34635802]

[81] Axsome Therapeutics receives FDA fast track designation for axs-05 for alzheimer's disease agitation.

[82] Lopez Lopez C, Tariot PN, Caputo A, *et al.* The Alzheimer's Prevention Initiative Generation Program: Study design of two randomized controlled trials for individuals at risk for clinical onset of Alzheimer's disease. Alzheimers Dement (N Y) 2019; 5(1): 216-27.
[http://dx.doi.org/10.1016/j.trci.2019.02.005] [PMID: 31211217]

[83] Lopez Lopez C, Caputo A, Liu F, *et al.* The Alzheimer's Prevention Initiative Generation Program: Evaluating CNP520 Efficacy in the Prevention of Alzheimer's Disease. J Prev Alzheimers Dis 2017; 4(4): 242-6.
[PMID: 29181489]

[84] Savica R, Grossardt BR, Bower JH, Ahlskog JE, Mielke MM, Rocca WA. Incidence and time trends of drug-induced parkinsonism: A 30-year population-based study. Mov Disord 2017; 32(2): 227-34.
[http://dx.doi.org/10.1002/mds.26839] [PMID: 27779780]

[85] Jankovic J, Tan EK. Parkinson's disease: etiopathogenesis and treatment. J Neurol Neurosurg Psychiatry 2020; 91(8): 795-808.
[http://dx.doi.org/10.1136/jnnp-2019-322338] [PMID: 32576618]

[86] Postuma RB, Berg D, Stern M, *et al.* MDS clinical diagnostic criteria for Parkinson's disease. Mov Disord 2015; 30(12): 1591-601.
[http://dx.doi.org/10.1002/mds.26424] [PMID: 26474316]

[87] Dorsey ER, Elbaz A, Nichols E, *et al.* Global, regional, and national burden of Parkinson's disease, 1990–2016: a systematic analysis for the Global Burden of Disease Study 2016. Lancet Neurol 2018; 17(11): 939-53.

[http://dx.doi.org/10.1016/S1474-4422(18)30295-3] [PMID: 30287051]

[88] Singleton AB, Farrer MJ, Bonifati V. The genetics of Parkinson's disease: Progress and therapeutic implications. Mov Disord 2013; 28(1): 14-23. [http://dx.doi.org/10.1002/mds.25249] [PMID: 23389780]

[89] Parkinson's disease in adults NICE guideline. (2017).

[90] Fox SH, Katzenschlager R, Lim SY, *et al.* International Parkinson and movement disorder society evidence-based medicine review: Update on treatments for the motor symptoms of Parkinson's disease. Mov Disord 2018; 33(8): 1248-66. [http://dx.doi.org/10.1002/mds.27372] [PMID: 29570866]

[91] Szökő É, Tábi T, Riederer P, Vécsei L, Magyar K. Pharmacological aspects of the neuroprotective effects of irreversible MAO-B inhibitors, selegiline and rasagiline, in Parkinson's disease. J Neural Transm (Vienna) 2018; 125(11): 1735-49. [http://dx.doi.org/10.1007/s00702-018-1853-9] [PMID: 29417334]

[92] Li C, Xue L, Liu Y, Yang Z, Chi S, Xie A. Zonisamide for the Treatment of Parkinson Disease: A Current Update. Front Neurosci 2020; 14: 574652. [http://dx.doi.org/10.3389/fnins.2020.574652] [PMID: 33408605]

[93] Seppi K, Ray Chaudhuri K, Coelho M, *et al.* Update on treatments for nonmotor symptoms of Parkinson's disease—an evidence-based medicine review. Mov Disord 2019; 34(2): 180-98. [http://dx.doi.org/10.1002/mds.27602] [PMID: 30653247]

[94] Mabley JG, Pacher P, Liaudet L, *et al.* Inosine reduces inflammation and improves survival in a murine model of colitis. Am J Physiol Gastrointest Liver Physiol 2003; 284(1): G138-44. [http://dx.doi.org/10.1152/ajpgi.00060.2002] [PMID: 12388199]

[95] Haskó G, Sitkovsky MV, Szabó C. Immunomodulatory and neuroprotective effects of inosine. Trends Pharmacol Sci 2004; 25(3): 152-7. [http://dx.doi.org/10.1016/j.tips.2004.01.006] [PMID: 15019271]

[96] Bluett B, Togasaki DM, Mihaila D, *et al.* Effect of Urate-Elevating Inosine on Early Parkinson Disease Progression. JAMA 2021; 326(10): 926-39. [http://dx.doi.org/10.1001/jama.2021.10207] [PMID: 34519802]

[97] Teixeira FC, Gutierres JM, Soares MSP, *et al.* Inosine protects against impairment of memory induced by experimental model of Alzheimer disease: a nucleoside with multitarget brain actions. Psychopharmacology (Berl) 2020; 237(3): 811-23. [http://dx.doi.org/10.1007/s00213-019-05419-5] [PMID: 31834453]

[98] Nicholson K, Chan J, Macklin EA, *et al.* Pilot trial of inosine to elevate urate levels in amyotrophic lateral sclerosis. Ann Clin Transl Neurol 2018; 5(12): 1522-33. [http://dx.doi.org/10.1002/acn3.671] [PMID: 30564619]

[99] Schmidt WJ, Lebsanft H, Heindl M, *et al.* Continuous *versus* pulsatile administration of rotigotine in 6-OHDA-lesioned rats: contralateral rotations and abnormal involuntary movements. J Neural Transm (Vienna) 2008; 115(10): 1385-92. [http://dx.doi.org/10.1007/s00702-008-0102-z] [PMID: 18726139]

[100] Li K, Zhang Y, Tian E, Liu Z, Wang T, Fu F. The Effect of Rotigotine Extended-Release Microspheres Alone or With Celecoxib on the Inflammatory Pain. Front Pharmacol 2020; 11: 594387. [http://dx.doi.org/10.3389/fphar.2020.594387] [PMID: 33192533]

[101] Woitalla D, Dunac A, Safavi A, *et al.* A noninterventional study evaluating the effectiveness of rotigotine and levodopa combination therapy in younger *versus* older patients with Parkinson's disease. Expert Opin Pharmacother 2018; 19(9): 937-45. [http://dx.doi.org/10.1080/14656566.2018.1480721] [PMID: 29916262]

[102] Soogrim V, Ruberto VL, Murrough J, Jha MK. Spotlight on pimavanserin tartrate and its therapeutic potential in the treatment of major depressive disorder: The evidence to date. Drug Des Devel Ther

2021; 15: 151-7.
[http://dx.doi.org/10.2147/DDDT.S240862] [PMID: 33469267]

[103] Kianirad Y, Simuni T. Pimavanserin, a novel antipsychotic for management of Parkinson's disease psychosis. Expert Rev Clin Pharmacol 2017; 10(11): 1161-8.
[http://dx.doi.org/10.1080/17512433.2017.1369405] [PMID: 28817967]

[104] Ballard C, Banister C, Khan Z, *et al.* Evaluation of the safety, tolerability, and efficacy of pimavanserin *versus* placebo in patients with Alzheimer's disease psychosis: a phase 2, randomised, placebo-controlled, double-blind study. Lancet Neurol 2018; 17(3): 213-22.
[http://dx.doi.org/10.1016/S1474-4422(18)30039-5] [PMID: 29452684]

[105] Kaufmann H, Norcliffe-Kaufmann L, Palma JA. Droxidopa in neurogenic orthostatic hypotension. Expert Rev Cardiovasc Ther 2015; 13(8): 875-91.
[http://dx.doi.org/10.1586/14779072.2015.1057504] [PMID: 26092297]

[106] Hauser RA, Heritier S, Rowse GJ, Hewitt LA, Isaacson SH. Droxidopa and Reduced Falls in a Trial of Parkinson Disease Patients With Neurogenic Orthostatic Hypotension. Clin Neuropharmacol 2016; 39(5): 220-6.
[http://dx.doi.org/10.1097/WNF.0000000000000168] [PMID: 27332626]

[107] Ruzieh M, Dasa O, Pacenta A, Karabin B, Grubb B. Droxidopa in the treatment of postural orthostatic tachycardia syndrome. Am J Ther 2017; 24(2): e157-61.
[http://dx.doi.org/10.1097/MJT.0000000000000468] [PMID: 27563801]

[108] Dooley M, Markham A. Pramipexole. Drugs Aging 1998; 12(6): 495-514.
[http://dx.doi.org/10.2165/00002512-199812060-00007] [PMID: 9638397]

[109] Pramipexole - StatPearls - NCBI Bookshelf. Available at: https://www.ncbi.nlm.nih.gov/books/NBK557539/

[110] Jiang DQ, Jiang LL, Wang Y, Li MX. The role of pramipexole in the treatment of patients with depression and Parkinson's disease: A meta-analysis of randomized controlled trials. Asian J Psychiatr 2021; 61: 102691.
[http://dx.doi.org/10.1016/j.ajp.2021.102691] [PMID: 33992852]

[111] Tundo A, Filippis R, De Crescenzo F. Pramipexole in the treatment of unipolar and bipolar depression. A systematic review and meta-analysis. Acta Psychiatr Scand 2019; 140(2): 116-25.
[http://dx.doi.org/10.1111/acps.13055] [PMID: 31111467]

[112] Ettcheto M, Busquets O, Sánchez-Lopez E, *et al.* The preclinical discovery and development of opicapone for the treatment of Parkinson's disease. Expert Opin Drug Discov 2020; 15(9): 993-1003.
[http://dx.doi.org/10.1080/17460441.2020.1767580] [PMID: 32450711]

[113] Fabbri M, Ferreira JJ, Lees A, *et al.* Opicapone for the treatment of Parkinson's disease: A review of a new licensed medicine. Mov Disord 2018; 33(10): 1528-39.
[http://dx.doi.org/10.1002/mds.27475] [PMID: 30264443]

[114] Teixeira FG, Gago MF, Marques P, *et al.* Safinamide: a new hope for Parkinson's disease? Drug Discov Today 2018; 23(3): 736-44.
[http://dx.doi.org/10.1016/j.drudis.2018.01.033] [PMID: 29339106]

[115] Caccia C, Maj R, Calabresi M, *et al.* Safinamide: From molecular targets to a new anti-Parkinson drug. Neurology 2006; 67(7, Supplement 2) (Suppl. 2): S18-23.
[http://dx.doi.org/10.1212/WNL.67.7_suppl_2.S18] [PMID: 17030736]

[116] Deeks ED. Safinamide: first global approval. Drugs 2015; 75(6): 705-11.
[http://dx.doi.org/10.1007/s40265-015-0389-7] [PMID: 25851099]

[117] Borgohain R, Szasz J, Stanzione P, *et al.* Randomized trial of safinamide add-on to levodopa in Parkinson's disease with motor fluctuations. Mov Disord 2014; 29(2): 229-37.
[http://dx.doi.org/10.1002/mds.25751] [PMID: 24323641]

[118] Choi JH, Lee JY, Cho JW, *et al.* Double-Blind, Randomized, Placebo-Controlled Trial of DA -9701 in Parkinson's Disease: PASS - GI Study. Mov Disord 2020; 35(11): 1966-76. [http://dx.doi.org/10.1002/mds.28219] [PMID: 32761955]

[119] Xu T, Sun R, Wei G, Kong S. The Protective Effect of Safinamide in Ischemic Stroke Mice and a Brain Endothelial Cell Line. Neurotox Res 2020; 38(3): 733-40. [http://dx.doi.org/10.1007/s12640-020-00246-5] [PMID: 32613602]

[120] Bautista-Aguilera OM, Esteban G, Bolea I, *et al.* Design, synthesis, pharmacological evaluation, QSAR analysis, molecular modeling and ADMET of novel donepezil–indolyl hybrids as multipotent cholinesterase/monoamine oxidase inhibitors for the potential treatment of Alzheimer's disease. Eur J Med Chem 2014; 75: 82-95. [http://dx.doi.org/10.1016/j.ejmech.2013.12.028] [PMID: 24530494]

[121] Kwon YS, Son M. DA-9701: A new multi-acting drug for the treatment of functional dyspepsia. Biomol Ther (Seoul) 2013; 21(3): 181-9. [http://dx.doi.org/10.4062/biomolther.2012.096] [PMID: 24265862]

[122] Durães F, Pinto M, Sousa E. Old drugs as new treatments for neurodegenerative diseases. Pharmaceuticals (Basel) 2018; 11(2): 44. [http://dx.doi.org/10.3390/ph11020044] [PMID: 29751602]

[123] Newman DJ, Cragg GM. Natural Products as Sources of New Drugs from 1981 to 2014. J Nat Prod 2016; 79(3): 629-61. [http://dx.doi.org/10.1021/acs.jnatprod.5b01055] [PMID: 26852623]

[124] Khan RA. Natural products chemistry: The emerging trends and prospective goals. Saudi Pharm J 2018; 26(5): 739-53. [http://dx.doi.org/10.1016/j.jsps.2018.02.015] [PMID: 29991919]

[125] Alves C, Silva J, Pinteus S, *et al.* From marine origin to therapeutics: The antitumor potential of marine algae-derived compounds. Front Pharmacol 2018; 9: 777. [http://dx.doi.org/10.3389/fphar.2018.00777] [PMID: 30127738]

[126] Dias DA, Urban S, Roessner U. A historical overview of natural products in drug discovery. Metabolites 2012; 2(2): 303-36. [http://dx.doi.org/10.3390/metabo2020303] [PMID: 24957513]

[127] Atanasov AG, Zotchev SB, Dirsch VM, Supuran CT. Natural products in drug discovery: advances and opportunities. Nat Rev Drug Discov 2021; 20(3): 200-16. [http://dx.doi.org/10.1038/s41573-020-00114-z] [PMID: 33510482]

[128] Xie A, Gao J, Xu L, Meng D. Shared mechanisms of neurodegeneration in Alzheimer's disease and Parkinson's disease. BioMed Res Int 2014; 2014: 1-8. [http://dx.doi.org/10.1155/2014/648740] [PMID: 24900975]

[129] Habtemariam S. Natural products in Alzheimer's disease therapy: Would old therapeutic approaches fix the broken promise of modern medicines? Molecules 2019; 24(8): 1519. [http://dx.doi.org/10.3390/molecules24081519] [PMID: 30999702]

[130] Li J, Larregieu CA, Benet LZ. Classification of natural products as sources of drugs according to the biopharmaceutics drug disposition classification system (BDDCS). Chin J Nat Med 2016; 14(12): 888-97. [http://dx.doi.org/10.1016/S1875-5364(17)30013-4] [PMID: 28262115]

[131] Zanni R, Garcia-Domenech R, Galvez-Llompart M, Galvez J. Alzheimer: A Decade of Drug Design. Why Molecular Topology can be an Extra Edge? Curr Neuropharmacol 2018; 16(6): 849-64. [http://dx.doi.org/10.2174/1570159X15666171129102042] [PMID: 29189164]

[132] Caraci F, Castellano S, Salomone S, Drago F, Bosco P, Nuovo S. Searching for disease-modifying drugs in AD: can we combine neuropsychological tools with biological markers? CNS Neurol Disord Drug Targets 2014; 13(1): 173-86.

[http://dx.doi.org/10.2174/1871527311312999010 3] [PMID: 24040795]

[133] Troncoso-Escudero P, Sepulveda D, Pérez-Arancibia R, *et al.* On the Right Track to Treat Movement Disorders: Promising Therapeutic Approaches for Parkinson's and Huntington's Disease. Front Aging Neurosci 2020; 12: 571185.
[http://dx.doi.org/10.3389/fnagi.2020.571185] [PMID: 33101007]

[134] Hampel H, Caraci F, Cuello AC, *et al.* A Path Toward Precision Medicine for Neuroinflammatory Mechanisms in Alzheimer's Disease. Front Immunol 2020; 11: 456.
[http://dx.doi.org/10.3389/fimmu.2020.00456] [PMID: 32296418]

[135] Carpenter KA, Huang X. Machine Learning-based Virtual Screening and Its Applications to Alzheimer's Drug Discovery: A Review. Curr Pharm Des 2018; 24(28): 3347-58.
[http://dx.doi.org/10.2174/1381612824666180607124038] [PMID: 29879881]

[136] Basile L. Virtual screening in the search of new and potent anti-alzheimer agents.Neuromethods132, 107–137. Humana Press Inc. 2018.
[http://dx.doi.org/10.1007/978-1-4939-7404-7_4]

[137] Van Bulck M, Sierra-Magro A, Alarcon-Gil J, Perez-Castillo A, Morales-Garcia J. Novel approaches for the treatment of alzheimer's and parkinson's disease. Int J Mol Sci 2019; 20(3): 719.
[http://dx.doi.org/10.3390/ijms20030719] [PMID: 30743990]

[138] Sonmez F, Zengin Kurt B, Gazioglu I, *et al.* Design, synthesis and docking study of novel coumarin ligands as potential selective acetylcholinesterase inhibitors. J Enzyme Inhib Med Chem 2017; 32(1): 285-97.
[http://dx.doi.org/10.1080/14756366.2016.1250753] [PMID: 28097911]

[139] de Almeida JR, Figueiro M, Almeida WP, de Paula da Silva CHT. Discovery of novel dual acetylcholinesterase inhibitors with antifibrillogenic activity related to Alzheimer's disease. Future Med Chem 2018; 10(9): 1037-53.
[http://dx.doi.org/10.4155/fmc-2017-0201] [PMID: 29676170]

[140] Marco-Contelles J, Unzeta M, Bolea I, *et al.* ASS234, As a New Multi-Target Directed Propargylamine for Alzheimer's Disease Therapy. Front Neurosci 2016; 10: 294.
[http://dx.doi.org/10.3389/fnins.2016.00294] [PMID: 27445665]

[141] Serrano MP, Herrero-Labrador R, Futch HS, *et al.* The proof-of-concept of ASS234: Peripherally administered ASS234 enters the central nervous system and reduces pathology in a male mouse model of Alzheimer disease. J Psychiatry Neurosci 2017; 42(1): 59-69.
[http://dx.doi.org/10.1503/jpn.150209] [PMID: 27636528]

[142] Wang J, Land D, Ono K, *et al.* Molecular topology as novel strategy for discovery of drugs with aβ lowering and anti-aggregation dual activities for Alzheimer's disease. PLoS One 2014; 9(3): e92750.
[http://dx.doi.org/10.1371/journal.pone.0092750] [PMID: 24671215]

[143] Hong W, Wang Z, Liu W, *et al.* Diffusible, highly bioactive oligomers represent a critical minority of soluble Aβ in Alzheimer's disease brain. Acta Neuropathol 2018; 136(1): 19-40.
[http://dx.doi.org/10.1007/s00401-018-1846-7] [PMID: 29687257]

[144] Sideris DI, Danial JSH, Emin D, *et al.* Soluble amyloid beta-containing aggregates are present throughout the brain at early stages of Alzheimer's disease. Brain Commun 2021; 3(3): fcab147.
[http://dx.doi.org/10.1093/braincomms/fcab147] [PMID: 34396107]

[145] Lin CH, Hsieh YS, Wu YR, *et al.* Identifying GSK-3β kinase inhibitors of Alzheimer's disease: Virtual screening, enzyme, and cell assays. Eur J Pharm Sci 2016; 89: 11-9.
[http://dx.doi.org/10.1016/j.ejps.2016.04.012] [PMID: 27094783]

[146] Martin L, Latypova X, Wilson CM, *et al.* Tau protein kinases: Involvement in Alzheimer's disease. Ageing Res Rev 2013; 12(1): 289-309.
[http://dx.doi.org/10.1016/j.arr.2012.06.003] [PMID: 22742992]

[147] Oizumi H, Yamasaki K, Suzuki H, *et al.* Fatty Acid-Binding Protein 3 Expression in the Brain and

Skin in Human Synucleinopathies. Front Aging Neurosci 2021; 13: 648982.
[http://dx.doi.org/10.3389/fnagi.2021.648982] [PMID: 33841128]

[148] Cheng A, Wang Y, Shinoda Y, *et al.* Fatty acid-binding protein 7 triggers α-synuclein oligomerization in glial cells and oligodendrocytes associated with oxidative stress. Acta Pharmacol Sin 2022; 43(3): 552-62.
[http://dx.doi.org/10.1038/s41401-021-00675-8] [PMID: 33935286]

[149] Kawahata I, Bousset L, Melki R, Fukunaga K. Fatty Acid-Binding Protein 3 is Critical for α-Synuclein Uptake and MPP^+-Induced Mitochondrial Dysfunction in Cultured Dopaminergic Neurons. Int J Mol Sci 2019; 20(21): 5358.
[http://dx.doi.org/10.3390/ijms20215358] [PMID: 31661838]

[150] Parsons CG. CNS repurposing - Potential new uses for old drugs: Examples of screens for Alzheimer's disease, Parkinson's disease and spasticity. Neuropharmacology 2019; 147: 4-10.
[http://dx.doi.org/10.1016/j.neuropharm.2018.08.027] [PMID: 30165077]

[151] Ameen AM, Elkazaz AY, Mohammad HMF, Barakat BM. Anti-inflammatory and neuroprotective activity of boswellic acids in rotenone parkinsonian rats. Can J Physiol Pharmacol 2017; 95(7): 819-29.
[http://dx.doi.org/10.1139/cjpp-2016-0158] [PMID: 28249117]

[152] Kim SM, Park YJ, Shin MS, *et al.* Acacetin inhibits neuronal cell death induced by 6-hydroxydopamine in cellular Parkinson's disease model. Bioorg Med Chem Lett 2017; 27(23): 5207-12.
[http://dx.doi.org/10.1016/j.bmcl.2017.10.048] [PMID: 29089232]

[153] Derf A, Sharma A, Bharate SB, Chaudhuri B. Aegeline, a natural product from the plant Aegle marmelos, mimics the yeast SNARE protein Sec22p in suppressing α-synuclein and Bax toxicity in yeast. Bioorg Med Chem Lett 2019; 29(3): 454-60.
[http://dx.doi.org/10.1016/j.bmcl.2018.12.028] [PMID: 30579794]

[154] Nataraj J, Manivasagam T, Justin Thenmozhi A, Essa MM. Neurotrophic Effect of Asiatic acid, a Triterpene of Centella asiatica Against Chronic 1-Methyl 4-Phenyl 1, 2, 3, 6-Tetrahydropyridine Hydrochloride/Probenecid Mouse Model of Parkinson's disease: The Role of MAPK, PI3K-Ak--GSK3β and mTOR Signalling Pathways. Neurochem Res 2017; 42(5): 1354-65.
[http://dx.doi.org/10.1007/s11064-017-2183-2] [PMID: 28181071]

[155] Ren Z, Zhao Y, Cao T, Zhen X. Dihydromyricetin protects neurons in an MPTP-induced model of Parkinson's disease by suppressing glycogen synthase kinase-3 beta activity. Acta Pharmacol Sin 2016; 37(10): 1315-24.
[http://dx.doi.org/10.1038/aps.2016.42] [PMID: 27374489]

[156] Ojha S, Javed H, Azimullah S, Haque ME. β-Caryophyllene, a phytocannabinoid attenuates oxidative stress, neuroinflammation, glial activation, and salvages dopaminergic neurons in a rat model of Parkinson disease. Mol Cell Biochem 2016; 418(1-2): 59-70.
[http://dx.doi.org/10.1007/s11010-016-2733-y] [PMID: 27316720]

[157] Hu J, Ferchmin PA, Hemmerle AM, Seroogy KB, Eterovic VA, Hao J. 4R-Cembranoid Improves Outcomes after 6-Hydroxydopamine Challenge in Both *In vitro* and *In vivo* Models of Parkinson's Disease. Front Neurosci 2017; 11: 272.
[http://dx.doi.org/10.3389/fnins.2017.00272] [PMID: 28611572]

[158] Alvariño R, Alonso E, Tribalat MA, Gegunde S, Thomas OP, Botana LM. Evaluation of the Protective Effects of Sarains on H_2O_2-Induced Mitochondrial Dysfunction and Oxidative Stress in SH-SY5Y Neuroblastoma Cells. Neurotox Res 2017; 32(3): 368-80.
[http://dx.doi.org/10.1007/s12640-017-9748-3] [PMID: 28478531]

[159] Leirós M, Alonso E, Rateb ME, *et al.* Bromoalkaloids protect primary cortical neurons from induced oxidative stress. ACS Chem Neurosci 2015; 6(2): 331-8.
[http://dx.doi.org/10.1021/cn500258c] [PMID: 25387680]

[160] Boccitto M, Lee N, Sakamoto S, *et al.* The neuroprotective marine compound psammaplysene A binds the RNA-binding protein HNRNPK. Mar Drugs 2017; 15(8): 246. [http://dx.doi.org/10.3390/md15080246] [PMID: 28783126]

[161] Wang D, Feng Y, Murtaza M, *et al.* A Grand Challenge: Unbiased Phenotypic Function of Metabolites from *Jaspis splendens* against Parkinson's Disease. J Nat Prod 2016; 79(2): 353-61. [http://dx.doi.org/10.1021/acs.jnatprod.5b00987] [PMID: 26883470]

[162] Kao CJ, Chen WF, Guo BL, *et al.* The 1-tosylpentan-3-one protects against 6-hydroxydopamin--induced neurotoxicity. Int J Mol Sci 2017; 18(5): 1096. [http://dx.doi.org/10.3390/ijms18051096] [PMID: 28534853]

[163] Wang X, Cong P, Liu Y, *et al.* Neuritogenic effect of sea cucumber glucocerebrosides on NGF-induced PC12 cells *via* activation of the TrkA/CREB/BDNF signalling pathway. J Funct Foods 2018; 46: 175-84. [http://dx.doi.org/10.1016/j.jff.2018.04.035]

[164] Lee HW, Choi H, Nam SJ, Fenical W, Kim H. Potent Inhibition of Monoamine Oxidase B by a Piloquinone from Marine-Derived *Streptomyces* sp. CNQ-027. J Microbiol Biotechnol 2017; 27(4): 785-90. [http://dx.doi.org/10.4014/jmb.1612.12025] [PMID: 28068665]

[165] Wang W, Song N, Jia F, Xie J, Zhang Q, Jiang H. Neuroprotective effects of porphyran derivatives against 6-hydroxydopamine-induced cytotoxicity is independent on mitochondria restoration. Ann Transl Med 2015; 3(3): 39. [PMID: 25815300]

[166] Souza RB, Frota AF, Silva J, *et al. In vitro* activities of kappa-carrageenan isolated from red marine alga Hypnea musciformis: Antimicrobial, anticancer and neuroprotective potential. Int J Biol Macromol 2018; 112: 1248-56. [http://dx.doi.org/10.1016/j.ijbiomac.2018.02.029] [PMID: 29427681]

[167] Souza RB, Frota AF, Sousa RS, *et al.* Neuroprotective Effects of Sulphated Agaran from Marine Alga *Gracilaria cornea* in Rat 6-Hydroxydopamine Parkinson's Disease Model: Behavioural, Neurochemical and Transcriptional Alterations. Basic Clin Pharmacol Toxicol 2017; 120(2): 159-70. [http://dx.doi.org/10.1111/bcpt.12669] [PMID: 27612165]

[168] Silva J, Alves C, Martins A, *et al.* Loliolide, a new therapeutic option for neurological diseases? *In vitro* neuroprotective and anti-inflammatory activities of a monoterpenoid lactone isolated from codium tomentosum. Int J Mol Sci 2021; 22(4): 1888. [http://dx.doi.org/10.3390/ijms22041888] [PMID: 33672866]

[169] Silva J, Alves C, Pinteus S, *et al.* Disclosing the potential of eleganolone for Parkinson's disease therapeutics: Neuroprotective and anti-inflammatory activities. Pharmacol Res 2021; 168: 105589. [http://dx.doi.org/10.1016/j.phrs.2021.105589] [PMID: 33812007]

[170] Ye Q, Wang W, Hao C, Mao X. Agaropentaose protects SH-SY5Y cells against 6-hydroxydopamin--induced neurotoxicity through modulating NF-κB and p38MAPK signaling pathways. J Funct Foods 2019; 57: 222-32. [http://dx.doi.org/10.1016/j.jff.2019.04.017]

[171] Shen Y, Zhang B, Pang X, *et al.* Network Pharmacology-Based Analysis of Xiao-Xu-Ming Decoction on the Treatment of Alzheimer's Disease. Front Pharmacol 2020; 11: 595254. [http://dx.doi.org/10.3389/fphar.2020.595254] [PMID: 33390981]

[172] Liu P, Chen X, Zhou H, *et al.* The isoquinoline alkaloid dauricine targets multiple molecular pathways to ameliorate Alzheimer-like pathological changes *in vitro*. Oxid Med Cell Longev 2018; 2018: 1-19. [http://dx.doi.org/10.1155/2018/2025914] [PMID: 30057671]

[173] Huang D S, *et al.* Protective Effects of Wogonin against Alzheimer's Disease by Inhibition of Amyloidogenic Pathway. Evidence-based Complement Altern Med 2017: 2017: 3545169.

[174] Cho HM, Ha TKQ, Doan TP, *et al.* Neuroprotective Effects of Triterpenoids from *Camellia japonica* against Amyloid β-Induced Neuronal Damage. J Nat Prod 2020; 83(7): 2076-86. [http://dx.doi.org/10.1021/acs.jnatprod.9b00964] [PMID: 32569471]

[175] Wang YM, Ming WZ, Liang H, Wang YJ, Zhang YH, Meng DL. Isoquinolines from national herb Corydalis tomentella and neuroprotective effect against lipopolysaccharide-induced BV2 microglia cells. Bioorg Chem 2020; 95: 103489. [http://dx.doi.org/10.1016/j.bioorg.2019.103489] [PMID: 31862456]

[176] Xia CL, Tang GH, Guo YQ, Xu YK, Huang ZS, Yin S. Mulberry Diels-Alder-type adducts from Morus alba as multi-targeted agents for Alzheimer's disease. Phytochemistry 2019; 157: 82-91. [http://dx.doi.org/10.1016/j.phytochem.2018.10.028] [PMID: 30390605]

[177] Zhang Q, Xia Y, Luo H, *et al.* Codonopsis pilosula polysaccharide attenuates tau hyperphosphorylation and cognitive impairments in htau infected mice. Front Mol Neurosci 2018; 11: 437. [http://dx.doi.org/10.3389/fnmol.2018.00437] [PMID: 30542264]

[178] Dhouafli Z, Leri M, Bucciantini M, *et al.* 1,2,4-trihydroxynaphthalene-2-O-β-D-glucopyranoside delays amyloid-β_{42} aggregation and reduces amyloid cytotoxicity. Biofactors 2018; 44(3): 272-80. [http://dx.doi.org/10.1002/biof.1422] [PMID: 29582494]

[179] Wang C, Cai X, Hu W, *et al.* Investigation of the neuroprotective effects of crocin *via* antioxidant activities in HT22 cells and in mice with Alzheimer's disease. Int J Mol Med 2019; 43(2): 956-66. [PMID: 30569175]

[180] Gan SY, Wong LZ, Wong JW, Tan EL. Fucosterol exerts protection against amyloid β-induced neurotoxicity, reduces intracellular levels of amyloid β and enhances the mRNA expression of neuroglobin in amyloid β-induced SH-SY5Y cells. Int J Biol Macromol 2019; 121: 207-13. [http://dx.doi.org/10.1016/j.ijbiomac.2018.10.021] [PMID: 30300695]

[181] Wang J, Zheng J, Huang C, *et al.* Eckmaxol, a Phlorotannin Extracted from *Ecklonia maxima*, Produces Anti-β-amyloid Oligomer Neuroprotective Effects Possibly *via* Directly Acting on Glycogen Synthase Kinase 3β. ACS Chem Neurosci 2018; 9(6): 1349-56. [http://dx.doi.org/10.1021/acschemneuro.7b00527] [PMID: 29608860]

[182] Oh J, Choi J, Nam TJ. Fucosterol from an edible brown alga ecklonia stolonifera prevents soluble amyloid beta-induced cognitive dysfunction in aging rats. Mar Drugs 2018; 16(10): 368. [http://dx.doi.org/10.3390/md16100368] [PMID: 30301140]

[183] Lin J, Yu J, Zhao J, *et al.* Fucoxanthin, a Marine Carotenoid, Attenuates *β* -Amyloid Oligomer-Induced Neurotoxicity Possibly *via* Regulating the PI3K/Akt and the ERK Pathways in SH-SY5Y Cells. Oxid Med Cell Longev 2017; 2017: 1-10. [http://dx.doi.org/10.1155/2017/6792543] [PMID: 28928905]

[184] Xiang S, Liu F, Lin J, *et al.* Fucoxanthin Inhibits β-Amyloid Assembly and Attenuates β-Amyloid Oligomer-Induced Cognitive Impairments. J Agric Food Chem 2017; 65(20): 4092-102. [http://dx.doi.org/10.1021/acs.jafc.7b00805] [PMID: 28478680]

[185] Wei H, Gao Z, Zheng L, *et al.* Protective effects of fucoidan on Aβ25-35 and D-gal-induced neurotoxicity in PC12 cells and D-gal-induced cognitive dysfunction in mice. Mar Drugs 2017; 15(3): 77. [http://dx.doi.org/10.3390/md15030077] [PMID: 28300775]

[186] Liu Y, Jiang L, Li X. κ-carrageenan-derived pentasaccharide attenuates Aβ25–35-induced apoptosis in SH-SY5Y cells *via* suppression of the JNK signaling pathway. Mol Med Rep 2017; 15(1): 285-90. [http://dx.doi.org/10.3892/mmr.2016.6006] [PMID: 27959440]

[187] Wang X, Yi K, Zhao Y. Fucoidan inhibits amyloid-β-induced toxicity in transgenic *Caenorhabditis elegans* by reducing the accumulation of amyloid-β and decreasing the production of reactive oxygen species. Food Funct 2018; 9(1): 552-60.

[http://dx.doi.org/10.1039/C7FO00662D] [PMID: 29260173]

[188] Yang EJ, Mahmood U, Kim H, *et al.* Phloroglucinol ameliorates cognitive impairments by reducing the amyloid β peptide burden and pro-inflammatory cytokines in the hippocampus of 5XFAD mice. Free Radic Biol Med 2018; 126: 221-34.
[http://dx.doi.org/10.1016/j.freeradbiomed.2018.08.016] [PMID: 30118828]

[189] Shanmuganathan B, Sathya S, Balasubramaniam B, Balamurugan K, Devi KP. Amyloid-β induced neuropathological actions are suppressed by Padina gymnospora (Phaeophyceae) and its active constituent α-bisabolol in Neuro2a cells and transgenic Caenorhabditis elegans Alzheimer's model. Nitric Oxide 2019; 91: 52-66.
[http://dx.doi.org/10.1016/j.niox.2019.07.009] [PMID: 31362072]

[190] Martens N, Schepers M, Zhan N, *et al.* 24(S)-Saringosterol Prevents Cognitive Decline in a Mouse Model for Alzheimer's Disease. Mar Drugs 2021; 19(4): 190.
[http://dx.doi.org/10.3390/md19040190] [PMID: 33801706]

[191] Cao M, Xu K, Yu X, *et al.* A chromosome-level genome assembly of *Pyropia haitanensis* (Bangiales, Rhodophyta). Mol Ecol Resour 2020; 20(1): 216-27.
[http://dx.doi.org/10.1111/1755-0998.13102] [PMID: 31600851]

[192] Li Q, Che HX, Wang CC, *et al.* Cerebrosides from Sea Cucumber Improved $A\beta_{1-42}$ -Induced Cognitive Deficiency in a Rat Model of Alzheimer's Disease. Mol Nutr Food Res 2019; 63(5): e1800707.
[PMID: 30512229]

[193] Wang X, Tao S, Cong P, Wang Y, Xu J, Xue C. Neuroprotection of Strongylocentrotus nudus gangliosides against Alzheimer's disease *via* regulation of neurite loss and mitochondrial apoptosis. J Funct Foods 2017; 33: 122-33.
[http://dx.doi.org/10.1016/j.jff.2017.03.030]

[194] Alvariño R, Alonso E, Lacret R, *et al.* Caniferolide A, a Macrolide from *Streptomyces caniferus*, Attenuates Neuroinflammation, Oxidative Stress, Amyloid-Beta, and Tau Pathology *In vitro*. Mol Pharm 2019; 16(4): 1456-66.
[http://dx.doi.org/10.1021/acs.molpharmaceut.8b01090] [PMID: 30821469]

[195] Alvariño R, Alonso E, Lacret R, *et al.* Streptocyclinones A and B ameliorate Alzheimer's disease pathological processes *in vitro*. Neuropharmacology 2018; 141: 283-95.
[http://dx.doi.org/10.1016/j.neuropharm.2018.09.008] [PMID: 30205103]

CHAPTER 2

Neurobiology of Placebo: Interpreting its Evolutionary Origin, Meaning, Mechanisms, Monitoring, and Implications in Therapeutics

Akash Marathakam[1], Vimal Mathew[2] and **MK Unnikrishnan[3,*]**

[1] *Department of Pharmaceutical Chemistry, National College of Pharmacy Kozhikode, Kerala 673602, India*

[2] *Department of Pharmaceutics, National College of Pharmacy Kozhikode, Kerala 673602, India*

[3] *Department of Pharmacy Practice, NGSM Institute of Pharmaceutical Sciences, Nitte (Deemed to be University), Deralakatte, 575018 Mangaluru, Karnataka, India*

Abstract: Placebo is defined as the therapeutic response to inert treatment. However, this is a bit simplistic because comprehending the biological basis of the placebo effect requires understanding the entire therapeutic context and the patient immersed in it. Placebo does not cure the disease but alleviates symptoms. The placebo impact must be seen in the context of the recipients' cultural milieu, psychosocial background, the tone and tenor of the accompanying verbal communication (caring, indifferent, unfriendly), therapeutic rituals (*e.g.*, tablet, injection, or a procedure, including diagnostic tests), symbols (white coat, syringe, the diagnostic paraphernalia), and its meanings to the patient (past experiences and personal hope). Placebo is the inert treatment juxtaposed against the broad context of the accompanying sensory and sociocultural inputs that signal benefit. It could also be the harm in the case of nocebo. A major objective of a standard clinical trial is to eliminate or at least minimise the influence of placebo. Many methods have been devised to measure and eliminate placebo responders in the trial populations. The neurological basis of the placebo effect is complex and must have an evolutionary basis because the susceptibility to placebos may be traced back to animals and birds. The placebo effect probably owes its evolutionary origin to signalling sickness and the ability to draw comfort from winning sympathetic attention and care from conspecifics. Pain being a complex sensory experience with a strong affective component, the neuronal pathways that reflect both sensory experience and the affective components have been explored in the study of the placebo effect. Placebo research, having expanded from psychology to neurology, presently involves research tools that include pharmacology, brain imaging, genetics, animal models, *etc.* This review will discuss multiple dimensions of the placebo effect, including evolutionary, cultural, psychosocial, and neurological aspects, in addition to providing cues for transformational implications in clinical trials and therapeutic modalities that benefit

[*] **Corresponding author MK Unnikrishnan:** Department of Pharmacy Practice, NGSM Institute of Pharmaceutical Sciences, Nitte (Deemed to be University), Deralakatte, 575018 Mangaluru, Karnataka, India; Tel: +918089007260; E-mail:unnikrishnan.mk@nitte.edu.in

Zareen Amtul (Ed.)

society. Contemporary medicine is demonising placebo because it is a confounder in clinical trials. It would be much more useful if the healthcare system can harness the therapeutic potential of the placebo effect by manipulating the therapeutic context.

Keywords: Cognition, Evolutionary biology, Neurobiology, Placebo.

INTRODUCTION

The urge to heal is intrinsic to all forms of life. The enduring recuperative drive, the *sine qua non* for survival, is mostly unconscious and innate, and begins at the molecular level -- even the DNA is equipped with the technology for self-repair. The most primitive parts of the brain handle the most fundamental life-sustaining functions, demonstrating its evolutionary antiquity as well as the minimal need for 'intelligent' interventions by the conscious mind. The healing power of physicians, until recently, rested mostly on the results of the placebo effect [1]. This is self-evident today because we know that many of the pre-World War remedies were either useless or positively harmful. There is a scandalous account of how Calvin Coolidge Jr died, quite possibly from the treatment he received for infection [2]. Despite so many ineffective and hazardous remedies, the medical profession has consistently enjoyed society's awe and trust for thousands of years. Voltaire probably had the placebo effect in mind when he famously said, "The art of medicine consists of amusing the patient while nature cures the disease" [3].

There is a growing tendency to attribute the combined role of mind and body in fighting and recovering from illnesses [4]. Studies suggest that subjective perceptual states such as mindsets and expectations influence human physiology. For instance, young adults could improve their vision when exposed to a mindset patterned to suit pilots and athletes. For instance, vision improved when subjects learned to read the Snellen Eye Charts in reversed progression [5]. Those who were scared of catching the flu in winter actually expressed more real flu symptoms. This is consistent with the idea that the brain influences metabolic and endocrine regulation based on environmental challenges, appropriate to the context of available resources. Such a scheme has an evolutionary advantage because expectations prepare the body to cope better with an anticipated event.

The many case records of miraculous cures reinforce the value of subjective interventions. For instance, at the famous healing shrine of Lourdes, literature reports 176 cases of cancer that regressed without treatment [6]. Considering that people of different religious faiths experienced such cures, success appears to have depended more on the patient's state of mind than religious faith. On the other hand, these cures do not violate the laws of nature. They merely hasten the normal reparative process [7]. 'No one has grown a new limb or an eye at Lourdes', quotes a review on placebo!

The Context of Health and Disease: The Mind-Body Continuum

A recent study by Harvard University threw up a surprise [8]. When diabetics were asked to drink the same sweetened beverage from different bottles, with labels declaring different sugar contents, the rise in blood sugar in each individual was proportionate to the labelled sugar content, not the actual sugar content in what was consumed. In another study by the same department, they employed manipulated 'fast' and 'slow' clocks to modify the perception of time in type 2 diabetics. Interestingly, a fall in blood sugar correlated with the perceived time interval, not the actual time interval. In other words, the metrics of the environment must be meaningful to the mind; subjective perception can control the metabolic trajectory.

The role of the mind in metabolic regulation is not limited to diabetic subjects [9]. Women who were conditioned to consider work as a kind of exercise achieved greater weight control and a fall in BMI (body mass index). In another study, the ghrelin levels in subjects consuming milkshakes depended on the labelled calorific value rather than the true calorific value [10]. Yet another study showed that just believing to follow a low-calorie diet (while actually on an energy-balanced diet) can help reduce body mass. Endocrine regulation, as well as its potential disruption, is influenced by the conscious mind [11].

There is an evolutionary advantage in such perceptual neuroendocrine controls. Consciously anticipated events can potentially prepare the body to become future-ready. Maintaining the time course of blood sugar must have been much more difficult in the food-insecure primitive environment. In corollary, an evolutionary mismatch can explain the recent surge in metabolic disorders.

Placebo and the Therapeutic Context: the Evolutionary Backdrop

While the placebo effect is the response to inert treatment, the placebo phenomenon, in its totality, is a lot more complex [12]. PE constitutes a number of diverse variables/ entities that create the therapeutic context as a whole. The words uttered by the physician during encounters at the hospital, the body language of physicians, nurses and bystanders, the therapeutic rituals of diagnosis and treatment, the seemingly trivial, but ominous symbols and signals, such as white coats, stethoscopes, the smells of antiseptics, signboards notifying say, 'Cancer Hospital', (possibly a 'death sentence'), *etc.*, combine to create the therapeutic context. Every accompanying sensory/social stimulus that signal benefit, along with all the procedural rituals of therapeutic interventions, merges to create the cumulative impact of the placebo. Depending on the kind of these stimuli, the placebo can raise the expectations of the patient or lower them [13]. A higher dose of pills, a sharper taste of medications, a higher price, a well-known brand label, and high-profile advertisements have been shown to add to the patient's hope and increase his/her expectation from the treatment. Invasive methods, such as injections and intravenous infusions further raise expectations towards positive outcomes [14].

Humphrey and Skoyles, in a guest editorial in *Current Biology,* wrote that humans are among the best in deploying strategies for self-recovery [15]. First, predicting the cost/benefit ratio from environmental cues, enables the optimal deployment of natural defenCes. Secondly, exploiting placebo cures by deploying "culturally-based 'medical disinformation'".

Natural selection has already granted a standard set of healing defences in the course of evolution [16]. These include, but are not limited to, spontaneous tissue repair, pain-induced mobility restriction for limiting injury aggravation, fighting infections, preventing and purging toxins, and 'sickness behaviours'.

A good part of placebo research looks for proximate mechanisms of placebo effect, on how it is mediated physiologically. There is limited research seeking answers to ultimate mechanisms, the more puzzling 'why' question. Patrick Wall, a pain expert, says that pain not only signals injury but also motivates attention and avoidance action. Pain urges the victim to withdraw from a hurtful situation, restrict movement, adopt a relieving posture and seek support. The act of help and provision of treatment fulfill the need, allowing the pain to fade. Placebo effect is thus an adaptation [17]. Unaware of the physiological futility of placebo, the patient believes that appropriate action has been taken, fulfilling the need!

Healing of tissue injury and immune responses against infections are not knee-jerk reactions because they are energy-intensive [18]. Body defences are calibrated to meet environmental constraints. A cost-benefit analysis at the unconscious level helps postpone the pain of injury in danger; say, an attacking lion. Endorphins (endogenous pain killers) kill pain during such emergencies. Likewise, fighting the flu can wait until there are immediate options for procreation. Zebra finches suppress flu symptoms in the presence of a female [19]. Lack of food and energy reserves can restrain a full-blown immune response; that is why flu lasted longer in winter when food used to be scarce. (Natural selection occurred in a primitive environment; that is why there is an evolutionary mismatch in nutritional adaptations). Sick hamsters have been shown to mount a more powerful immune response in lab-simulated summer, but not winter, resting on the premise that food is more easily available in summer [20]

Self-healing processes are deployed based on the costs, opportunity costs, and potential benefits. Such cost-benefit analysis is performed by the subjective judgement of the environment. Placebo treatment modifies this cost-benefit analysis by creating an illusion that the immune system is ready to deploy resources for rapid recovery. Circumstances have changed for the better, allowing for a full-blown immune response regardless of the season.

Recovering from injury and infection has been a crucial adaptive challenge throughout evolutionary history. The immune system developed as the adaptive answer to such challenges [21]. Top-down neuronal control of immunity occurs even in nematodes [22]. The 'mammalian health governor' has a 'knowledge base' about direct costs and opportunity Costs associated with mounting a defence. The body consumes about 50% more energy during immune activation. There is a 20% increase in energy expenditure with a 2-degree rise in body temperature. The potential for tissue damage due to high fever becomes an associated opportunity cost.

Likewise, sickness behaviour is also adapted to hasten the healing process. The victim feels sleepy. Apathy and social withdrawal accompany a motivational shift away from social, sexual, and aggressive behaviour. The corresponding opportunity costs are decreased prospects for mating, bonding, and fighting for dominance hierarchies. Avoidance behaviour of conspecifics adds to the social costs of sickness behaviour, but prevents infections from spreading within the group.

"The body is evolved to survive, not to be comfortable," says Randolph Nesse, a pioneer in evolutionary medicine. Bodily instincts prefer to err on the side of caution. Better to mistake a rope for a snake than the snake for a rope. Bodily instinct prepares for the worst, while the mental state is goaded with optimism that hopes for the best possible outcome [23]. Hedonism and optimism provided the *raison d'être* for the evolutionary struggle in sentient creatures.

Optimism, a key promoter of the placebo effect, is the complementary mental state that enhances survival, compensating for the pessimistic prudence and parsimony in bodily instincts. Research has shown that optimism is a 'dispositional factor' in animal behaviour, from 'rats to honeybees'. In humans, optimism is specifically linked to health outcomes [24]. Sleep improved when the patient was provided with an optimistic account of the placebo being administered. Optimism is a strong predictor of the rate of recovery. Optimistic patients were faster at reaching "behavioural milestones of recovery," such as sitting in bed, walking around, *etc.* Optimistic patients were also rated by staff as having a more favourable physical recovery. Optimists were faster to resume normal routine activities six months later. Optimism is also associated with longevity. For example, a Boston-based study published in PNAS in 2019 [25] reported that adjusting for demographics and health conditions, women in the highest quartile in optimism scores lived about 15% longer than cohorts in the lowest quartile. Results in men were similar. Those with the highest optimism levels had 1.5 (women) and 1.7 (men) better chances of living up to 85 years, compared to their least optimistic cohorts, when adjusted for health behaviours such as smoking, alcohol and diet. Optimistic anticipation of a forthcoming festive occasion has been associated with a drop in the death rate, followed by a peak soon after. For example, Mortality among Chinese (n = 1288) fell by about 35% during the week preceding the Harvest Moon Festival, and rose by a similar amount in the week following the festival [26]. Similar results have been reported for Jewish members before and after Passover, the Jewish holiday. Such a drop in mortality was not observed among non-Jewish cohorts like blacks, orientals and Jewish infants, during the 18-year study period between 1966 to 84 ($p = 0{\cdot}045$). ($p = 0{\cdot}003$).

What is the evolutionary purpose of sickness behaviour? Most importantly, sickness behaviours help to advertise illness and attract sympathetic support from conspecifics. In the struggle for survival, the opportunity costs of immune activation must have been compensated by the support of conspecifics. The optimistic hope of support and its fulfillment by conspecifics is fundamental to sickness behaviour. This is not just peculiar to humans but pervasive across different species. In house sparrows that breed in couples and feed offspring together, the male collects more food when the female shows symptoms of sickness. Female chimpanzees reduce travel speed to let a sick member catch up with the group. Chimpanzees have also been found to take care of a sick mother's child while she was sleeping. An injured member in a group of wild mongooses was found to be groomed and fed by mates. Feeding sick members is observed only in some non-human animals [27]. Support for the sick or injured is crucial among hunter-gatherer societies. Social support could be a reason for human longevity. Skeletal remains of severely handicapped individuals who survived more than 100000 years ago imply that social support for the sick increased survival. Care also promoted a "belief in medicine"; where "mumbo-jumbo and snake-oil" worked nearly as well as the practical cures [28]. Medical care must have started with nursing care, well before doctors ascended to occupy the top slot in the healthcare hierarchy.

Signaling Theory of Symptoms: Leander Steinkopf has put forward a very interesting theory, namely the Signaling Theory of Symptoms (STS), explaining how sickness behaviour augments the placebo effect [29]. The immune reaction has discernible symptoms such as fever, swelling, apathy, and signs of pain. STS, therefore, not only serves to defend the body but also employs the same signals to advertise for help and receive supportive treatment from potential conspecifics. Stronger the symptoms, the greater their signaling efficacy. Therefore, it is likely that symptoms are exaggerated to heighten the likelihood of mobilizing help and treatment. Once help and treatment are received, the signaling function is fulfilled, potentially diminishing the symptoms.

In a study that used an ultrasound device to reduce swellings after dental treatment [30], the swelling was found to be reduced, even if the device was turned off, demonstrating a placebo effect. However, the swelling came down only when the experimenter administered the treatment, not when the subject used it himself. Similarly, intranasal oxytocin increased the analgesic effect of a placebo ointment applied to the forearm [31]. The rise in analgesia is likely the result of oxytocin's ability to increase the subject's trust in the doctor who applied the placebo ointment [32]. Presenting facial expressions alters the magnitude of the placebo effect in placebo analgesia [33]. Sad faces enhanced the placebo effect more than neutral faces. However, a happy face enhances the placebo effect even more. A sad face signals empathy with the sufferer, a precondition for helping. A happy face also indicated that help is very likely. On the other hand, a neutral face probably suggested indifference [34].

In contemporary medicine, treatments rarely include touching. Even slight physical contact in medical settings can soothe pain. There is great demand for treatments that involve greater physical contact: many alternative and complementary treatments feature touching [35]. Acupuncture involves much physical contact and a strong placebo effect [36]. So do Ayurvedic massages in which experts apply medicated oils for treating chronic conditions [37].

From the perspective of STS, these treatments satisfy the need for help and its fulfillment through touch. Contemporary medicine has become entirely materialistic, objective, and impersonal. Physical contact with the patient is a subject in itself, where informed policy interventions can bring about transformative changes in contemporary medical practice.

STS also suggests a flexible component of immune regulation adapted to the culture. Patients' faith in a given treatment increases the placebo effect. Advertising of a drug (a form of clever education) enhances its placebo effect [38]. Patients experience a greater placebo effect when the prescribing physician is convinced of the medicine's efficacy [39]. Physicians' authority tells the patient which treatment is effective and, in turn, diminishes signaling symptoms upon treatment. Patients also learn from their own experiences: Classical conditioning is reported as the main mechanism of the placebo effect [40, 41].

The manipulative power of advertisement on patient expectations and the consequent augmentation of the placebo effect has been demonstrated over a period of thirty years. Placebo effect of antidepressants increased [42, 43], possibly on account of large-scale direct-to-patient advertising on antidepressants, especially in the USA, where it has become a part of popular medical culture and standard therapy [44, 45]. STS predicts that exaggerated immune reactions occur in the presence of potential helpers. Support systems possibly precipitate a full-blown response whenever the primitive society facilitates survival by guaranteeing rest and security in full measure.

Measuring the Placebo Effect

Measuring the placebo effect is problematic. Patient responses may be misattributed to placebo because of methodological biases. Biased reports by the patient and the experimenter are particularly significant when subjective symptoms are assessed as therapeutic outcomes. Patients' positive expectations diminish anxiety and activate reward mechanisms. That is why placebo effect should be compared with real treatment. Regardless of the intervention, the observed improvement can be due to spontaneous remission and/or methodological biases and/or the patient's expectations.

Placebo effect has been a problem in the reliable measurement of clinical trial (CT) results. Eliminating the placebo effect in CTs is a priority. Many new CT designs are envisaged in depression [46] and different placebo components are described within the CT setting [47].

Therefore, the mechanistic aspects of the placebo effect are key to CT design [48 - 50]. Placebo research has steadily expanded from Psychology to Neurology [51] and currently employs a diverse variety of strategies and research tools in pharmacology, brain imaging genetics, animal models, *etc.* Together, these multiple tools address complex issues in the neurobiological domain, with translational implications in clinical trials, medical practice, and for society at large.

In a comprehensive, high-profile review, Wager and coworkers (*Nature Reviews Neuroscience* 2015, 16, 403–418) describe four models for assessing the placebo effect [52]. The first model consists of employing three groups in parallel, where one group receives a placebo, the second the active treatment, and the third, neither intervention. The difference between the outcomes of the active treatment and placebo group will indicate treatment effectiveness. Likewise, the difference between the outcomes of the placebo group and the nil-intervention group would indicate the placebo response. The second paradigm is to compare an open *vs.* hidden study. While one group knowingly receives the active treatment, the second group receives the same treatment without any knowledge about it. The difference in the outcomes would indicate placebo effect.

The third paradigm employs 'response conditioning', in which two skin sites are marked for testing analgesic potency. One site receives a placebo in the guise of a potent analgesic, and the second site receives the same placebo dubbed as control. Both sites are then subjected to painful skin stimuli. The pain stimuli to both placebo and control sites are stated to be identical, but the placebo-treated patch of skin (surreptitiously) receives a milder stimulus, reinforcing the placebo illusion. In the next stage, when painful stimuli of equal intensity are applied to both sites, the reported difference in responsiveness will reflect placebo conditioning. This paradigm is useful in neuroimaging studies where placebo and control treatments can be compared in a 'within-person', crossover design.

The fourth paradigm involves pharmacological conditioning. This method combines instructions and cues paired with active drugs during a conditioning phase. The conditioning phase is repeated over multiple days. Subsequently, the placebo effect is determined by presenting cues alone and comparing outcomes in drug-paired *versus* non-drug-paired groups. Response conditioning and pharmacological conditioning designs have been used in both humans and non-human animals.

The Physiology of Placebo: Mechanistic Aspects

Neurobiological Mechanisms of Placebo Response [53]: While the placebo effect has many dimensions, pain relief is well studied. Pain perception is tricky, a double-edged sword. The inability to feel pain would instantly compromise survival because there would be no incentive to avoid injury. On the other hand, the problem of inappropriate pain has been the greatest source of misery and the most recurrent reason for seeking treatment. The evolutionary course in optimising and reconciling the life-saving potential of pain perception, along with the unavoidable suffering associated with it, has resulted in many layers of controls and counter-checks, together creating a very complex physiological scaffolding. Among the most important aspects of pain sensation are the following:

Opioid Pathway: Antinociceptive mechanism, operating *via* mu-opioid receptors, is activated during placebo analgesia. The role of this crucial opioid pathway in placebo response has been demonstrated by its blockade by Naloxone, a nonselective competitive antagonist of opioid receptors [54]. Naloxone blocks the placebo effect by blocking the mu receptors, which have been implicated in addiction behaviour [55].

Cholecystokinin: The pro-nociceptive system (nocebo) operates *via* cholecystokinin, which antagonizes the opioid system. Pro-nociceptive CCK system is activated by anticipatory anxiety in nocebo hyperalgesia. The involvement of CCK-2 receptors (more important) has been demonstrated by the blockade of cholecystokinin-induced hyperalgesia by proglumide, a CCK-2 antagonist [56].

Endocannabinoids: In addition to cholecystokinin and opioid receptors, placebos also activate CB1 cannabinoid receptors that inhibit prostaglandin (PG) synthesis [57]. Nocebos, on the other hand, increase PG synthesis. The enzyme Fatty acid amide hydrolase (FAAH), a serine hydrolase enzyme, was first shown to break down anandamide (the endogenous cannabinoid principle). Anandamide mediates pain perception *via* cannabinoid receptors [58]. Its breakdown by FAAH is, therefore, critical in the placebo effect. Different genetic variants of FAAH affect the magnitude of placebo analgesia. Pro129Thr, a functional missense variant of the gene coding FAAH affects placebo analgesia and placebo-induced mu-opioid neurotransmission. Likewise, rimonabant, a CB1 receptor blocker, can block placebo analgesia.

Interestingly, endocannabinoids mediate non-opiate analgesia. When Ketorolac is administered for 2 days, and when a placebo is administered on the third day, the placebo analgesia was found to operate through CB1 receptors. Therefore, it is not reversed by naloxone. On the other hand, rimonabant, a CB1 cannabinoid receptor blocker, blocks this placebo analgesia [51]. Thus, the multiple pathways in placebo add redundancy to a critical biological function. Multiple mechanisms create a surplus of self-limiting equilibria by crosstalk. Such complexity is pervasive in critically important biological pathways because the body cannot take chances in survival.

While previous exposure to non-opioids stimulates CB1 receptor-mediated placebo analgesia, opioid analgesics mediate placebo analgesia through the opioid pathway. The larger pathways that modulate pain sensation involve arachidonic acid, endogenous cannabinoid ligands (*e.g.*, anandamide), prostaglandins and thromboxane. Fatty Acid Amide Hydrolase (FAAH) is the major endocannabinoid-degrading enzyme. Functional missense variant (Pro129Thr) of gene coding FAAH affects analgesic responses to placebo as well as placebo-induced m opioid neurotransmission [59].

Placebo response in Parkinson's disease (PD) involves the activation of D2-D3 dopamine receptors in the striatum. D2-D3 receptors & m opioid receptors in the nucleus accumbens mediate PE in pain [60]. Quite, by contrast, deactivation of D2-D3 and m receptors occurs in hyperalgesia. Placebo in PD can decrease the firing rate and bursting activity of subthalamic nucleus neurons, and substantia nigra pars reticulata [61].

Placebo Effect (PE): A Distraction in Clinical Trials [62]

Placebos are employed in double-blind CTs to compensate for PE, and to minimise PE when evaluating the therapeutic value of a test drug. Sample sizes should anticipate and neutralise placebo responders. PE can confound CT results and lead to uninterpretable results. PE is a common reason for negative results in CT.

Negative and ambiguous CT outcomes have increased since the 1980s. Negative results indicate failure of assay sensitivity in CT. CT is judged not by design but by outcomes. In other words, a faulty design can in itself lead to misleading outcomes. NDA submissions sometimes do not include positive CT results. FDA requires only 2 trials that demonstrate superiority over placebo, thereby admitting a lot of medicines with doubtful clinical benefit. In 5 NDA submissions to FDA for depression in the 1990s, only 14% of trials (out of a total of 39 trials) showed superiority to placebo. Paroxetine needed 9 trials to get 2 positive outcomes. Fluoxetine with 13 efficacy trials had only 5 positive outcomes.

A study by Walsh *et al.* reported highly variable but substantial PE (averaging 30%, Range 12– 52%) in 75 CTs for major depressive disorder (MDD), published between 1981 and 2000. PE has been rising significantly in recent years, possibly because of the greater degree of anticipation created by direct-to-patient advertisements. Manufacturers' websites prominently provide more positive drug information to patients.

Experts say that PE has increased alarmingly across psychiatric disorders such as generalized anxiety disorder, panic disorder, *etc.* The rise in PE has dissuaded private players from pursuing drug discovery in psychiatric disorders [63].

To consider another class of drugs, take smoking cessation. Quit rates in the ninth week were 31%, which was similar for both placebo and nicotine patches. However, a meta-analysis of 17 trials (n = 5,098) showed that a nicotine patch was significantly more effective than a placebo. In another study on 989 smokers, the quit rate was not different between placebo and fluoxetine at 2 doses.

The problem lies deeper. Most negative trials are unpublished, suggesting that the average placebo response in MDD trials is even higher, say 35– 45% (very high!). There are multiple factors behind the high placebo effect. Diagnostic misclassification; inclusion/ exclusion criteria; outcome measures' lack of sensitivity to change; measurement errors; poor quality of data entry and verification; waxing and waning of the natural course of illness; regression toward the mean; patient and clinician expectations; study design issues; non-specific therapeutic effects; high attrition (drop-outs) are among the important factors. Minimising these individually can bring down PE, but going into further details would be going beyond the scope of this review. However, those working in the clinical trial design should pay special attention to such topics.

Placebo Responders *vs* Nonresponders: Placebo responses differ from PE! Placebo responses include all health changes that follow inactive treatment, including natural history and regression to the mean. PE, on the other hand, refers to changes that may be attributed specifically to placebo mechanisms. Since PE is mediated by central nervous system mechanisms (expectations, learning processes, *etc.*), specific metrics of brain function might help in locating and quantifying the brain function changes that accompany placebo responses. Thus, such measurements could also be employed to understand the underlying neuropsychological attributes that distinguish responders from non-responders [64].

Neuroimaging: Neuroimaging helps understand placebo, particularly analgesia. Prefrontal circuitry in placebo analgesia is seen in chronic back pain [3], fibromyalgia [4], and chronic knee osteoarthritis [5]. The right midfrontal gyrus connectivity has been identified as a specific biomarker to predict placebo response in osteoarthritis [5]. Right midfrontal gyrus influences decision-making, memory, and planning. Therefore, placebo analgesia is a complex top-down modulation. Amygdala, nucleus accumbens (NAc), and ventral striatum (VS) are closely connected to medial prefrontal circuitry. Strong placebo analgesia is predicted by NAc–VS activity, including stronger placebo-related opioid & fMRI activity responses during pain, increased grey matter volume, and stronger fMRI responses in a reward pursuit task unrelated to pain [6].

In addition to neuroimaging techniques, pharmacological approaches can also be employed to understand the placebo mechanism. Analgesia can be boosted by both vasopressin and oxytocin agonists and oxytocin's modulation of PE is dose-dependent [7, 8]. In a randomized trial on healthy participants, intranasal vasopressin (40 IU) increased placebo analgesia significantly. However, lower doses of intranasal oxytocin (24 IU) and intranasal saline (0.4 ml) showed effects only in women. Subjects who inhaled a higher dose of oxytocin (40 IU) showed no effect at lower doses of 24 IU intranasal oxytocin. Women with lower dispositional anxiety and low cortisol levels showed largest vasopressin-induced modulation of PE. This suggests that pre-existing psychological factors and cortisol changes have a role in modifying PE [7].

Leveraging the Placebo Effect

Harnessing/Leveraging Placebo: While PE is accepted as a reality and counted as a major component of therapeutic success in most situations, the clinical trials see PE as a nuisance and a distraction. Not many consider PE as something to be harnessed for the benefit of the patient. This is probably because of the excessive objectivity and reductionism in modern medicine. The healthcare curriculum considers the body as a machine with many parts. Modern medicine has developed therapeutic practices that require minimal bodily contact or dialogue. In fact, most of the ward accessories are expressly designed to avoid personal contact, whatever the situation. With the advent of COVID-19, this tendency to isolate doctors from patients has become a statutory need. Modern medical practices and guidelines, including the medical curriculum, ignore PE, which contributes a substantial portion of therapeutic benefit at minimal cost in infrastructure or human resources. The emphasis on measurability has undermined the less measurable subjective realm. Training modules in the healthcare curriculum, except in physiotherapy, do not consider the importance of bodily contact, except during invasive procedures. Ethical norms generally forbid placebos.

The Need to Instill Hope in Patients and Meet Their Expectations: During World War II, injured soldiers were being treated with morphine injections to control pain. When the morphine supplies ran out, Henry Beecher, to comfort wounded soldiers, injected them with saline. The results were compellingly positive. He concluded that the saline placebo injections showed 40% effect.

Patient expectations grow from the physician's authority. In a double-blind trial, two groups of asthmatics were asked to inhale a bronchodilator or constrictor. One group was told that the inhalant would produce its true pharmacological effect. The other group was told that it would produce the opposite effect. The effect on airway resistance of the dilator was twofold greater when the physician's version matched the pharmacological effect of the bronchodilator. In other words, the bronchodilator produced twice the effect in those who were told it was a bronchodilator. Likewise, the constrictor too produced greater constriction when the physician told patients that it was a bronchoconstrictor [65]. In other words, a reassuring physician can deliver a substantially greater therapeutic impact on a trusting patient.

Psychological effects of surgery are strongest in cardiac surgery. The heart is most susceptible to emotions. Literature has described the heart as the "end-organ of anxiety". Placebo effects of heart surgery were first demonstrated by ligating the internal mammary artery, based on the logic that it would divert more blood into the coronary circulation [66]. Notably, 40% patients showed they marked symptomatic relief. But, it was subsequently demonstrated that even a mock surgery (just opening the skin under anesthesia, suture it back; nothing else) was found as effective. Patients who underwent such mock surgery demonstrated a reduced need for nitroglycerin and better exercise tolerance [67].

Placebo effect has also been demonstrated in surgery for retinal detachment. It was shown that the patients' level of 'acceptance' can influence the speed of healing after surgical repair of retinal detachment [68]. Before the surgery, 98 patients were interviewed to evaluate the level of "acceptance" based on four parameters, namely (1) trust in the surgeon, (2) optimism about outcomes, (3) confidence about coping with outcomes, and [4] readiness to accept the bad with the good. There was a significant correlation ($p < .001$) between acceptance scores and speed of healing.

There is no ethical dilemma in promoting a placebo indirectly, say, by increasing patient expectations. Expectations can be raised by creating a patient-friendly healthcare setting, which can indirectly enhance the placebo effect. Health Technology Assessment (HTA) can study various facets of healthcare towards facilitating the placebo effect in patients. Such a strategy would be particularly cost-effective in resource-poor settings. For instance, in an HTA study commissioned by the Secretary of State for Health, UK, researchers extensively reviewed both published literature and opinions from narrative interviews on patient expectations and their fulfilment [69]. Reviewing a selection of 211 articles on site-specific data, the study listed and compared the pre-visit expectations and post-visit experiences, focusing on what patients expected and which expectations were fulfilled. The patients' major previsit expectations were cleanliness, information about where to go, convenient and punctual appointments, and helpful reception staff. Patients' expectations from the doctor were knowledge, clarity, easy to understand, and involvement in treatment decisions. The study revealed that expectations least likely to be met were being seen on time, the ability to choose the hospital and doctor, the helpfulness of reception staff, and the doctor dealing with the patient respectfully and with dignity. Expectations were less likely to be met in hospitals than with GP visits, demonstrating the importance of personalized approaches. Similarly, patients also found a lack of reassurance, advice about health/condition, information about cause management of the condition, and information about benefits and side effects of treatment. Patients also felt that the opportunity to discuss problems fell below their expectations. Previous consultations/experiences commonly influenced expectations.

Anticipation of an adverse drug reaction is a strong predictor of experiencing an ADR. It has been shown that among patients undergoing cancer chemotherapy, those who anticipate vomiting actually vomit more than those who don't. Chemotherapy-induced vomiting has a very contextual connection. Anecdotal references suggest that patients vomit even when they happen to see the doctor who prescribed the medicine. Being warned about an ADR actually aggravates its occurrence. Sexual dysfunction for anti-prostate medicines increased by 28.3% when patients were informed about it. Likewise, gastrointestinal side effects for antianginal medicines increased 6 times after patients were informed. A review of 27 studies reported an increased incidence of ADRs when patients were warned against them. This poses an ethical dilemma. Not informing a patient about ADRs amounts to withholding information, while warnings potentially worsen patients' chances of an event free recovery. In this context, it is worth emulating a recently published novel method to leverage the placebo effect by telling patients that experiencing an ADR is actually a sign that the drug is working [70, 71].

In a study on healthy volunteers receiving diclofenac, they reported fewer intense side effects when they were told *via* video that ADR is a signal that the drug is working [72]. Further, there was a positive correlation between side effects and pain reduction. In another study, adolescents on oral immunotherapy for peanut allergies, when told that side effects mean that treatment is working [73], experienced less anxiety about side effects and were less likely to contact staff about ADR concerns. They were also less likely to skip or reduce doses due to anxiety. Even the biomarker of allergic desensitisation (IgG4 levels) was higher in those who were warned about ADRs. Thus, there is already evidence that the therapeutic context can be cleverly manipulated to create optimism in a patient [74]. Rephrasing the same message in different words can lighten ethical embarrassment and improve therapeutic outcomes.

CONCLUSION

Practices, such as mock cardiac surgery, *etc.*, have disappeared. Even contemplating such remedies would be considered outrageous today because there are effective interventions, such as CABG, *etc.* Similarly, there is no need to inject saline because morphine products are readily available on prescription. Mindboggling material progress, pharmaceutical innovation, profusion of technology, path-breaking diagnostics, and a plethora of gadgetry have redefined the physicians' role. Placebo has become a distraction (even worse, a villain) in the luxury of choice proffered by contemporary medicine. However, from a public health point of view, placebo therapy might still not be inappropriate for a large number of impoverished sections of the society, that have no hospitals, physicians, or systems for preventive care.

Unfortunately, employing placebos has become an ethical transgression. Ethical propriety and its codification are invariably verbalized by the developed countries, who also can take credit for most of the innovation. The rules are framed in an environment that takes for granted that patients have access to expensive medicines, well-trained healthcare experts and supportive gadgetry. Likewise, the innovators can also invoke ethical propriety to deny essential medicines to millions, as has already happened in Africa, where anti-HIV medicines were virtually unavailable until an Indian pharma company made them affordable. In the end, one can see that socio-political and economic considerations overwhelm the logic of science in healthcare.

In the wake of chronic diseases overwhelming the healthcare system, patients need to build a life-long partnership of trust with physicians and healthcare professionals in general. In this context, the lessons from the positive impact of placebos can be rightfully exploited by training physicians to contribute better towards enriching the trust between patients and healthcare functionaries. When measurable outcomes have been experienced from placebos, it is rather puzzling how an understanding of the placebo mechanism has never been a part of the healthcare curriculum. Research can help advance the methodology for exploiting the placebo effect to greater advantage, without investment in infrastructure or human resources. Training healthcare professionals to engage with patients more reassuringly, providing opportunities to patients and bystanders to discuss therapy with physicians, etc., can bring about significantly better outcomes at a minimal cost. Placebo effect has an evolutionary basis and deserves greater attention from public health departments. It is high time professional bodies and universities consider investing in the placebo potential. A policy can work towards creating a therapeutic ambiance for optimising the placebo effect.

REFERENCES

[1] Frank JD. Biofeedback and the placebo effect. Biofeedback Self Regul 1982; 7(4): 449-60. [http://dx.doi.org/10.1007/BF00998885] [PMID: 7165779]

[2] Hager T. The Demon Under the Microscope: From Battlefield Hospitals to Nazi Labs, One Doctor's Heroic Search for the World's First Miracle Drug. Harmony Books 2006.

[3] Monleón Getino T, Canela i Soler J. Causality in medicine and its relationship with the role of statistics. Biomedical Statistics and Informatics 2017; 2(2): 61-8.

[4] Cousins N. 3. Anatomy of an Illness (As Perceived by the Patient). Columbia University Press; 1991 Mar 2.

[5] Johnson AT, Dooly CR, Simpson CR. Generating the Snellen chart by computer. Comput Methods Programs Biomed 1998; 57(3): 161-6. [http://dx.doi.org/10.1016/S0169-2607(98)00036-4] [PMID: 9822853]

[6] Everson TC, Cole WH. Spontaneous regression of cancer: Preliminary report. Ann Surg 1956; 144(3): 366-83. [http://dx.doi.org/10.1097/00000658-195609000-00007] [PMID: 13363274]

[7] Cranston WI, Pepper MC, Ross DN. Carbon dioxide and control of respiration during hypothermia. J Physiol 1955; 127(2): 380-9.
[http://dx.doi.org/10.1113/jphysiol.1955.sp005264] [PMID: 14354679]

[8] Park C, Pagnini F, Langer E. Glucose metabolism responds to perceived sugar intake more than actual sugar intake. Sci Rep 2020; 10(1): 15633.
[http://dx.doi.org/10.1038/s41598-020-72501-w] [PMID: 32973226]

[9] Lee SH, Zabolotny JM, Huang H, Lee H, Kim YB. Insulin in the nervous system and the mind: Functions in metabolism, memory, and mood. Mol Metab 2016; 5(8): 589-601.
[http://dx.doi.org/10.1016/j.molmet.2016.06.011] [PMID: 27656397]

[10] Hogenkamp PS, Cedernaes J, Chapman CD, *et al.* Calorie anticipation alters food intake after low□caloric not high□caloric preloads. Obesity (Silver Spring) 2013; 21(8): 1548-53.
[http://dx.doi.org/10.1002/oby.20293] [PMID: 23585292]

[11] Hill JO, Wyatt HR, Peters JC. Energy balance and obesity. Circulation 2012; 126(1): 126-32.
[http://dx.doi.org/10.1161/CIRCULATIONAHA.111.087213] [PMID: 22753534]

[12] Busse WW, Lemanske RF Jr. The placebo effect in asthma: Far more complex than simply "I shall please". J Allergy Clin Immunol 2009; 124(3): 445-6.
[http://dx.doi.org/10.1016/j.jaci.2009.07.042] [PMID: 19733293]

[13] Planès S, Villier C, Mallaret M. The nocebo effect of drugs. Pharmacol Res Perspect 2016; 4(2): e00208.
[http://dx.doi.org/10.1002/prp2.208] [PMID: 27069627]

[14] Matza L, Cong Z, Chung K, *et al.* Utilities associated with subcutaneous injections and intravenous infusions for treatment of patients with bone metastases. Patient Prefer Adherence 2013; 7: 855-65.
[http://dx.doi.org/10.2147/PPA.S44947] [PMID: 24039408]

[15] Humphrey N, Skoyles J. The evolutionary psychology of healing: A human success story. Curr Biol 2012; 22(17): R695-8.
[http://dx.doi.org/10.1016/j.cub.2012.06.018] [PMID: 22975000]

[16] Hunt T. The middle way of evolution. Commun Integr Biol 2012; 5(5): 408-21.
[http://dx.doi.org/10.4161/cib.20581] [PMID: 23181154]

[17] Apkarian VA, Hashmi JA, Baliki MN. Pain and the brain: Specificity and plasticity of the brain in clinical chronic pain. Pain 2011; 152(3) (Suppl.): S49-64.
[http://dx.doi.org/10.1016/j.pain.2010.11.010] [PMID: 21146929]

[18] Ahluwalia A, Tarnawski AS. Critical role of hypoxia sensor--HIF-1α in VEGF gene activation. Implications for angiogenesis and tissue injury healing. Curr Med Chem 2012; 19(1): 90-7.
[http://dx.doi.org/10.2174/092986712803413944] [PMID: 22300081]

[19] Huber-Lang M, Lambris JD, Ward PA. Innate immune responses to trauma. Nat Immunol 2018; 19(4): 327-41.
[http://dx.doi.org/10.1038/s41590-018-0064-8] [PMID: 29507356]

[20] Bonneaud C, Mazuc J, Gonzalez G, *et al.* Assessing the cost of mounting an immune response. Am Nat 2003; 161(3): 367-79.
[http://dx.doi.org/10.1086/346134] [PMID: 12703483]

[21] Snyder JM, Molk DM, Treuting PM. Increased mortality in a colony of zebra finches exposed to continuous light. J Am Assoc Lab Anim Sci 2013; 52(3): 301-7.
[PMID: 23849414]

[22] Ohio State University. Symptoms of illness less severe in hamsters during winter, Study Finds. ScienceDaily. 2002. Available from: www.sciencedaily.com/releases/2002/02/0202260 75131.htm

[23] Scheier MF, Carver CS. Effects of optimism on psychological and physical well-being: Theoretical overview and empirical update. Cognit Ther Res 1992; 16(2): 201-28.

[http://dx.doi.org/10.1007/BF01173489]

[24] Nesse RM, Stearns SC. The great opportunity: Evolutionary applications to medicine and public health. Evol Appl 2008; 1(1): 28-48.
[http://dx.doi.org/10.1111/j.1752-4571.2007.00006.x] [PMID: 25567489]

[25] Peterson C, Park N, Kim ES. Can optimism decrease the risk of illness and disease among the elderly? Aging Health 2012; 8(1): 5-8.
[http://dx.doi.org/10.2217/ahe.11.81]

[26] Lee LO, James P, Zevon ES, *et al.* Optimism is associated with exceptional longevity in 2 epidemiologic cohorts of men and women. Proc Natl Acad Sci USA 2019; 116(37): 18357-62. [http://dx.doi.org/10.1073/pnas.1900712116] [PMID: 31451635]

[27] Panesar NS, Goggins W. Postponement of death around Chinese holidays: a Hong Kong perspective. Singapore Med J 2009; 50(10): 990-6.
[PMID: 19907890]

[28] Rasa OAE. A case of invalid care in wild dwarf mongooses. Z Tierpsychol 1983; 62(3): 235-40.
[http://dx.doi.org/10.1111/j.1439-0310.1983.tb02153.x]

[29] Steinkopf L. The signaling theory of symptoms: an evolutionary explanation of the placebo effect. Evol Psychol 2015; 13: 3.
[http://dx.doi.org/10.1177/1474704915600559]

[30] Ho KH, Hashish I, Salmon P, Freeman R, Harvey W. Reduction of post-operative swelling by a placebo effect. J Psychosom Res 1988; 32(2): 197-205.
[http://dx.doi.org/10.1016/0022-3999(88)90055-4] [PMID: 3404502]

[31] Steinkopf L. The signaling theory of symptoms. Evol Psychol 2015; 13: 3.
[http://dx.doi.org/10.1177/1474704915600559]

[32] Hashish I, Haia HK, Harvey W, Feinmann C, Harris M. Reduction of postoperative pain and swelling by ultrasound treatment: A placebo effect. Pain 1988; 33(3): 303-11.
[http://dx.doi.org/10.1016/0304-3959(88)90289-8] [PMID: 3419838]

[33] Kessner S, Sprenger C, Wrobel N, Wiech K, Bingel U. Effect of oxytocin on placebo analgesia: A randomized study. JAMA 2013; 310(16): 1733-5.
[http://dx.doi.org/10.1001/jama.2013.277446] [PMID: 24150470]

[34] Kosfeld M, Heinrichs M, Zak PJ, Fischbacher U, Fehr E. Oxytocin increases trust in humans. Nature 2005; 435(7042): 673-6.
[http://dx.doi.org/10.1038/nature03701] [PMID: 15931222]

[35] Valentini E, Martini M, Lee M, Aglioti SM, Iannetti G. Seeing facial expressions enhances placebo analgesia. Pain 2014; 155(4): 666-73.
[http://dx.doi.org/10.1016/j.pain.2013.11.021] [PMID: 24315986]

[36] Fishman E, Turkheimer E, DeGood DE. Touch relieves stress and pain. J Behav Med 1995; 18(1): 69-79.
[http://dx.doi.org/10.1007/BF01857706] [PMID: 7595953]

[37] Kerr CE, Jones SR, Wan Q, *et al.* Effects of mindfulness meditation training on anticipatory alpha modulation in primary somatosensory cortex. Brain Res Bull 2011; 85(3-4): 96-103.
[http://dx.doi.org/10.1016/j.brainresbull.2011.03.026] [PMID: 21501665]

[38] Kaptchuk TJ, Kelley JM, Conboy LA, *et al.* Components of placebo effect: Randomised controlled trial in patients with irritable bowel syndrome. BMJ 2008; 336(7651): 999-1003.
[http://dx.doi.org/10.1136/bmj.39524.439618.25] [PMID: 18390493]

[39] Madhukar LS, Nivrutti BA, Bhatngar V, Bhatnagar S. Physio-anatomical explanation of abhyanga: An ayurvedic massage technique for healthy life. Journal of Traditional Medicine & Clinical Naturopathy 2018; 7(1): 252.

[http://dx.doi.org/10.4172/2573-4555.1000252]

[40] Kamenica E, Naclerio R, Malani A. Advertisements impact the physiological efficacy of a branded drug. Proc Natl Acad Sci USA 2013; 110(32): 12931-5. [http://dx.doi.org/10.1073/pnas.1012818110] [PMID: 23878212]

[41] Gracely R, Dubner R, Deeter W, Wolskee P. Clinicians' expectations influence placebo analgesia. Lancet 1985; 325(8419): 43. [http://dx.doi.org/10.1016/S0140-6736(85)90984-5] [PMID: 2856960]

[42] Benedetti F, Pollo A, Lopiano L, Lanotte M, Vighetti S, Rainero I. Conscious expectation and unconscious conditioning in analgesic, motor, and hormonal placebo/nocebo responses. J Neurosci 2003; 23(10): 4315-23. [http://dx.doi.org/10.1523/JNEUROSCI.23-10-04315.2003] [PMID: 12764120]

[43] Stewart-Williams S, Podd J. The placebo effect: dissolving the expectancy *versus* conditioning debate. Psychol Bull 2004; 130(2): 324-40. [http://dx.doi.org/10.1037/0033-2909.130.2.324] [PMID: 14979775]

[44] Undurraga J, Baldessarini RJ. Randomized, placebo-controlled trials of antidepressants for acute major depression: thirty-year meta-analytic review. Neuropsychopharmacology 2012; 37(4): 851-64. [http://dx.doi.org/10.1038/npp.2011.306] [PMID: 22169941]

[45] Walsh BT, Seidman SN, Sysko R, Gould M. Placebo response in studies of major depression: variable, substantial, and growing. JAMA 2002; 287(14): 1840-7. [http://dx.doi.org/10.1001/jama.287.14.1840] [PMID: 11939870]

[46] Kirsch I. Antidepressants and the placebo effect. Z Psychol 2014; 222(3): 128-34. [http://dx.doi.org/10.1027/2151-2604/a000176] [PMID: 25279271]

[47] Moerman Daniel. Against 'Placebo.' The Case for Changing our Language, and for the Meaning Response. Placebo and Pain: From Bench to Bedside 2013; 183-8. [http://dx.doi.org/10.1016/B978-0-12-397928-5.00018-0]

[48] Fava M, Evins AE, Dorer DJ, Schoenfeld DA. The problem of the placebo response in clinical trials for psychiatric disorders: culprits, possible remedies, and a novel study design approach. Psychother Psychosom 2003; 72(3): 115-27. [http://dx.doi.org/10.1159/000069738] [PMID: 12707478]

[49] Kaptchuk TJ, Friedlander E, Kelley JM, *et al.* Placebos without deception: a randomized controlled trial in irritable bowel syndrome. PLoS One 2010; 5(12): e15591. [http://dx.doi.org/10.1371/journal.pone.0015591] [PMID: 21203519]

[50] Enck P, Klosterhalfen S, Weimer K, Horing B, Zipfel S. The placebo response in clinical trials: more questions than answers. Philos Trans R Soc Lond B Biol Sci 2011; 366(1572): 1889-95. [http://dx.doi.org/10.1098/rstb.2010.0384] [PMID: 21576146]

[51] Enck P, Bingel U, Schedlowski M, Rief W. The placebo response in medicine: minimize, maximize or personalize? Nat Rev Drug Discov 2013; 12(3): 191-204. [http://dx.doi.org/10.1038/nrd3923] [PMID: 23449306]

[52] Benedetti F. Placebo and the new physiology of the doctor-patient relationship. Physiol Rev 2013; 93(3): 1207-46. [http://dx.doi.org/10.1152/physrev.00043.2012] [PMID: 23899563]

[53] Wager TD, Atlas LY. The neuroscience of placebo effects: connecting context, learning and health. Nat Rev Neurosci 2015; 16(7): 403-18. [http://dx.doi.org/10.1038/nrn3976] [PMID: 26087681]

[54] Benedetti F. Placebo effects: from the neurobiological paradigm to translational implications. Neuron 2014; 84(3): 623-37. [http://dx.doi.org/10.1016/j.neuron.2014.10.023] [PMID: 25442940]

[55] Zubieta JK, Stohler CS. Neurobiological mechanisms of placebo responses. Ann N Y Acad Sci 2009; 1156(1): 198-210.
[http://dx.doi.org/10.1111/j.1749-6632.2009.04424.x] [PMID: 19338509]

[56] Colloca L, Klinger R, Flor H, Bingel U. Placebo analgesia: Psychological and neurobiological mechanisms. Pain 2013; 154(4): 511-4.
[http://dx.doi.org/10.1016/j.pain.2013.02.002] [PMID: 23473783]

[57] Prudic J, Fitzsimons L, Nobler MS, Sackeim HA. Naloxone in the prevention of the adverse cognitive effects of ECT: a within-subject, placebo controlled study. Neuropsychopharmacology 1999; 21(2): 285-93.
[http://dx.doi.org/10.1016/S0893-133X(99)00015-9] [PMID: 10432476]

[58] Ballaz S. The unappreciated roles of the cholecystokinin receptor CCK(1) in brain functioning. Rev Neurosci 2017; 28(6): 573-85.
[http://dx.doi.org/10.1515/revneuro-2016-0088] [PMID: 28343167]

[59] Benedetti F, Amanzio M, Rosato R, Blanchard C. Nonopioid placebo analgesia is mediated by CB1 cannabinoid receptors. Nat Med 2011; 17(10): 1228-30.
[http://dx.doi.org/10.1038/nm.2435] [PMID: 21963514]

[60] Bracey MH, Hanson MA, Masuda KR, Stevens RC, Cravatt BF. Structural adaptations in a membrane enzyme that terminates endocannabinoid signaling. Science 2002; 298(5599): 1793-6.
[http://dx.doi.org/10.1126/science.1076535] [PMID: 12459591]

[61] Peciña M, Martínez-Jauand M, Hodgkinson C, Stohler CS, Goldman D, Zubieta JK. FAAH selectively influences placebo effects. Mol Psychiatry 2014; 19(3): 385-91.
[http://dx.doi.org/10.1038/mp.2013.124] [PMID: 24042479]

[62] Scott DJ, Stohler CS, Egnatuk CM, Wang H, Koeppe RA, Zubieta JK. Placebo and nocebo effects are defined by opposite opioid and dopaminergic responses. Arch Gen Psychiatry 2008; 65(2): 220-31.
[http://dx.doi.org/10.1001/archgenpsychiatry.2007.34] [PMID: 18250260]

[63] Fava M, Evins AE, Dorer DJ, Schoenfeld DA. The problem of the placebo response in clinical trials for psychiatric disorders: culprits, possible remedies, and a novel study design approach.. Psychother Psychosom 2003; 72(3): 115-27.
[http://dx.doi.org/10.1159/000069738] [PMID: 12707478]

[64] Huneke NTM, van der Wee N, Garner M, Baldwin DS. Why we need more research into the placebo response in psychiatry. Psychol Med 2020; 50(14): 2317-23.
[http://dx.doi.org/10.1017/S0033291720003633] [PMID: 33028433]

[65] Frisaldi E, Shaibani A, Benedetti F. Placebo responders and nonresponders: what's new? 2018; 8: 6.
[http://dx.doi.org/10.2217/pmt-2018-0054]

[66] Luparello TJ, Leist N, Lourie CH, Sweet P. The interaction of psychologic stimuli and pharmacologic agents on airway reactivity in asthmatic subjects. Psychosom Med 1970; 32(5): 509-14.
[http://dx.doi.org/10.1097/00006842-197009000-00009] [PMID: 4097491]

[67] Miller FG. Homage to Henry Beecher (1904-1976). Perspect Biol Med 2012; 55(2): 218-29.
[http://dx.doi.org/10.1353/pbm.2012.0013] [PMID: 22643759]

[68] Tufo HM, Ostfeld AM, Shekelle R. Central nervous system dysfunction following open-heart surgery. JAMA 1970; 212(8): 1333-40.
[http://dx.doi.org/10.1001/jama.1970.03170210039006] [PMID: 5467674]

[69] Mason RC Jr, Clark G, Reeves RB Jr, Wagner SB. Acceptance and healing. J Relig Health 1969; 8(2): 123-42.
[http://dx.doi.org/10.1007/BF01533141] [PMID: 24419988]

[70] Bowling A, Rowe G, Lambert N, *et al.* The measurement of patients' expectations for health care: a review and psychometric testing of a measure of patients' expectations. Health Technol Assess 2012;

16(30): i-xii, 1-509.
[http://dx.doi.org/10.3310/hta16300] [PMID: 22747798]

[71] Roscoe JA, Morrow GR, Aapro MS, Molassiotis A, Olver I. Anticipatory nausea and vomiting. Support Care Cancer 2011; 19(10): 1533-8.
[http://dx.doi.org/10.1007/s00520-010-0980-0] [PMID: 20803345]

[72] Leibowitz KA, Howe LC, Crum AJ. Changing mindsets about side effects. BMJ Open 2021; 11(2): e040134.
[http://dx.doi.org/10.1136/bmjopen-2020-040134] [PMID: 33526496]

[73] Fernandez A, Kirsch I, Noël L, *et al.* A test of positive suggestions about side effects as a way of enhancing the analgesic response to NSAIDs. PLoS One 2019; 14(1): e0209851.
[http://dx.doi.org/10.1371/journal.pone.0209851] [PMID: 30605458]

[74] Howe LC, Leibowitz KA, Perry MA, *et al.* Changing patient Mindsets about Non–Life-Threatening symptoms during oral immunotherapy: a randomized clinical trial. J Allergy Clin Immunol Pract 2019; 7(5): 1550-9.
[http://dx.doi.org/10.1016/j.jaip.2019.01.022] [PMID: 30682576]

CHAPTER 3

Role of Gut Microbiota in Neuroinflammation and Neurological Disorders

Khadga Raj[1,*], **Navneet Arora**[2], **Rohit**[2], **Anupam Awasthi**[1], **Mayank Patel**[1], **Ankit Chaudhary**[1], **Shamsher Singh**[1] and **G.D. Gupta**[3]

[1] *Department of Pharmacology, ISF College of Pharmacy, Moga, Punjab-142001, India*

[2] *Department of Pharmacy Practice, ISF College of Pharmacy, Moga, Punjab-142001, India*

[3] *Department of Pharmaceutics, ISF College of Pharmacy, Moga, Punjab-142001, India*

Abstract: The prevalence of neurological diseases such as Alzheimer's disease (AD), Parkinson's disease (PD), and Multiple sclerosis (MS) are growing in the world, but their pathogenesis is unclear and effective treatment does not exist. Neuroinflammation is associated with many neurodegenerative mechanisms involved in neurodegenerative diseases. The human gut microbiota is an aggregate of microorganisms that live in the gastrointestinal tract (GIT) that plays a crucial role in maintaining human health and the pathogenesis disease condition. The microbiota can affect neuronal function through neurotransmitters, vitamins, and neuroactive microbial metabolites like short-chain fatty acids. The change in gut microbiota architecture causes increased permeability of the intestine and immune system activation, contributing to systemic inflammation, neurological injury, and eventually neurodegeneration. Available data suggest that the microbiota send signals to the central nervous system (CNS) by activating afferent neurons of the vagus nerve *via* neuroendocrine and neuroimmune pathways. The molecular interaction between the gut/microbiome and CNS is complex and bidirectional, ensuring gut homeostasis and proper digestion. Evidence suggests that dysfunction of the gut-brain axis could be a significant factor leading to many disorders of CNS. In this chapter, we explore how the gut microbiome may affect brain function and the development of neurological disorders. In addition, we are also trying to highlight the recent advances in improving neurological disease by supplemental probiotics and faecal microbiota transplantation *via* the concept of the gut-brain axis to combat brain-related dysfunction.

Keywords: Alzheimer's disease, Central nervous system, Parkinson's disease, Gastrointestinal tract, Microbiota, Multiple sclerosis, Neuroinflammation.

* **Corresponding author Khadga Raj:** Department of Pharmacology, ISF College of Pharmacy, Moga, Punjab-142001, India; Tel: +91-8288979902; E-mail: bishalarann@gmail.com

Zareen Amtul (Ed.)

INTRODUCTION

The human microbiome comprises bacteria, viruses, archaea and eukaryotic organisms that live in and on our bodies. These microbes potentially affect human physiology, both in health and illness conditions [1]. The human microbiota is estimated to include 10^{13}–10^{14} microbial cells, with a microbial cell-to-human cell ratio of about 1:1. The varied gut microbiota comprises bacteria classified as Firmicutes, Bacteroidetes, and Actinobacteria [2]. This diverse and complex microbiome functions as a functional extension of host genomes, with an estimated 50–100-fold increase in gene count compared to the host. These additional genes have resulted in various enzymatic proteins that were not previously encoded by the host and serve a vital function in aiding host metabolism and therefore contribute to the control of host physiology [3]. The microbiota helps the host in many ways, including improving gut integrity and shaping the intestinal epithelium, harvesting energy, protecting against pathogens, and regulating host immunity. However, these processes may be disturbed due to changed microbial composition, a condition known as dysbiosis. However, these processes may be concerned due to altered microbial composition, a condition known as dysbiosis. With the development of more sophisticated techniques for profiling and characterizing complex ecosystems, it has become increasingly clear that the microbiota plays a role in a wide variety of intestinal and extra-intestinal illnesses [4].

COMPOSITION AND STRUCTURE OF THE HUMAN GI MICROBIOTA

The human gastrointestinal tract is a complicated system that starts at the oesophagus and ends at the anus, with most data collected to date from the distal colonic microbiota owing to specimen collecting practicalities. Significant physiological factors like as pH, bile content, and transit time change throughout the GI tract, contributing to the existence of different microbial populations in the upper and lower GI tracts [5]. The gut microbiota is not as varied as bacteria found in other body parts and has a high degree of functional redundancy. A comprehensive inventory of the human gut microbiome's functional capability was recently acquired, with 9879896 genes discovered using a combination of 249 newly sequenced and 1018 previously published samples. The study identified country-specific microbial signatures, suggesting that environmental factors, such as diet, shape gut microbiota composition, and possibly also by host genetics [6]. However, it is noted that microbiotas with varying designs may exhibit some functional redundancy, resulting in comparable protein or metabolite profiles. This knowledge is crucial for developing treatment therapeutic strategies to modify and shape the microbial community in disease. It is critical to evaluate cognition, hearing, vision, and speech first to interpret examination results accurately.

INTERACTION OF GUT AND NERVOUS SYSTEM, GUT-BRAIN-AXIS

The gut-brain connection was identified for the first time by Gershon American physicist at the end of the 19th century. The gut-brain axis communicates in two directions between the brain and the gut. For instance, the CNS influences gut function in reaction to psychological and physical stresses, altering motility, secretion, and immunological reactivity, while alterations in the gut microbiota may result in behavioral and neurochemical changes [7]. The microbiota-gut-brain axis control the GI tract and CNS through the vagus nerve, hypothalamic–pituitary–adrenal axis, and various cytokines.

The vagus nerve has been a primary focus in recent studies of the gut-brain axis because it represents a major bidirectional connection between the body and the brain. The vagus nerve is the 10th cranial nerve and serves several GI organs such as the oesophagus, stomach, small intestine, and colon, as well as other digestive organs such as the liver, pancreas, and gallbladder, as well as cardiopulmonary organs such as the heart, lung, trachea, and aortic arch [8]. Sensory (afferent) and motor (efferent) fibres coexist in the vagus nerve, but their cell bodies are located in different places: sensory neurons in the nodose/jugular ganglia adjacent to the jugular foramen and motor neurons in the dorsal motor nucleus of the vagus (DMV) and the nucleus ambiguus in the brainstem. Sensory neurons of the vagus nerve are genetically diverse, with a wide range of molecular machines for sensing stretch, tension, and other chemical signals, as well as connecting with other sensory cells such enteroendocrine cells, neuroepithelia bodies, taste buds, and enteric neurons. The vagus nerve's genetically unique sensory neurons are expected to encode different body information [8].

The gut microbiota controls the gut-brain axis, where brain communication to the GI tract is linked to permeability, motility, secretion, and immunological modulation in the brain's function. The gut-brain axis comprises the immune system, neuroendocrine system, GIT, ENS, ANS, and CNS, controlling afferent and efferent regulation. The GI tract regulates endocrine and exocrine secretions, motility, and microcirculation and is also involved in immunological and inflammatory processes. Numerous brain regions, including the hypothalamus, pituitary adrenal (HPA), and vagus nerve system, regulate metabolic, immunological, and homeostatic functions [9].

Changing the composition of the gut microbiota may disrupt gut homeostasis and affect the CNS and many metabolic diseases such as inflammatory bowel disease (IBD). The health-promoting bacteria concentration is reduced when the microbiota is altered in the gut by *lactobacilli* and *Bifidobacteria* [10]. Gram-negative bacteria contribute to dysbiosis by producing immunogenic neurotoxins such as LPS, which increases intestinal permeability. Thus, it is necessary to balance the immune response and colonization of the intestinal microbiota. Due to the complex relationship between gut microorganisms and the host population, the authors proposed a new concept: the microbiota-gut-brain axis [11]. Over the last two decades, research has shown that changes in the gut microbiota linked with ageing lead to an increased susceptibility of older adults to various illnesses, including cardiovascular disease, cancer, obesity, cancer, diabetes, and neurological disorders. Ageing is a complex process that affects organisms' physiological, metabolic, and immunological functions, resulting in inflammation and metabolic dysfunction. The intestinal barrier's permeability increases with age in various animal species, including worms, rats, and mice [12]. It has been suggested that age-related deterioration of the intestinal barrier function results in the leaking of gut microorganisms into the systemic circulation, resulting in increased antigenic load and systemic immunological activation [13]. For example, it has been shown that age-associated remodelling of the gut microbiome results in increased production of pro-inflammatory cytokines and intestinal barrier failure in mice. Gammaproteobacteria have been linked to increased intestinal permeability, inflammation, and death in Drosophila. The research shows that intestinal dysbiosis, independent of chronological age, indicates age-onset end in flies. Short-chain fatty acids (SCFAs) produced by the microbiome, such as butyrate, propionate, acetate, and valerate, are an essential energy source for the epithelium and affect the epithelium hypoxia-inducible factor-mediated fortification of the epithelial barrier [14].

Interestingly, elderly people demonstrated a decrease in SCFA levels, including butyrate, while centenarians demonstrated a population rearrangement of particular butyrate-producing bacteria. Metabolites produced from the microbiota have also been implicated in intestinal epithelial stem cell growth. For instance, butyrate and nicotinic acid, both by-products of the gut microbiota, are involved in the colon's inhibition and stimulation of stem cell proliferation, respectively. Additionally, neurostimulators produced from the microbiota, such as serotonin, glutamate, and gamma-aminobutyric acid, have been shown to control intestinal epithelial stem cell proliferation through the enteric nervous system. Other microbiota-derived compounds have been found to impact various host systems directly, but their role in host ageing remains unknown [15].

MICROBIAL METABOLITES AND CELLULAR COMPONENTS ON CNS

The gut microbiota converts macro-and micronutrients, fiber, and polyphenols into various metabolites, including short-chain fatty acids, trimethylamines, amino acid derivatives, and vitamins [16]. These metabolites and dietary components generated from microbes perform critical metabolic and signalling activities that affect host homeostasis, including BBB integrity and brain function. Microorganisms' influence has been recognized on behaviour and cognition. Microbial signals may control critical processes in healthy human bodies, and mounting evidence indicates that gut bacteria disruptions cause many illnesses. Microbial signals may control vital processes in healthy human bodies, and mounting evidence suggests that gut bacteria disruptions cause many diseases. Microbial signals may control critical functions in healthy human bodies, and mounting evidence suggests that gut bacteria disruptions cause many illnesses [17]. The GIT is a co-dominant organ of the CNS, ANS, and ENS. Enteric nerve control consists of four distinct levels of neurological regulation. The first level of law is that imposed by the ENS on a local level. The ENS comprises two nerve plexuses, the myenteric and submucosal, and its motor and sensory neurons communicate with one another to conduct separate information integration and processing functions comparable to the brain and spinal cord [18]. The second level is the prevertebral ganglia, which receives input from both the ENS and CNS nerves. After integrating data from various brain and spinal cord centres in response to signals about internal or external environmental changes, the CNS transmits regulatory information to the ENS or acts directly on gastrointestinal effector cells *via* the autonomic nervous system and the neuroendocrine system to regulate glands, smooth muscle, glands, and blood vessels. The fourth level comprises sophisticated brain centres; information from the cortex and subcortical area converges downhill to particular basal ganglia brain stem nuclei [19]. This kind of neuroendocrine network, which links the gastrointestinal tract and the CNS at various levels, provides the anatomical foundation for microbiota-gu--brain axis activity. The vagus nerve connects the gut directly to the brain, and bacteria may activate the afferent neurons of ENS. Disorders of the microbiota-gut-brain axis have been linked to depression, anxiety, and IBD. Microorganisms may also alter the neurophysiology of the host by generating chemical compounds, which bind to receptors both within and outside the stomach. The SCFAs have been shown in animal models to enhance the neurodevelopment and cognitive function of animals suffering from neurodegenerative diseases [20]. On the other hand, injected a particular type of SCFA [propionic acid (PPA)] into rats showed autism-related features and neurochemical alterations. Neuroinflammation, increased oxidative stress, and antioxidant depletion were all seen due to these neurochemical alterations, resulting in mitochondrial dysfunction. Other molecules produced from microorganisms, such as serotonin, melatonin, histamine, and acetylcholine, also play a role in the gut microbiome. -axis of the brain. Studies indicate that the microbiota may affect the CNS by altering adult hippocampal neurogenesis [21]. Ogbonnaya *et al.* recently found a difference between the hippocampal neurons of sterile and normal mice and that colonization with the microbiota after weaning did not affect the amount of hippocampal neurogenesis, implying that microorganisms play a critical role in hippocampal neurogenesis early in life. Early animal experiments showed that introducing a single, unique flora may develop anxiety-like behaviour, followed by the activation of brain neurons dependent on gut signals sent to the brain through the vagus nerve [22]. Early animal experiments show that the introduction of a single, unique flora may result in the development of anxiety-like behaviour; this shift was accompanied by the activation of brain neurons that were dependent on signals supplied from the gut through the vagus nerve [23]. Later research discovered that colonization of faecal bacteria from one kind of mice with another type might result in the receptor animals exhibiting behaviour identical to that of the donor mice. When microorganisms from specific-pathoge--free (SPF) Swiss mice were transferred to the bodies of sterile BALB/c mice, the exploration behaviour of sterile BALB/c mice increased; however, when the microorganisms of SPF BALB/c mice were transferred to the body of sterile Swiss mice, the exploration behaviour of sterile Swiss mice became lower than that of normal Swiss mice, thus indicating that behaviour type may be directly associated with microorganisms [24].

MICROBIOTA AND NEURODEVELOPMENT

Similar to gut microbiota development, brain maturation goes through a critical developmental stage throughout infancy and adolescence. Adolescence is believed to be the most important time for the development and beginning of various brain diseases. These include synaptic pruning, which eliminates excess synapses, reducing cortical grey matter as the brain develops [25]. This is accompanied by new neural connections, resulting in a period of great plasticity in most brains. The high level of susceptibility to pathological insults such as stress, drug misuse, and nutritional deficiency results from this massive neural rewiring throughout adolescence. This developmental period is also the peak time for the onset of numerous psychiatric disorders, including schizophrenia, substance abuse, and mood disorders. As a result, the teenage brain's susceptibility to pathological damage, along with the volatility and immaturity of the gut microbiota throughout adolescence, renders the brain particularly vulnerable to abnormal alterations that predict the development of brain diseases during this period [26].

MICROBIOTA AND AGEING

One potentially productive approach to the roles of the gut microbiota in human ageing is to compile age-related changes in the gut microbiota and examine whether these changes have any biological relevance [27]. The adult gut microbiota is specific to the person and generally stays constant over time. It can withstand negative environmental factors like antibiotic usage and stress by recovering its varied and stable 'normal' core microbiota. However, it is worth mentioning that new research challenges the idea that the gut microbiota is stable throughout adulthood, indicating that short-term dietary may significantly alter gut microorganisms. At the moment, it is unknown if such alterations may potentially have a rapid effect on CNS health [28].

Similarly, to how the gut microbiota matures and stabilizes, ongoing brain development happens throughout this stage of adulthood. Although the brain achieves its maximum weight around the age of 20, white matter volume continues to grow until the mid-forties, corresponding with the peak of myelination around 50. While adulthood does not seem to be a crucial or sensitive age, it is still a period during which changes in the microbiota may affect the brain and behaviour. While adulthood does not seem to be a crucial or sensitive age, it is still a period during which changes in the microbiota may affect the brain and behaviour. While adulthood does not seem to be a crucial or sensitive age, it is still a period during which changes in the microbiota may affect the brain and behaviour. Ageing has been shown to harm the composition of the gut microbiota, which may affect health outcomes throughout this stage of life [29]. While the gut microbiome develops throughout life, the diversity and stability of the microbiota diminish with age. Recently, it was shown that the microbial composition of older adults was linked to and affected by their residential environment, dietary regimen, and overall health condition. A variety of medicines used by the elderly reduced digestion and motility processes, resulting in nutritional malabsorption, and a weaker immune system contribute to a decreased diversity and stability of the gut microbiota composition. In the aged, reduced stability and variety of the gut microbiota are associated with reduced brain volume and cognitive performance [30]. Indeed, at the age of 55, the brain starts to lose weight. Gray and white matter tissue loss occurs in an age-dependent way in most dorsal brain areas between the ages of 50 and 70, with a more progressive decrease afterwards. These age-related changes in brain morphology occur concurrently with a compromised immune system, increased oxidative stress, and amyloid plaque accumulation in the brain, resulting in diminished cognitive and behavioural function and may manifest as a variety of age-related memory impairments and disorders Alzheimer's disease [31]. Interestingly, the study consortium research results that concentrate on the future description of the gut microbiota of older adults indicate how important diet is for the composition of microbiota in the elderly. These results encourage the idea that dietary/microbiota modification treatments may modulate the microbiota of the elderly to restore variety and thus enhance general and potentially mental health, particularly in such a crucial context as ageing [32].

MICROBIOTA AND NEUROINFLAMMATION

Neuroinflammation in the nervous system is now known as a cardinal feature of various neurological diseases. In CNS autoimmune disorders such as multiple sclerosis, there is an influx of immune cells from the periphery into the CNS, leading to demyelination and neuronal destruction [33]. However, there is also a high level of inflammatory mediators or moderate immune cell infiltration in neurodegenerative conditions PD, AD, acute ischemia, traumatic brain injury, mood diseases, and autism spectrum disorders. Chronic inflammation may have undesirable consequences for the health status of the nervous system. The GI tract is considered one of the primary immune bodies and has similar neurological functions for the brain. Several data have suggested that intestinal microbiota mainly mediates immune activities and alterations in the intestinal milieu that causes inflammation. Such inflammatory modulation is related to the pathogenesis of neurodegenerative disease, including AD. The intestinal bacterial population is affected by age, prolonged antibiotic therapy, unlimited calorie diet, and lifestyle changes that lead to intestinal contamination. Lowe *et al.* have shown that elevated bacterial change increases the inflammatory cytokine (TNF-α, IL-17, MCP-1, IL-23) level in the mouse model of alcohol-induced neuroinflammation. Microbiota altered by 5xFAD compared to wild-type mice, Alzheimer's mouse models had a higher number of precursors and a Firmicutes/Bacteroidetes genus ratio, resulting in inflammation and neurophysiology. Based on these findings, it is hypothesized that a high calorie diet modifies the framework of the gut and is linked with a variety of pathological changes in the brain.

Additionally, probiotic supplements and a low-calorie diet are sure to have therapeutic benefits by restoring a healthy gut microenvironment. Additionally, probiotic supplements and a restricted diet have therapeutic benefits by restoring a healthy gut microenvironment. Thus, altering the composition of the gut may delay or prevention of the progression of AD. Additionally, vagus nerve-derived molecular pathways targeted by cholinergic agonists are helpful in brain diseases by reducing inflammation and disrupting the BBB. Antibiotic treatment of genetically engineered (APP/PS1) animals produced faulty cognitive outcomes. Antibiotic-treated animals exhibited higher levels of pro-inflammatory cytokines, altered microbiota composition, and microglial morphology. As a result of these results, many medicines targeting neuroinflammatory components are now undergoing clinical trials in AD.

Alzheimer's Disease

Alzheimer's disease (AD) was first diagnosed in 1901 by German psychiatrist "Dr. Alois Alzheimer" in Auguste D, a 50-year-old woman, and they give the Alzheimer's name on his surname. AD is a progressive neurological disease marked by memory loss and motor impairment, disrupting learning, perceptions, language, thinking, and planning. AD may be divided into two types based on its start: early-onset (typically affects individuals between 30 and 60) and late-onset (usually involving people over 65 years). The prevalence of Alzheimer's disease increases with age, and the global majority of AD is about 0.7 to 1.8% in individuals aged 60 to 65 years, while in the population aged 85 years and older, AD is about 32%. The pathological hallmarks of AD include the deposition of Aβ and hyperphosphorylated tau proteins in the brain and the formation of neurofibrillary tangles (NFTs) [34]. These structural abnormalities and inflammatory lesions result in the gradual degeneration of neurons in the brain, which further contributes to AD development. There are various molecular mechanisms of neurodegeneration in AD includes neuroinflammation, mitochondrial dysfunction, disruption of Ca^{2+} homeostasis, reduced cytoskeletal integrity, impaired cell stress response, lipid metabolism, oxidative stress, apoptosis, epigenetic changes, deregulation of enzymes (proteases, phosphatases, and kinases) and most importantly neurotransmitter pathways failures [35].

Impact of Gut Microbiota on Brain

The CNS and gut microbiota are directly connected through the vagus nerve, which exchanges neurotransmitters, the neurological system, and chemical substances that cross the BBB. The gut microbiota may generate various bioactive compounds during metabolic processes, reaching the bloodstream through enterohepatic circulation [36]. We can study compounds associated with disease phenotypes in bodily fluids such as urine, plasma, or faeces using nuclear magnetic resonance (NMR) and mass spectrometry-based metabolomics. The GM metabolizes a broad variety of neurotransmitters and neuromodulators (short-chain fatty acids), including glutamate, acetylcholine, gamma-aminobutyric acid, dopamine, and serotonin. Bacterial treatments can alter the levels of neurotransmitters involved in synaptic plasticity and control the activation of serotonin and N-methyl-d-aspartate (NMDA) receptors [37]. Impairment in the composition of gut microbiota and their metabolites can alter the gut-brain axis, which controls memory, mood, cognition, and social behaviour. However, dysbiosis may produce harmful misfolded proteins, promoting cell malfunction, neurodegeneration, and synapse loss [38]. The major neurotransmitter and their role in the brain are described in Table **1**.

Table 1. The major neurotransmitter and their role in the brain.

Gut Microbiota	Neurotransmitters	Function	References
Lactobacillus, Lactobacillus plantarum, Bacillus	Acetylcholine	Cognition, arousal, emotional, memory, motivation, attention, self-care ability.	[39]
Lactobacillus brevis, Streptococcus genera, Bifidobacterium infantis, Bifidobacterium dentium, Bifidobacterium adolescentis	GABA	Regulates feelings of fear, anxiety, cognitive functions, depression, behavioral	[40]
Coryneform, Brevibacterium lactofermentum, Brevibacterium flavum, Bacteroides vulgatus, Campylobacter jejuni, Lactobacillus plantarum, Lactobacillus paracasei, and Lactococcus lactis	Glutamate	Synaptic plasticity, motor functions, memory, learning	[41]
Streptococcus, Enterococcus Lactobacillus plantarum, Akkermansia muciniphila	Serotonin	Regulates mood, learning, cognition, memory, feelings of well-being, and happiness.	[42]
Lactobacillus, Escherichia, Lactococcus, Bacillus, Streptococcus	Dopamine	Protects neuron loss, improves motor deficits, cognition, reduced anxiety and stress, decision making, attention, motivation and reward	[43]

Role of Gut Microbiota in AD

The microorganisms that live in the human gut are complex and dynamic species. These gut bacteria coexist with the host in a symbiotic relationship and are evolved to coordinate and integrate conserved metabolic signals, regulate homeostasis, and enhance immunity, thus ensuring the host's survival [44]. The microbiome is critical for regulating neural functioning, behaviour, and brain development. Researchers continuously to investigate the specific role of the gut microbiota and how its disruption results in ongoing neurodegeneration, neuroinflammation, and cognitive impairment in AD. Along with neuronal abnormalities, changes in the makeup of the gut microbiota are linked with the development of various illnesses, including gastrointestinal problems, obesity, diabetes, metabolic diseases, and cardiovascular disturbances [45]. Alterations in the gut microbiota and dysbiosis occur due to continuous use of antibacterial agents, non-steroidal anti-inflammatory medications, and changes in dietary pattern and lifestyle [46].

Additionally, prior clinical research discovered many gut species, including *Bacillus subtilis*, *Escherichia coli*, *Klebsiella pneumonia*, *Mycobacterium spp*., *Salmonella spp*., *Staphylococcus aureus*, and *Streptococcus spp*., are linked with amyloid fiber formation [47]. Once produced in the intestine, Aβ increases the intestinal wall's permeability and can quickly enter the systemic circulation and accumulate in the brain. This process aggregation of Aβ is facilitated by a malfunction in subjects with AD of some carrier (ApoE and ApoJ) that regulates the influx of Aβ by crossing the BBB and causes impairment in cognition & memory. Additionally, amyloid produced from bacteria increased the amounts of pro-inflammatory cytokines (IL-1, IL-6, IL-17, IL-10, and IL-22) in the blood that travels from the gut to the brain, where they initiate immunogenic responses that lead to neurodegeneration. As a result of these cascades, phagocytosis is impaired, leading to A1-42 aggregation [48].

The intestinal microbes play an essential role in the process of neurogenesis. The symbionts of the gut microbiota support the host, and a healthy gut is vital to developing neural and behavioural functions. A stable gut is prone to changes in intestinal microbiota diversity, which leads to the production of neurodegenerative markers in AD pathology, including Aβ, NFT, and inflammatory cytokines [49]. Feeding a high-fat diet (HFD) to pregnant mothers compromises the intestinal microflora and shows harmful effects on the behaviour of offspring and the diversity of their intestines. However, the intake of HFD generates the reactive oxygen species, which further induces oxidative stress and increases the Aβ aggregation, permeability of BBB, and abnormal morphology of the microglial cells in the hippocampus [50, 51].

In addition, ageing, chronic antibiotic therapy, an unregulated calorie intake, and lifestyle modifications affect the intestinal bacterial population, resulting in gut contamination. In AD patients, increase the level of *Bacteroidetes* and decreases the population of *Firmicutes* and *Actinobacteria* [52]. In particular, an increase in the intestinal population of gram-negative bacteria leads to a systemic increase in LPS concentration. This may impact AD pathology by causing systemic inflammation (by activating NFκB) or co-localization with amyloid plaques. In addition, gut alterations are responsible for the progression of neuroinflammation and gut intestinal disorders [53].

Moreover, microbial pathogens and infections such as *E. coli*, *Helicobacter pylori*, *Citrobacter rodentium*, *Salmonella enterica*, *Salmonella typhimurium*, *Mycobacterium tuberculosis*, *Chlamydia pneumoniae*, and others may cause microbial dysbiosis throughout the gut. This may lead to gut and systemic inflammation and an increase in the amyloid plaques in the brain. Further, chronic alcohol consumption also promotes neuroinflammation through the expression of toll-like receptor-4 (TLR4) in the brain, and an increased population of bacterias in the gut also induces the levels of inflammatory cytokines (IL-10, Il-22, TNF-α, IL-17, MCP-1, IL-23) and causes neuroinflammation [54].

Probiotics

Probiotics are the microorganisms that promote mucosal colonization and prevent pathogenic bacteria from adhering. The probiotics bacterial strain belongs to the Lactobacillus and Bifidobacterium, where lowers the luminal pH and inhibits the abundance of gut microbiota in the human intestine [55]. The amelioration of cognitive dysfunction through prebiotics treatment with the strain of *Bifiobacterium breve* in AD. The administration of probiotics includes *L. fermentum*, *L. acidophilus*, *B. langum* and *B. lactis* may improve AD's pathological features, including learning, spatial memory, and oxidative stress [56]. They are dietary fibre that induces the growth or activity of beneficial microorganisms, and they cannot be digested in the human intestine. Prebiotics, in particular, acts as fermentation substrates for probiotic genera that produce short-chain fatty acids (SCFAs), such as Lactobacilli and Bifidobacteria, resulting in a rise in the anti-inflammatory metabolites acetate, propionate, and butyrate [57] (Fig. **1**).

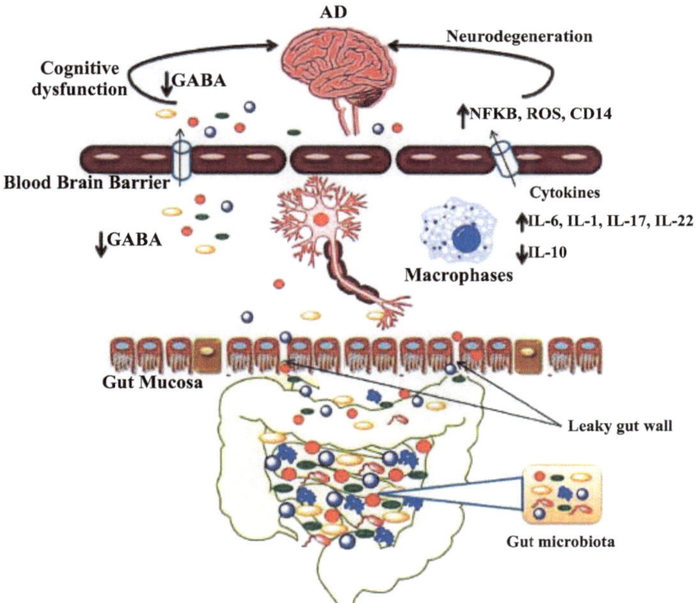

Fig. (1). MGB axis and its role in AD. Alteration of gut microbiota may initiate the pathogenesis *via* modulation of ENS and the formation of amyloid and bacterial metabolites in the gut. This pathological change induces systemic and localized inflammation *via* activation of immune cells and further increases permeability, and contributes to developing AD's pathogenesis. SCFA – short-chain fatty acids, HFD – high-fat diet.

Depression

Clinical depression is a complex condition that includes underlying temperament and personality factors, traumatic and stressful life experiences, and biological sensitivity. Depression is a "phasic" illness that can be unipolar or bipolar. Stressful life events trigger depressive episodes, however, their influence appears to lessen as the illness progresses. This shows that depression is linked to a progression of stress response abnormalities related to structural plasticity and cellular resilience deficits [58]. Thus, to avoid morphological and functional defects, it appears that proper treatment of depression in the early phases of the illness is critical. Depression is a mental disorder characterized by an aversion to action and can substantially impact daily life. It is classified as a cognitive and behavioural condition by medical professionals [59].

Although the specific aetiology of depression is unknown, many factors such as biological differences, brain chemistry, hormones, and inheritance are involved, shown in Fig. (**2**).

Fig. (2). Clinical features of Depression.

Major depression was classified as the third leading cause of illness burden by the World Health Organization (WHO) in 2008, and the disease is expected to rise to first place by 2020 [60]. Because of its diverse presentations, unexpected course and prognosis, and variable response to prognosis, its identification, diagnosis, and therapy often provide issues for clinicians in practices. Because of the high morbidity rate, researchers in India have traditionally focused on depression as a condition. As a result, it's a one-of-a-kind problem to figure out the pathophysiology of depression [61].

There is also a differentiation drawn between depression in those who have or do not have manic episodes. Both types of depression can be chronic (meaning they last for a long time) and have relapses, especially if left untreated. Depression is caused by a complex interplay of social, psychological, and biological variables. People who have faced adversity (such as job loss, grief, or psychological trauma) are more likely to develop depression. Increased stress and dysfunction can accompany depression, affecting both the affected person's situation and the depression itself. Depression is a life-threatening disorder that affects hundreds of millions of people all over the world. It can affect anyone at any age, from childhood to old age, and it costs society a lot of money since it causes so much pain and disturbance in people's lives, and it can be fatal if left untreated [62]. The psychopathological condition is characterized by low or depressed mood, anhedonia, and low energy or fatigue. Sleep and psychomotor difficulties, guilt, low self-esteem, suicidal impulses, and autonomic and gastrointestinal abnormalities are other signs and symptoms. Rather than being a particular ailment, depression is a complex phenomenon with various types and possibly multiple etiologies. It includes a predisposition to episodic and often progressive mood irregularities, symptomatology ranging from mild to severe symptoms with or without psychotic elements, and links to various mental and somatic disorders. The cause of Major Depressive Disorder (MDD), one of the most common and devastating mental diseases, is unknown [63].

Alterations in neurotransmission, HPA axis abnormalities associated with chronic stress, inflammation, diminished neuroplasticity, and network dysfunction are plausible pathophysiological reasons for depression. All of the hypothesized mechanisms are intertwined and have bidirectional interactions. Furthermore, psychological factors have been proven to directly impact neurodevelopment, resulting in a biological predisposition to depression [64]. While data suggests that severely depressed patients require antidepressant drug therapy and that non-severely depressed patients may benefit from alternative approaches (*i.e.*, "non-biological"), little research on the efficacy of other depression treatments has been done. In addition, antidepressant medication therapeutic response and tolerance may vary. Prevention initiatives have been shown to help people with depression. Effective community techniques for avoiding depression include school-based programs that help children and adolescents develop a habit of optimistic thinking. Exercise programs for the elderly can help to keep depression at bay [65].

Clinical Features

Symptoms of depression can range from moderate to severe. These symptoms include changes in appetite, difficulties sleeping, increased weariness, lack of interest in activities, feelings of worthlessness or guilt, suicidal ideation, slower speech, and difficulty thinking and concentrating in daily activities [66]. Not only are depressive syndromes different in terms of etiologies, but symptoms like guilt and suicidality are impossible to replicate in animal models. On the other hand, other symptoms have been accurately predicted, and these, along with clinical data, reveal information about the neurobiology of depression. Depressed mood, anhedonia (diminished ability to perceive pleasure from natural rewards), irritability, difficulty concentrating, and anomalies in eating and sleep (together known as "neurovegetative syndrome") are also common symptoms [67]. Depending on the number and severity of symptoms, a depressive episode can be classed as mild, moderate, or severe. Inflammation-induced depressed symptoms such as anhedonia, weariness, and interior attention are commonly referred to as "illness behaviours [68]."

Classification, Prevalence, and Course of Depression

According to the Diagnostic and Statistical Manual of Menial Health, Fourth Edition, the essence of major depressive disorder is a clinical course marked by one or more major depressive episodes without a history of manic, mixed, or hypomanic episodes (DSM.-IV) [69]. In addition, five of the following nine DSM-IV symptoms must be present for at least two weeks to make an accurate diagnosis: I a depressed mood; (ii) a loss of interest or pleasure; (iii) a significant change in weight or appetite; (iv) insomnia or hyposomnia; (v) fatigue or loss of energy; (vi) psychomotor agitation or retardation; (vii) diminished ability to think or concentrate or indecisiveness; (viii) feelings of worthlessness; and (ix) suicidal ideation.

The origins and classification of depression have been the subject of much debate in the past. Emil Kraepelin's understanding of depression as a disease and Sigmund Freud's understanding of depression as an expression of repressed wrath and loss were the two competing points of view around the turn of the century Sir Martin Roth, and the Newcastle Group made a significant contribution to our understanding of depression by categorizing clinical symptoms of depression (ranging from mild to severe psychosis) and dividing them into discrete groupings of "endogenous" and "reactive" subtypes of depression [70]. This notion has been employed in biological psychiatric research for decades to discover different etiological subtypes of the condition. The most current DSM-IV 1 and the International Classification of Diseases, 10th Revision (ICD-10) 3 editions distinguish unipolar (depression) from bipolar (manic depressive) illness, based on findings from collaborative projects"-5 in the United States and the United Kingdom.

Risk Factor for Depression

The Impact of Life Events

The existence of events triggering the beginning depression, whether endogenous or nonendogenous, are linked to the symptom pattern, implying that there is no clear-cut difference between the presence of events that cause endogenous and nonendogenous depression [72]. The role of events on depressing outcome is still being debated since pleasant events have been proven to improve outcomes while stressful situations have been shown to decrease progress and raise the risk of relapse. Females are more prone than males to suffer from severe depression cannot be explained by differences in rates or sensitivity to stressful life events [73].

Although women reported more interpersonal stressors than men, men reported more legal or work-related stressors; this cannot be connected to females experiencing more severe depression.

Genetic Influences

According to evidence from family, twin, and adoption studies, genetic characteristics play a critical role in the etiology of emotional disorders. In addition, there is strong epidemiological evidence for a genetic component, notably in bipolar disorders, with heritability estimates as high as 80% [74]. However, the inheritance does not follow the typical mendelian pattern, meaning that a single significant gene locus may not, or only in a few families, explain the increased intrafamilial risk of the illness [75].

Biochemical Basis of Depression

Neuroscience has evolved into an interdisciplinary field that encompasses a wide range of biological investigations, from molecular studies of cell and gene function to brain-imaging techniques, expanding our understanding of the cellular and molecular machinery that controls behaviour. It began with neuroanatomy and electrophysiology at the turn of the twentieth century [76].

Synaptic Transmission

Pre- and postsynaptic events are well-known for being highly controlled and serving as the foundation for central nervous system plasticity and learning

(CNS). Several steps in the signal transduction cascade are involved in the chemical transmission, including neurotransmitter synthesis, storage in secretory vesicles, and regulated release into the synaptic cleft between pre-and postsynaptic neurons, as well as the termination of neurotransmitter action and induction of final cellular responses [76]. The facilitated transport of amino acids from the bloodstream to the brain is the first step in the synthesis process, where precursors are converted into transmitters *via* enzymatic reactions and then stored in synaptic vesicles before being released into the synaptic cleft *via* a Ca^{2+}-dependent process. Somatodendritic autoreceptors are another significant regulator of release because binding of released transmitter molecules reduces synthesis or prevents future release from the presynapse [78]. Neurotransmitter molecules do not pass the postsynaptic membrane, but their first binding to surface receptors within the postsynaptic membrane, frequently associated with guanine nucleotide-binding proteins, triggers a cascade of processes (G-proteins). These G-proteins are important early regulators in transmembrane signalling because they regulate various effector systems within cells, such as adenylyl cyclases, phospholipases, and the phosphoinositide-mediated system [79]. The early cellular events of the signal transduction cascade (*i.e.*, an increase in intracellular calcium ions or second messengers such as cyclic adenosine monophosphate cAMP) initiate a pathway *via* protein kinases, which regulates many biological responses and controls short- and long-term brain functions by regulating neuronal ion channels, receptor, and other neuronal structures [80]. A fundamental mechanism driving depression could be disrupted in one or more steps of this chemical transfer. On the other hand, it is now widely accepted that these processes are ineffective [81].

Monoamine Hypothesis

Around 30 years ago, the first significant hypothesis of depression was proposed, claiming that depression is caused by a functional deficiency of the brain monoaminergic transmitters norepinephrine (NE), 5-HT, and/or dopamine (DA), whereas mania is caused by a valuable excess of monoamines at critical synapses in the brain [82]. Animal research and clinical observations supported this theory, showing that the antihypertensive drug reserpine, which depletes presynaptic stores of NE, 5-HT, and DA, causes depression-like symptoms. However, in contrast to reserpine, some patients taking iproniazid, a tuberculosis treatment that increased NE and 5-HT levels in the brain by inhibiting the metabolic enzyme MAO, experienced euphoria and hyperactivity [83].

Transporters for Neurotransmitter Reuptake

Because transport proteins limit the availability of neurotransmitters in the synaptic cleft, they prevent neurotransmitters from activating on pre-and postsynaptic receptors; they are vital in monoaminergic transmission. The 5-HT transport system is prevalent in human platelets, even though much of our knowledge about transporter failure comes from animal and postmortem brain investigations. This enables us to investigate its action *in vivo* and in other depressive stages [84].

Impact of Gut Microbiota in Depression

In the human body, the gut microbiota plays a crucial function. According to reports, one of these microflora's benefits is improved mental wellness. Multiple factors contribute to depression, which is a severe mental condition. Low emotional disposition, loss of confidence, and indifference are all symptoms [85]. Depression is thought to be caused by intricate connections between a person's genetics and environment. According to the World Health Organization (WHO), major depressive disorder (MDD) is the most significant cause of worldwide disease burden. According to WHO estimates, almost 350 million people are affected by depression [86]. Under steady or stressful conditions, healthy gut microbiota conveys brain signals through pathways involved in the neuronal transmission, neurogenesis, behavioural regulation, and microglial activation; as a result of this process, various studies have recognized the role of microbiomes in the treatment of mental health problems. The neuroendocrine and neuroimmune pathways are also dysregulated in depression [87]. Sleep difficulties and sad behaviours affect more than 20% of patients with inflammatory bowel disease (IBD). Treatments targeting this phenomenon have risen in popularity in IBD patients and healthy people because inflammation impacts the brain and how one thinks. The pathophysiology of gut dysbiosis has been linked to several risk factors [88]. Antibiotics have been linked to alterations in the gut microbiome's makeup in the short and long term. At various periods of life, environmental factors are hypothesized to play a role in developing gut dysbiosis. The link between gut microbiota and mental health is a relatively new area of research that has gained popularity in recent years. However, there are still several issues that need to be researched and understood more [89]. Activation of intestinal transmission *via* gut permeability, regional gut motility, and secretion changes the function community pattern of the gut microbiota throughout the autonomic nervous system (ANS). The degree of changes in function and composition of gastrointestinal microflora leading to depression and the causal association of bacterial commensals and depression must be carefully explored to establish the importance of the gut microbiome in depression. Microbial lipopolysaccharides

(LPS) are produced in response to changes in the gut microbiome composition [90].

As a result, inflammatory responses are triggered. Cytokines deliver signals to the vagus nerve, connecting the process to the hypothalamic-pituitary-adrenal axis, resulting in behavioural consequences. Another school of thinking contends that inflammation in the gastrointestinal (GI) tract causes neuroinflammation. The kynurenine pathway is then triggered, which drives microglial activity [91]. However, gut microbiota may alter brain stress by increasing neuroactive chemicals like noradrenaline, serotonin, glutamate, and dopamine, which all have excitatory effects on post-synaptic neurons. Supplemental probiotics have demonstrated potential results in the treatment of brain-related issues. Although promising, additional research is needed to determine probiotics' mode of action and negative effects [92]. Under stable and stressful conditions, gut commensals may send signals to the central nervous system (CNS) *via* various pathways, including neurogenesis, neurotransmission, and others. It has been discovered that these bacterial commensals can affect stress levels and that stressful experiences can control our gut microbiome makeup. However, the gut microbiota may play a role in the amounts of these neurotransmitters in the brain and stomach [93].

Furthermore, alterations in immunological and inflammatory states may affect brain function indirectly as a result of gut dysbiosis. Paying attention to what we eat, avoid medicines that have a detrimental impact on our gut commensals, and integrate probiotic supplements into our diet regimen after discovering how microbiomes affect human behaviour [94]. The importance of gut microbiota to the body is still controversial in society, despite its well-known scientific world. Anti-inflammatory management is being used to target depressive behaviours of reducing IBD exacerbations. It will be a medical breakthrough if the inflammatory pathway is a part of the pathophysiology of psychiatric diseases. More doctors will understand that stressful experiences and inherited genes aren't the only things that lead to a patient's mental illness; the inflammatory process also plays a role [95]. Effective psychiatric illness care would entail addressing both the behavioural part of the illness and the patient's inflammatory process, resulting in overall health. Early exposure to stressful events, early drug use that eliminates gut commensals, and poor childhood food choices have all been linked to the development of depression. This information is essential for everyone who works with children [96].

Gut Dysbiosis Association with Antidepressant Drugs

The development or worsening of mental diseases, such as major depressive disorder, has been linked to dysbiosis of the gut microbiota (MDD). The aetiology of major depressive disorder (MDD) is multifactorial, and disturbed gut microbiota has been linked to it [97]. Therefore, it's crucial to figure out any differences in the gut microbiome in MDD patients. Antidepressants drugs are used for the treatment of depression. Still, they alter the composition of gut microbiota through the modulation of microbiota, and their antidepressant effects are exerted, at least in part. However, some of the antidepressant's drugs were shown antimicrobial activity in various microorganisms, and they inhibit the number of microorganisms processes, such as bacterial motility and slime production [98].

Furthermore, the gut contains high levels of norepinephrine and serotonin, which stimulate the proliferation and virulence of bacteria that operate as interkingdom signalling molecules. Impairing the serotonin transporter also disturbed gut bacteria balance, increasing early life stress effects [99]. Antidepressants' impact on the microbiota, however, may influence depressive behaviour through neurotransmitter regulation. Microbiota can create neuroactive substances, such as neurotransmitters, affecting the host's physiology and behaviour. Furthermore, through the enterochrome, the host-microbiota can impact serotonin synthesis. Furthermore, the enterochromaffin cells in the host gut can regulate serotonin synthesis by the microbiota [100].

How Gut Microbiota Alters in Depression

Physiological effects such as behaviour, nausea, satiety, and pain may be mediated *via* bidirectional communication between the gut and the brain. GI motility and secretions are altered in stressful situations. Communications may begin as sensory information from the GI tract and evolve into neuronal, hormonal, and immunological signals as a result [101]. The central nervous system is linked to these signals both cooperatively and independently. Both neurological and immunological pathways more deeply relay the progression of changed gut microbiota [102]. Gut peptides and signalling molecules like serotonin are released from enteroendocrine cells (EECs) along the GI tract, making up the gut endocrine system EECs makeup about 1% of epithelial cells in the gut lumen, and they're split into subtypes based on where they're found in the gut and the types of compounds they release, such as peptides [103]. The EECs govern the release of serotonin on 5-HT3 receptor-expressing primary afferent nerves that extend into the intestinal villi, allowing them to transmit information from the gut to the neurological system [104]. Peptides are short chains of amino-acid monomers linked by amide bonds (around 50 amino-acid residues) that can be released from neurons in the brain and bind to G-protein-coupled receptors (GPCR) and processed by endogenous enzymes, and reach receptors far from the release site [105].

Various peptides found in EECs and the CNS regulate the GI tract. Calcitonin gene-related peptide, for example, is expressed in intrinsic sensory neurons [106]. Vasoactive intestinal peptides relate to the adjacent postsynaptic targets to regulates the circadian rhythm and absorption from the intestinal lumen and inhibit gastric acid secretion [107]. Furthermore, pituitary adenylate cyclase-activating polypeptide, the vasoactive intestinal peptide, stimulates the enterochromaffin-like cell [108]. Oxyntomodulin, Insulin, GLP-2, enteroglucagon, glucagon, calcitonin gene-related peptide, gastrin inhibitory polypeptide and somatostatin are examples of other peptides. Because of their preferred location in the GI tract, those peptides are probably influenced by changes in the gut microbiota and modulate upstream communication to the brain (directly or indirectly). X/A-like cells in the intestine release orexigenic and ghrelin peptides. Gut peptides diffuse across the lamina propria, lined with immune cells, to eventually reach the blood, stimulating sensory neurons and may be acting on the vagus nerve, forming an intersection for gut peptidergic communication to the brain [109].

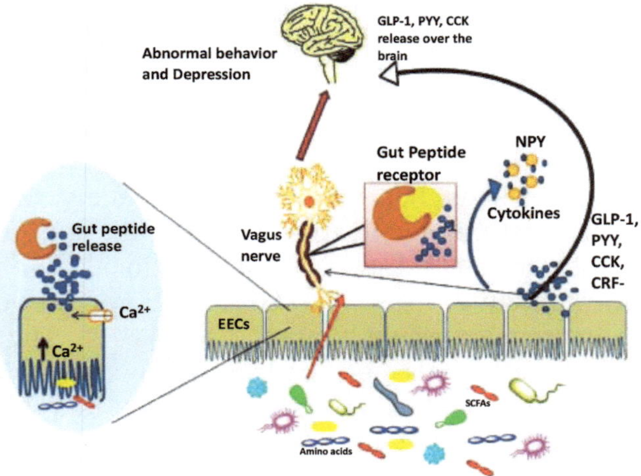

Fig. (3). Gut microbiota and brain axis communication routes the gut bacteria may communicate with the brain *via* several pathways that control physiological functions. Enteroendocrine cells (EECs) emit gut peptides, which activate immune system gut peptide receptors and vagus terminals in the gut. Under immunological stress, cytokines can also release gut peptides including neuropeptide Y (NPY). This diagram shows examples of the stomach–brain route and gut peptides. GLP-1 - glucagon-like peptide; SCFA - short-chain fatty acid; PYY - peptide YY; CRF - corticotropin-releasing factor; GABA – Gamma-aminobutyric acid; CCK = cholecystokinin.

Similarly, the BBB selectively promotes specific peptides from the stomach to the brain, proposing a direct conduit for particular gut peptides that aren't quickly degraded. Furthermore, because peptides are widely expressed in both the gut and the bloodstream, it appears plausible to suppose that they can act in ways other than their core signalling function, implying a role (most likely indirect) for gut peptides in neuropsychiatric disorders [110]. Finally, the dynamic character of gut peptides, particularly their direct link to mood disorders and their extensive role in the gut, implies that psychobiotics could target the microbiota gut peptide and brain communication axis to prevent and treat such illnesses shown in Fig. (**3**) [111].

Parkinson's Disease

In 1817, 'James Parkinson' published "An Assay on Shaking Palsy," which established the movement abnormality that is today recognised as Parkinson's disease (PD). Parkinson's disease (PD) is a multifactorial degenerative disease that causes tremors, gait rigidity, and hypokinesia, making it difficult to live usually [112]. Unfortunately, due to a lack of early detection methods, this disease is typically discovered in its later stages, when neurons have entirely degraded. As a result, the cure is on hold, and death is certain. As a result, biomarkers are required to diagnose the condition early on when it is still possible to avoid it [113].

Clinical Features of PD

Symptoms of Parkinson's disease usually appear gradually. Resting tremors, muscular stiffness, akinesia/bradykinesia, and postural instability are the four main clinical hallmarks of Parkinson's disease (PD). However, these symptoms can be accompanied by nonmotor symptoms such as olfactory deficiencies, sleep disturbances, and neuropsychiatric abnormalities [114]. Tremors are usually unilateral at the start, present at rest, and fade away with physical activity. Rigidity or elevated muscular tone, which is characterised by jerky resistance comparable to that of a cogwheel (cogwheel rigidity), can also indicate a motor control problem [115]. Bradykinesia, or excessive slowness of movement, is the most disabling symptom because it affects all of the body's motor systems. Bradykinesia is characterised by a hunched posture when standing or walking, as well as a shuffling gait with no regular arm winging motions. The lack of facial muscle function leads to the absence of facial expressiveness (mask-like face). Drooling is caused by an inability to swallow, whereas changes in voice quality are caused by bradykinesia of the larynx muscles. In Parkinson's disease patients, dementia and cognitive impairment are also common [116].

Pathology/Etiology

The loss of dopaminergic neurons in the substantia nigra pars compacta (SNpc), which give dopaminergic innervations throughout the basal ganglia, is a pathological feature of Parkinson's disease (particularly striatum). However, while the steady death of these dopaminergic neurons is a natural part of ageing, symptoms of Parkinson's disease (PD) usually appear after the loss of more than 70–80 percent of neurons [117].

While mutations in several genes, including PINK1, parkin, DJ1, and LRRK2, as well as environmental contaminants (such as MPTP and 6-OHDA), pesticides/herbicides (such as rotenone and paraquat), illicit medications (antipsychotics), other toxins produced by manufacturing plants, or air pollution, have been linked to the in Parkinson's disease psychosis, minor phenomena (illusions, passage hallucinations, and presence hallucinations), visual and nonvisual hallucinations, and delusions are all included (PDP) [118]. In addition, deterioration in function and quality of life has been related to PDP. Therefore, the initial management strategy should include the identification and treatment of any contributing medical factors, the reduction or discontinuation of medications that have the potential to induce or worsen psychosis, nonpharmacological strategies, and consideration of acetylcholinesterase inhibitor treatment in the setting of dementia. Treatment options for PDP include quetiapine, clozapine and pimavanserin [119].

Mechanisms of Neurodegeneration in PD

The mechanisms causing dopamine neuron degeneration are unknown; however, oxidative stress and excitotoxicity have been suggested as possible. The inability of neurons to eliminate oxidative stress could lead to a self-perpetuating cycle of oxidative damage and neuronal death. Thus, dopamine metabolism could be one source of oxidative damage. In addition, excessive excitatory activity in the substantia nigra produced by a lack of dopamine activity in the striatum may contribute to glutamate-mediated excitotoxicity by increasing Ca^{+2} overloads *via* NMDA receptors [120].

Role of Gut Microbiota in PD

The gut microbiota (GM) is well-known for its role in gut function, energy cycles, metabolism, and it also has a role in Parkinson's disease (PD) shown in Figs. (**4** and **5**). In addition, the GM influences neurological outcomes through various pathways, including metabolite synthesis and the gut-brain axis. According to mounting data, GM dysbiosis appears to play a role in various neurological illnesses, including Parkinson's disease (PD), AD, depression, and multiple sclerosis [121].

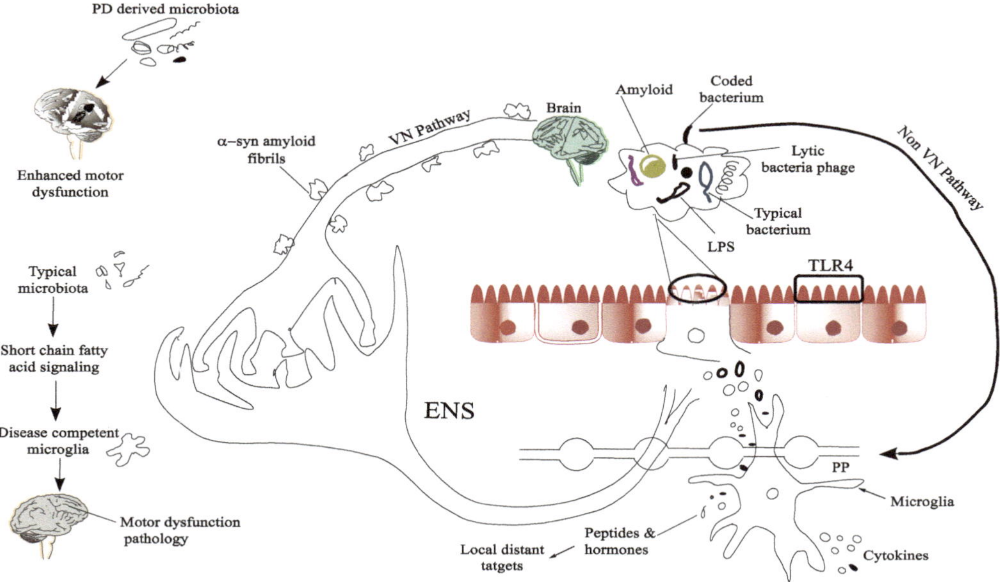

Fig. (4). The epithelium of the intestine performs several tasks. The bidirectional link between the brain and the gut is mediated by the vagus nerve (VN-gateway) and humoral channels, such as lymphatic tissue and circulation (Non-VN gateways). A monolayer of epithelial cells separates the intestinal lumen and varied gut bacteria from the underlying lymphoid and enteric neuronal systems. The structure of alpha-synuclein amyloid fibrils (PDB 2N0A) was built using atomic-resolution molecular. Microfold cells allow microbial antigens to get through the intestinal epithelium, where they are involved in localised inflammatory reactions. Microbe-sensing proteins found in intestinal epithelial cells called toll-like receptors help discriminate between good and harmful bacteria. TLR4 stands for Toll-like receptor 4; VN stands for vagus nerve; ENS stands for the enteric nervous system; M stands for microfold cells; NP stands for neuropods; PP stands for Peyer's patches; TLR4 stands for Toll-like receptor 4.

Certain strains of microbiota are helpful to humans because they aid in the generation of important neurotransmitters. Parkinson's disease (PD) is a progressive neurodegenerative sickness that affects the central nervous system and, eventually, the motor system. It is the second most common neurodegenerative ailment after AD [122]. Tremors, bradykinesia, muscle stiffness, and poor gait are common symptoms of PD, which affects an estimated 3 million individuals globally (around the age of 60). The accumulation of the protein alpha-synuclein (-synuclein [-syn]), also known as Lewy bodies and Lewy neurites, and cell death in the brain's basal ganglia, where up to 70% of the dopamine-secreting neurons in the substantia nigra pars compacta are affected by the end of life, are the key pathological characteristics of PD [123]. The unique movement problem and vagal nerve dysfunction associated with PD are caused by damage to dopaminergic neurons. Although Parkinson's disease is often thought of as a movement illness, it has increasingly been evident that people with the disease frequently experience non-motor symptoms such as rapid eye movement (REM), sleep impairments and hyposomia, cognitive impairment, orthostatic hypotension, and, most typically, intestinal dysfunction, with constipation affecting 80% of PD patients [124]. Some of these symptoms can occur years before clinical motor symptoms and aetiology (Lewy bodies) do. The transfer of - syn forms from the GI tract to the brain has been observed in rats, supporting the concept that the gut predominantly mediates PD pathogenesis [125]. Some gut microbiota which helps in the synthesis of various neurotransmitters that regulates brain functioning has been mention in Table **2**.

Table 2. List of gut microbiota helps in the synthesis of various neurotransmitters that regulates brain functioning.

Sr. No	Neurotransmitters	GUT Microbiotas	References
1.	Dopamine	*Bacillus spp.*, (*Bacillus cereus, Bacillus mycoides, Bacillus subtilis*), *Escherichia coli, Klebsiella pneumoniae, Morganella morganii* (*Proteus vulgaris, Serratia marcescens*, and *Staphylococcus aureus*)	[126, 127]
2.	GABA	*Lactobacillus spp., Bifidobacterium spp.*, and *Streptococcus salivarius* subsp. *Thermophilus*	[128]
3.	Serotonin	*Streptococcus spp., Candida spp., Enterococcus spp.*, and *Escherichia spp*	[129]
4.	Noradrenaline	*Escherichia spp., Saccharomyces spp.*, and *Bacillus spp.*	[129]
5.	Acetylcholine	*Lactobacillus spp.*	[126]

Other critical compounds generated in the colon by Prevotellaceace family members under physiologic conditions include short-chain fatty acids (SCFAs) such as butyrate, acetate, and propionate, which regulate intestinal inflammation immunological function. Increased intestinal permeability and bacterial endotoxin generation are caused by decreased *Prevotellaceae* and increased *Enterobacter spp*. abundance under pathological conditions [130]. Lipopolysaccharides (LPS) are large molecules consisting of a lipid and a polysaccharide that are bacterial toxins. LPS is derived from the cell walls of gram-negative bacteria, which crosses the intestinal wall and enters the bloodstream, where it induces the production of pro-inflammatory cytokines such as tumour necrosis factor (TNF-α), interleukin (IL-1), and IL-6, which disrupts the BBB and promotes -synuclein overexpression and misfolding to form Lewy bodies (LW), which causes DA neurons in the SN to be Pathological abnormalities in the structure and quantity of the gut flora have been associated to PD, according to clinical study [131].

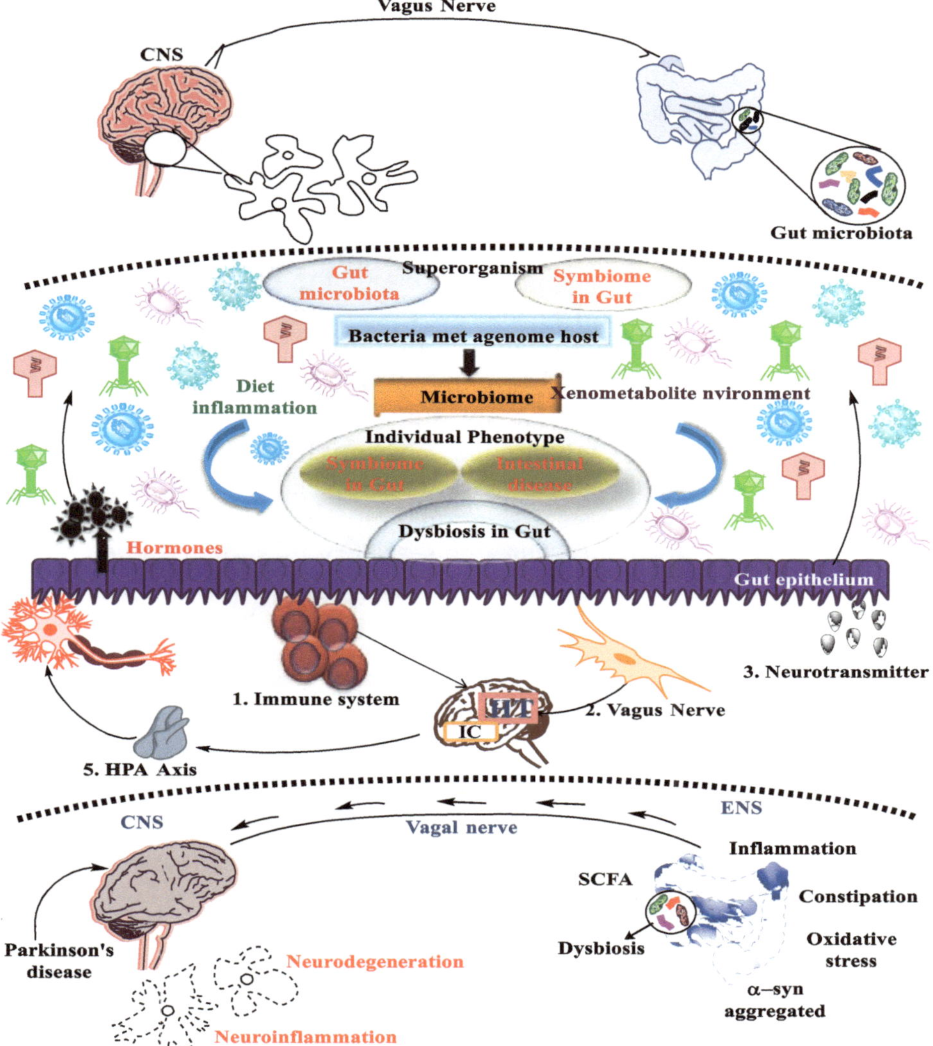

Fig. (5). The function of the gut microbiota in the progression of Parkinson's disease. (A) The vagus nerve is essential in the brain's and gut's healthy bidirectional connection. (B) The brain-gut axis and non-motor symptoms of Parkinson's disease (PD) include central and gastrointestinal dysfunction. (C) Environmental factors, such as gut microbiota, may trigger a pathogenic process inside the enteric nerve plexus, resulting in mucosal inflammation and oxidative stress, hence alpha-synuclein (-syn) accumulation. The vagal nerve may serve as a conduit for -Syn to travel from the Enteric Nervous System to the brain, passing through the brainstem, the substantia nigra, the basal forebrain, and finally the cortical areas, where the neurodegeneration and neuroinflammation process seen in Parkinson's disease is triggered.

For example, increased intestinal permeability is caused by an overgrowth of *Helicobacter pylori* (*H. pylori*), *Escherichia coli* (*E. coli*), and a decrease in the quantity of *Prevotella* sp., which promotes a leaky gut by triggering neurochemical and neuroanatomical changes. This imbalance in the host's microbiota can lead to inflammation, cellular degeneration, and an imbalance of cellular energy, which leads to increased oxidative stress, which disrupts the blood-brain barrier (BBB) and causes motor dysfunctions by decreasing dopamine levels in the brain (Fig. **6**). Synuclein aggregation and trans-synaptic cell-to-cell transmission from the intestine to the brain *via* a trans-synaptic cell-to-cell communication can produce motor dysfunctions by reducing dopamine levels [132]. The internal and external innervations of the GI tract, the dorsal motor nucleus of the vagus nerve (DMV), and the ENS of the vagus nerve have all been shown to be impacted to varying degrees, implying that PD pathogenesis in the gut occurred considerably earlier than in the substantia nigra. Endoscopic biopsies of the GI tract have confirmed that intestinal disease is connected to PD and provide ideal settings for studying the ENS in living individuals. Probiotic treatment for Parkinson's disease may have a mechanism [133]. GM has three main effects on Parkinson's disease (neuronal mechanism, endocrine mechanism, and immunological mechanism). Through secretory ECs, the GM can both create and stimulate specific neurotransmitters. Certain neuroactive factors, such as PYY, Trp, and His, can also be produced by ECs. These two types of components enter the BBB and affect the CNS. Ghrelin and IPA, two gut hormones induced by neuroactive components, can have dual effects on the CNS. The GM can send electrical impulses to the DMV straight from the ENS *via* the vagus nerve. Cortisol and ferulic acid, for example, are endocrine components that have various functions in PD pathways [134]. Specific GM members could suppress both chronic and pathological inflammation. Microbe-associated molecular patterns on the surface of GM members trigger immune cell receptors, such as DCs, and upregulate or inhibit inflammatory cytokines. Finally, GM has an effect on mucin synthesis in the stomach [135].

Current Pharmacotherapy to Treat PD

Since there is no permanent cure for PD, current treatment objectives are to either increase DA or prevent its breakdown, which significantly provides symptomatic relief. Commonly used pharmacotherapy, mode of action, dose ranges, and side effects are summarized in Table **3**.

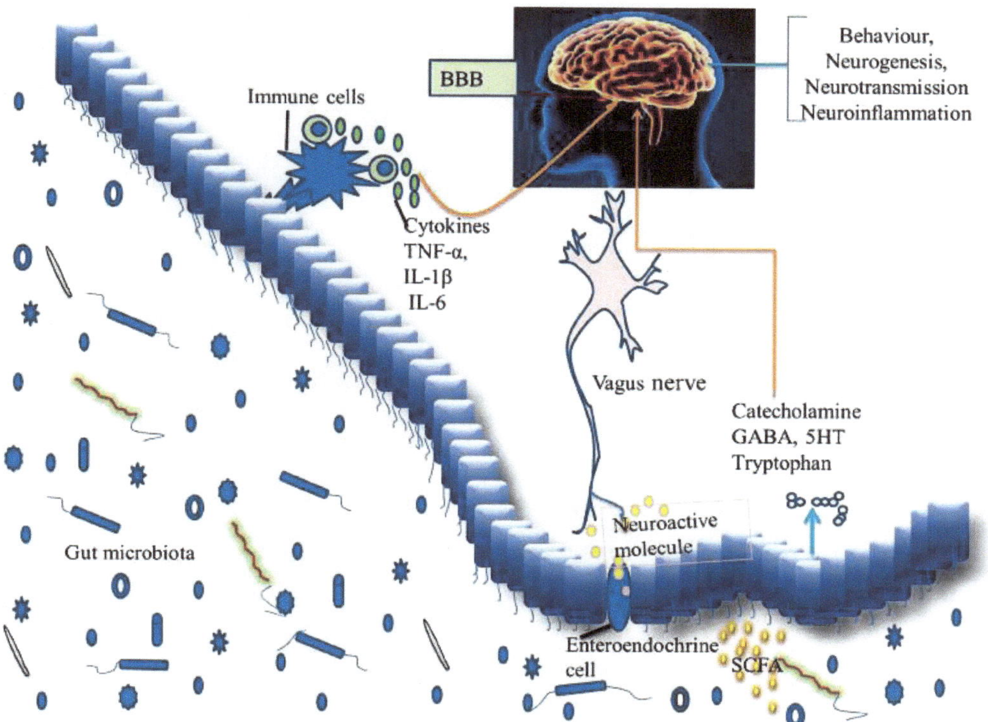

Fig. (6). According to new research, the gut microbiome regulates motor deficits and neurodegeneration in PD. The stomach and the brain are linked through neuronal, immunological, and metabolite-mediated communication routes. The signs of gut dysbiosis include increased inflammation, abnormal neurotransmitter levels, and bacterial metabolites. These elements could play a role in abnormal vagus nerve signalling. When the gastrointestinal barrier is breached, bacterial migration (leaky gut) and inflammation are generated. Inflammatory cytokines compromise the integrity of the BBB.
Abbreviations: BBB- Blood-Brain Barrier, DA- dopamine, 5HT- 5 hydroxy tryptophan (serotonin), GABA- gamma amino butyric acid.

Table 3. Current pharmacotherapy for PD.

Drugs Name	Dose Range	Mode of Action	Side Effects	References
Dopamine precursor				
Carbidopa/levodopa	25+100 mg, 50+200 mg BID	Levodopa converted to dopamine in dopaminergic terminals by dopa-decarboxylase, carbidopa protect its peripheral breakdown.	Mild headedness, nausea, dyskinesia	[136]
Dopamine agonists				

(Table 3) cont.....

Drugs Name	Dose Range	Mode of Action	Side Effects	References
Pramipexole	1.5-4.5 mg/day	Provide similar effects as dopamine by stimulating D2, D3, and D4 receptors	Hallucinations, sleepiness, and compulsive behaviors	[137]
Ropinirole	0.25 mg TID			[138]
Bromocriptine	1.25 mg BID	Partial D1 receptor agonist, and stimulates D2, 5-HT1, HT2 receptors		[137]
Pergolide	0.05 mg TID			
MAO-B inhibitors				
Selegiline	5 mg BID	Selective irreversible inhibitor of MAO-B, and thus inhibits presynaptic dopamine Receptors, and dopamine uptake	Nausea, vomiting, and insomnia	[139]
Rasagiline	10mg BID			
COMT inhibitors				
Entacapone	200 mg/day	Reversible inhibition of COMT (catechol-o-methyl transferase), and thus prevent dopamine breakdown	Risk of serious liver damage, diarrhea, and dyskinesia	[140]
Tolcapone	100mg BID			
NMDA receptor inhibitor				
Amantadine	100mg BID	Blocks NMDA glutaminergic receptor, and increase synthesis, and release of dopamine	Ankle swelling, skin purple mottling, and hallucinations	[141]
Anti-cholinergic				
Trihexyphenidyl HCl	1mg BID	Blockade of muscarinic receptors, and possible inhibition of cholinergic transmission in striatal interneuron's	Impaired memory, hallucinations, dry mouth, and impaired urination	[142]

Microbiota Targeted Strategies to Control PD

1. **Probiotics-** Probiotics are living bacteria that colonise the mucosa and prevent harmful germs from sticking. Many probiotic bacterium strains, such as Lactobacillus and Bifidobacterium species, lower luminal pH, as do bacteriocins, which reduce pathogenic bacteria abundance in the human intestine. Bacillus spp. may also convert L-tyrosine to L-DOPA, which is sufficient for dopamine production [143].

2. **Prebiotics**- Prebiotics are substances that encourage the growth or activity of helpful bacteria. Butyrate is most likely an intriguing prebiotic drug that promotes

synuclein breakdown and stimulates Atg5 and PI3K/Akt/mTOR-related autophagy. Increased butyrate-producing bacteria in the gut may aid in the prevention of intestinal barrier failure and boost striatal DA levels [144].

3. **Antibiotics**- Following broad-spectrum antibiotic administration, Firmicutes spp. were lower, whereas levels of Proteobacteria, Verrucomicrobia, Bacteroidetes, and Cyanobacteria spp. were found to be more significant. Treatment with antibiotics can also lower TNF- and IL-1 levels in the striatum, preserve dopaminergic neuron cell death, and ameliorate motor impairments in Parkinson's disease patients [145].

4. The most plausible conclusion is that GM dysfunction in Parkinson's disease patients exacerbates -syn deposition in the disease through various mechanisms,

resulting in neurodegeneration and, as a result, PD-related symptoms like movement difficulties [146].

5. In a few areas, however, there are still some unanswered questions and doubts. The nature of the connection between GM malfunction and PD is uncertain. According to current research, the unknown outside pathogen that causes PD first enters the GI tract and produces GM imbalance before breaking past the intestinal epithelial barrier and reaching the ENS [147]. In Parkinson's disease, -syn deposition may start in the ENS, increase to a certain level, and then travel to the CNS *via* trans-synaptic cell-to-cell transmission. The pathogen's GM translocation could create a pro-inflammatory milieu in the gastrointestinal system. Those impulses would be conveyed systemically to a specific part of the brain due to a defective blood-brain barrier construction. In investigations, alpha-syn was found in both the gut and the brain, suggesting that the gut may not be the first source of -syn illness in PD. As a result, we can't draw any firm conclusions about the origins of PD in the intestine just yet. More research into the GMBA and the effects of changing GM and microbial metabolites on PD is needed to establish a cause-and-effect relationship between GM dysbiosis and PD [147].

Multiple Sclerosis

Although multiple sclerosis is a complex disease, some of the most fundamental questions about its genesis and susceptibility have been answered. It typically affects young adults and has substantial functional, economic, and quality-of-life implications. The expenditures are high, and they continue to rise as the severity of the disability increases [148]. The focus of this session will be on recent developments in our understanding of multiple sclerosis progression and treatment. The genetic (*e.g.*, HLA DRB1*15:01), environmental (*e.g.*, vitamin D), and lifestyle (*e.g.*, cigarette smoking) factors that contribute to the disease's development are better understood, with environmental factors playing a more significant role in vulnerability than genetic factors [149]. Both the innate and adaptive immune systems, as well as their effector cells (microglia, activated macrophages, B and T lymphocytes), are known to play a role in the pathogenesis of multiple sclerosis, and the revelation that B cells play a key role in the disease has led to new therapeutic targets. MS is a chronic, autoimmune inflammatory illness of the CNS with unknown aetiology, characterised by demyelination, inflammation, and different degrees of axonal damage even in the early stages of the disease, and affecting predominantly women aged 20 to 40 [150]. Individual individuals can experience a wide range of symptoms, and it's believed that around 2 million people are affected worldwide, with the majority of data coming from North America and Europe [151]. Multiple sclerosis is a tough and disabling condition, but researchers are learning more about the genetic and environmental factors contributing to its development, such as low vitamin D levels, cigarettes, smoking and obesity. Diagnostic criteria for people presenting with a clinically isolated illness include imaging and spinal fluid abnormalities, allowing for an early and reliable diagnosis. Patients with the relapsing-remitting form of the disease, in particular, have a wide range of treatment choices, both oral and intravenous. Care must be used when selecting the proper treatment, balancing the side-effect profile with efficacy, and climbing as clinically necessary. This shift toward more personalised therapy is supported by a clinical guideline published in 2018. Finally, to improve health-related quality of life, all patients with multiple sclerosis should follow a complete management plan that emphasises wellness, addresses aggravating factors, and manages comorbidities. The greatest challenge is developing therapies, such as neuroprotection and remyelination, to treat and eventually prevent multiple sclerosis's devastating, progressive forms [152].

Clinical Features

MS has a wide range of clinical symptoms, including sensory, visual, motor, and brainstem pathways, acute onset of partial transverse myelitis, urine retention, and cognitive deterioration. Symptoms might be categorised as unilateral or bilateral depending on the degree of the lesion. MS usually affects the cervical cord, causing motor, bladder, and bowel function impairments, but acute full transverse myelitis with paraplegia is uncommon in MS patients [153]. The majority of MS patients, on the other hand, are diagnosed with relapsing-remitting episodes of new or recurring neurological symptoms. Clinically isolated syndrome (CIS) is the first clinical sign in these patients and is defined by optic neuritis, incomplete myelitis, or brainstem syndrome. Because the presence of typical demyelination lesions on the brain baseline or in the spinal cord can be seen by MRI, brain MRI is thought to be a good predictor of MS. The majority of MS patients, on the other hand, are diagnosed with relapsing-remitting neurological symptoms that come and go. The first clinical sign in these patients is a CIS, defined by optic neuritis, incomplete myelitis, or brainstem syndrome. Brain MRI is a valuable predictor of MS because it can detect typical demyelination lesions on the brain baseline or spinal cord [154].

Pathology

Multiple discrete areas of myelin breakdown inside the CNS, referred to as plaques or lesions, are a pathologic feature of multiple sclerosis. Variable gliosis and inflammation accompany demyelination, as well as axonal preservation. the subpial spinal cord, optic nerves, cerebellum, juxtacortical, brainstem, and periventricular white matter regions are among the areas where lesions are most common [155, 156]. Despite the fact that MS has long been assumed to be a disease affecting only the white matter of the CNS, new pathologic and imaging studies have identified demyelinated lesions in MS patients' cortical grey matter. MS is a complex illness in which many plaques accumulate in the grey and white matter of the brain and spinal cord, causing neuronal demyelination, axonal destruction, and neurological dysfunctions. According to evidence gathered from various research, genetic, environmental, infectious, and immunological variables may all affect illness progression [156].

Mechanisms of Demyelination and Axonal Dysfunction in MS

Primary demyelination with partial axon preservation is the most common brain tissue injury in MS. In RRMS, actively demyelinating plaques are caused by immune cells moving from the periphery into the CNS, linked to a disturbance of the blood-brain barrier (BBB). On the other hand, progressive disease entails the emergence of compartmentalised pathogenic processes within the brain, primarily mediated by resident CNS cells [157]. Demyelination and axonal dysfunction are thought to be caused by a breakdown of self-tolerance to myelin and other CNS antigens, resulting in continual peripheral activation of auto-reactive CD4+ myelin reactive T cells. On the other hand, this loss of self-tolerance may be triggered by an environmental antigen, probably a virus, in genetically sensitive individuals. As a result of cellular damage, the infection can activate T cells or cause the generation of autoantigens, which can then activate T cells *via* cross-reactivity between an endogenous protein (*e.g.*, myelin essential protein) and a pathogenic exogenous protein (viral or bacterial antigen) [158]. Once activated in the periphery, myelin-reactive T lymphocytes can traverse the blood-brain barrier (BBB). During the transmigration process, T cells' very late antigen-4 (VLA-4) interacts with each other. The synthesis and overexpression of various adhesion molecules, chemokines, and matrix metalloproteinases (MMPs) increase interaction between T lymphocytes' very late antigen-4 (VLA-4) and capillary endothelial cells' vascular cell adhesion molecule-1 (VCAM-1). When autoreactive peripherally activated T cells enter the CNS, they can be reactivated when they come into contact with autoantigenic peptides in the brain parenchyma in the context of MHC class II molecules expressed by local antigen-presenting cells (dendritic cells, macrophages, and B cells), triggering an inflammatory cascade that results in the release of chemokines and cytokines, including the recruitment of other immune cells [159].

Role of Gut Microbiota in MS

MS is a demyelinating illness that is accompanied by inflammatory symptoms. Is there a link between an imbalance in intestinal epithelial cells, immune system cells and intestinal microbiota in the intestinal mucosa and immune system overstimulation. Changes in BBB permeability, oligodendrocyte death, and microglial activation are the first to detect, but demyelination is still present [160]. According to studies, individuals with RR-MS have a microbiota that contains more, *Flavobacterium, Pedobacteria Pseudomonas, Acinetobacter, Mycoplasma, Eggerthella, Blautia, Akkermansia Streptococcus* and *Dorea*, and then healthy controls. *Bacteroides, Prevotella, Parabacteroides, Sutterella, Adlercreutzia, Coprobacillus, Haemophilus, Lactobacillus, Anaerostipes, Faecalibacterium*, and *Clostridium*, on the other hand, have a microbiome with depleted microbial populations in MS patients [161]. Finally, in patients with RR-MS, restoring the microbial community appears to minimise inflammatory episodes and immune system reactivation. SCFAs from the *Clostridium perfringens* bacteria are absorbed through the intestine, enter the bloodstream, and cause MS-like symptoms (such as blurred vision, lack of coordination, and spastic paralysis) when they bind to receptors in the brain's vascular system, as well as in both myelinated and demyelinated brain regions, such as the corpus callosum, causing inflammation and also pro-inflammatory effects. This can affect Bacteroides fragilis, causing Th1 T cell differentiation and the generation of IL-10. treg cells, a unique population of CD4+ T cells produced in the peripheral organs of the immune system and thymus, regulate the suppression of excessive immunological stimulation (Fig. **7**). Treg dysfunction is common in inflammatory disorders such as rheumatoid arthritis, systemic lupus erythematosus (SLE), and MS [162].

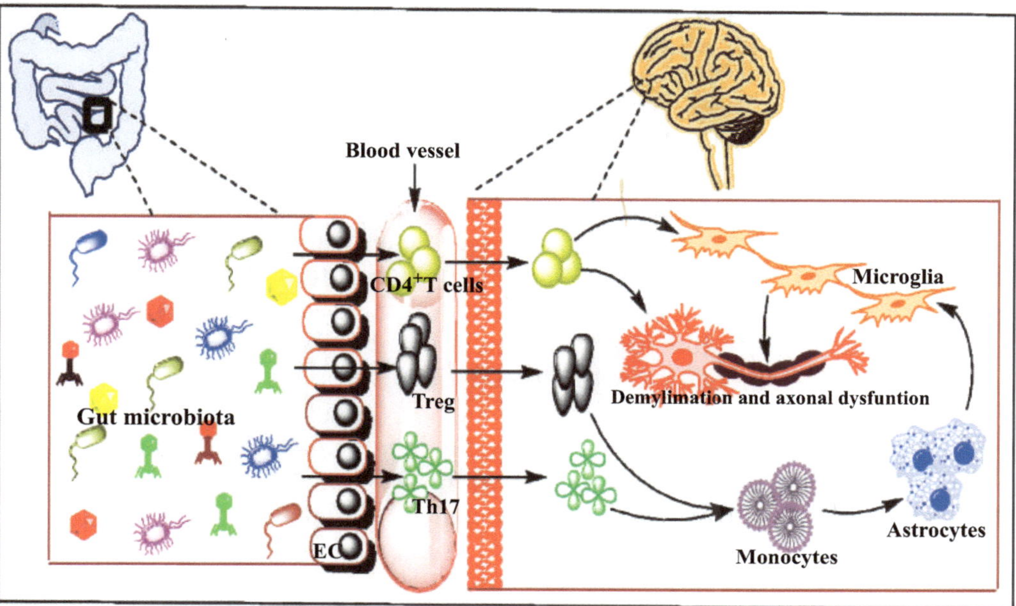

Fig. (7). Recent research has discovered major changes in the gut microbiota and links to pro-inflammatory pathways that can accelerate disease progression. CD4+ T cells are polarised into proinflammatory phenotypes (Th17) in MS, allowing them to bypass regulatory mechanisms like Treg repression, go to the CNS, and induce inflammation. Antigen-presenting cells (APCs) support the development of memory CD4+ T cells, which circulate throughout the body and eventually reach the CNS, activating astrocytes, microglia, and macrophages, causing inflammation altering disease pathogenesis.

Diet and Gut Microbiota in Multiple Sclerosis

The aetiology of Multiple Sclerosis is determined by a variety of factors, both environmental and genetic, that influence disease risk. The course of this disease can be influenced by lifestyle, eating habits, and some environmental variables. The composition of the intestinal bacterial flora can be controlled by diet, which can indirectly support the development of autoimmune inflammatory illnesses like MS51 [163]. Obesity is a risk factor, and several recent studies have found that it increases the risk of MS, especially in children and adolescents. Turnbaugh *et al.* conducted a significant study demonstrating how gut microbiota differs between obese and normal-weight patients. Obese people, like MS patients, exhibited lower levels of Bacteroidetes than healthy people [164]. This phylum produces short-chain fatty acids that protect regulatory T cells from inflammation. As a result, obese people, like MS patients, have a higher susceptibility for inflammatory processes due to their decrease. Finally, the altered gut microbiota has a higher presence of pathogenic Firmicutes than Bacteroidetes, disrupting the microbial balance between microbiota and host. Minor endotoxemia is encouraged by this syndrome, which contributes to chronisis. Mild endotoxemia, which promotes persistent and systemic gastrointestinal inflammation and increases the risk of immune-mediated illnesses like MS, thrives in this environment [165].

Vitamin D Insufficiency has been identified as a possible risk factor for MS. Identifying these risk variables as prospective therapeutic targets may offer a way to improve current MS treatments. The distribution of MS prevalence is affected by latitude, which is linked to both sunshine intensity and vitamin D serum levels [166]. Vitamin D is a potent immunomodulator that affects various immunological activities in innate and adaptive immune systems. T cells are also affected, both directly and indirectly. It has been discovered that 25-(OH)-D serum levels are linked to the likelihood of developing MS. Vitamin D also appears to be able to reduce the production of IFN, it also aids in the maintenance of gastrointestinal homeostasis by inducing innate immune responses and the generation of regulatory T cells. Vitamin D has also been linked to the maintenance of a healthy gut microbiome and epithelial cell integrity. Vitamin D is a powerful antioxidant that inhibits nitric oxide synthase and gamma-glutamyl transpeptidase, which produce free radicals [167].

The Role of Microbiota in Multiple Sclerosis: Clinical Trials

Several preclinical studies have suggested that the intestinal microbiota has played a significant role in treating neurodegenerative illnesses in recent decades. Hereditary and environmental factors both influence MS progression. The commensal microbiota is one of the environmental risk factors connected to the development of MS [168].

The most frequent species in the microbiota are Bacteroidetes, Proteobacteria, Actinobacteria, and Firmicutes. During the study of the genera Sutturella and Haemophilus (Phylum Proteobacteria), Aldercreutzia and Collinella (Phylum Actinobacteria), and Sutturella and Haemophilus (Phylum Proteobacteria), a dysbiosis of the gut was revealed as a continuous feature of MS. On the other hand, Blautista, Dorea, and *Streptococcus thermophilus* (phylum Firmicutes), Pedobacteria and Flavobacterium (phylum Bacteroidetes), Mycoplana (phylum Proteobacteria), *Eggerthella lenta* (phylum Proteobacteria) and Pseudomonas and all showed a rise (phylum Actinobacteria). In human T cells, this dysbiosis in MS patients has a pro-inflammatory and regulatory effect [169].

An intriguing element shows that MS patients treated with FMT and a modified diet normalised some of these microbial communities. In addition, FMT and a changed diet alter the gut microbiota, encouraging the development of anti-inflammatory bacteria such as Lactobacilli, Bacteroides, and Prevotella [170].

Some microbial populations differ in MS patients compared to healthy people, which are shown in Table **4**.

It is feasible to explore lowering the beginning of clinical relapses of disease in MS patients by using therapeutic interventions that alter the gut microbiota to encourage the formation of "good" anti-inflammatory microorganisms, including Lactobacilli, Bacteroides, and Prevotella [171]. As a result, a healthy and balanced diet can help slow the progression of the disease by altering the intestinal microbiota composition. As a result, it's critical to improve the quality of the fats consumed by minimising saturated and hydrogenated fats and favouring mono and polyunsaturated lipids. Furthermore, proper vitamin, mineral, and antioxidant intake must be ensured, particularly in fruits and vegetables. Vitamin D appears to be particularly significant among vitamins, as persons with MS are frequently low. Obesity, which can exacerbate impairment, should also be avoided [172].

Table 4. Some Microbial Populations Differ in MS Patients Compared to Healthy People. Increased (↑) or Reduced (↓) Bacterial Genera in the Microbiota Gut of MS Patients.

Phylum of Bacteria	Genus	Microbiota of MS Patients *vs* Healthy Individuals	References
Bacteroidetes	prevotella	decreased	[169 - 172]
-	pedobacteria	increased	
-	bacteroides	decreased	
-	parabacteroides	decreased	
Firmicutes	dorea	increased	
-	blautia	increased	
-	faecalibacterium	decreased	
-	anaerostipes	decreased	
Proteobacteria	pseudomonas	increased	
-	acinetobacter	increased	
-	sutterella	decreased	
-	mycoplana	increased	
-	haemophilus	decreased	
Actinobacteria	collinsella	decreased	
-	eggerthella	increased	
-	adlercreutzia	decreased	
Verrucomicrobia	akkermansia	increased	

Current Pharmacotherapy to Treat MS

Current MS pharmacotherapy comprises immunomodulatory drugs that attempt to delay the beginning of the disease, as well as symptomatic care aimed at alleviating specific symptoms such as stiffness, fatigue, pain and bladder dysfunction, Steroids such as (methylprednisolone) and adrenocorticotropic hormone (ACTH) have anti-inflammatory and immunomodulatory properties and are commonly used to speed up recovery and treat acute relapses. As a result, Table **5** summarises the current medication utilised in MS treatment [173].

Table 5. Summarises the current medication utilised in MS treatment.

Name of Drug	Dose	MOA	Side Effects	Reference
First-line Therapies				
(1β) Beta-interferon	250 µg	Immunomodulator that maintains the balance between Proinflammatory and Anti-inflammatory cytokines and prevents inflammation that causes nerves to damage in the brain	Flu-like symptoms, thyroid abnormalities, increase in liver enzyme, leucopenia	[174, 175]
Beta-interferon (1α)	30 µg			
Glatiramer acetate	20 mg	Act as T cell receptor antagonist and inhibits the T cell response to several myelin antigens	chest tightness, flushing, palpitation, and rare lipoatrophy with prolonged use and dyspnea	[176]
Second-line therapies				
Natalizumab	300 mg	A monoclonal antibody acts as an α-4 integrin antagonist to prevent immune cells (leukocytes) transmission into the brain.	Progressive multifocal encephalopathy (PML). Hepatotoxicity, infusion reactions	[177]
Mitoxantrone	Weight-based dose	It suppresses the proliferation of T cells, B cells, and macrophages and decreases the secretion of proinflammatory cytokines.	Cardiotoxicity, secondary leukaemia	[178]
Teriflunomide	7 or 14 mg	Reduces the proliferation of activated T and B lymphocytes by reversibly inhibiting dihydro-orotate dehydrogenase, a crucial mitochondrial enzyme in the *de-novo* pyrimidine production pathway.	Hair loss, diarrhoea, hepatotoxicity, teratogenicity, headache, increased risk of infections due to lymphopenia	[179]

(Table 5) cont.....

Name of Drug	Dose	MOA	Side Effects	Reference
Alemtuzumab	12 or 24 mg	Rapid and long-lasting predominantly depletion of B, and T cells	Serious infusion reactions, increased risk of malignancies- thyroid cancer, melanoma. Secondary autoimmune diseases- thyroiditis	[180]

Targeting Gut Microbiome to Treat MS

In humans, the gut microbiota is a crucial regulator of autoimmunity. Treatments based on the gut microbiome could help with disease progression as well as symptom management. The following are some of the therapeutics that target gut microbiota:

i. **Bacteriophage Therapy-** Phage treatment is another name for it. Phages are bacterium-specific viruses that can change the populations of bacteria of interest while leaving others alone. Oral phage delivery is safe and capable of bypassing intestinal epithelia and transporting it to the bloodstream. Phage therapy is a particularly appealing choice for dealing with bacterial translocation that can lead to disease because of its ability to permeate the bloodstream. Phages, which make up most of the intestinal virome, are critical in shaping the gut microbiome. On the other hand, Phages can be pathogenic because they cause intestinal dysbiosis or the uncontrolled killing of beneficial bacteria, disrupting the gut microbiome's general structure and function. Finally, there are many outstanding questions and concerns about the safety and efficacy of phage therapy in terms of unintended microbiome consequences [181].

ii. **Fecal Microbiota Transplantation-**Faecal microbiota transplantation (FMT) is a procedure that involves replacing the entire gut microbiome in the hopes of repairing abnormal gut microbiome architecture and functions. Because of the high success of FMTs, faecal transplantation is a common therapeutic strategy for C. difficile patients. The diseased state will theoretically be corrected by removing the host's abnormal gut microbiota and replacing it with a healthy gut microbiome [182]. FMT's efficacy has also been effective in other conditions, such as autoimmune, such as neurological problems and inflammatory bowel disease (IBD). However, only a few types of research have examined the beneficial benefits of FMT on neurological deficiency.

iii. **Role of Diet-** Diet is the most important component in reducing negative effects and chronic inflammatory states and promoting gut flora structure and function. On the other hand, Western diets tend to be high in carbs and saturated fats, leading to chronic inflammation. As a result of detecting the effects of the Western diet on the microbiota composition, the excess of short-chain fatty acids was discovered (SCFAs). SCFAs generated from carbohydrate catabolism include acetate, propionate, and butyrate [183].

iv. **Vitamin D**- Lowered vitamin D levels are common in MS patients, which may be due to reduced solar exposure at higher geographical latitudes, low vitamin D receptor expression levels, the host's metabolism, and the impact on the gut microbiota [178]. It regulates the immune system, lowers intestinal permeability, and influences immunomodulatory metabolites such as butyrate. As a result, vitamin D supplementation is a crucial therapeutic technique for treating MS, although the proper dosage and combination with other supplements has also been regarded as a helpful therapeutic approach for MS. Vitamin D insufficiency has been identified as a possible risk factor for MS. Identifying these risk variables as prospective therapeutic targets may offer a way to improve current MS treatments. The distribution of MS prevalence is affected by latitude, linked to both sunshine intensity and vitamin D serum levels. Vitamin D is a potent immunomodulator that affects various immunological activities in innate and adaptive immune systems. T cells are also affected, both directly and indirectly.

The presence of 25-(OH)-D in the blood has been connected to the risk of developing MS48. Vitamin D also appears to be able to reduce the production of IFN-. It also aids in the maintenance of gastrointestinal homeostasis by inducing innate immune responses and the generation of regulatory T cells. Vitamin D has also been linked to the maintenance of a healthy gut microbiome and epithelial cell integrity. Vitamin D is a powerful antioxidant that inhibits gamma-glutamyl transpeptidase and nitric oxide synthase, which produce free radicals.

Autism Spectrum Disorder

ASD refers to a group of early-onset social communication deficits and repetitive sensory-motor behaviours caused by hereditary and environmental causes. ASD is a complex neurodevelopmental behavioural disease characterised by impaired social interaction, intellectual impairment, language and speech disorders, and motor dysfunctions and learning. It primarily affects children under the age of three years and affects all stages of normal mental development. In the overall population, the prevalence rate is roughly 1%. Men are more likely than women to be affected by autism, and comorbidity is expected (more than 70 percent of people have multiple illnesses). Unusual cognitive profiles are found in people with autism, including impaired social cognition and social perception, atypical perceptual, executive dysfunction, and information processing. Atypical brain development at the systems level underpins these characteristics. Genetics has a significant role in the aetiology of autism, in addition to developmentally early environmental factors shown in Fig. (**8**).

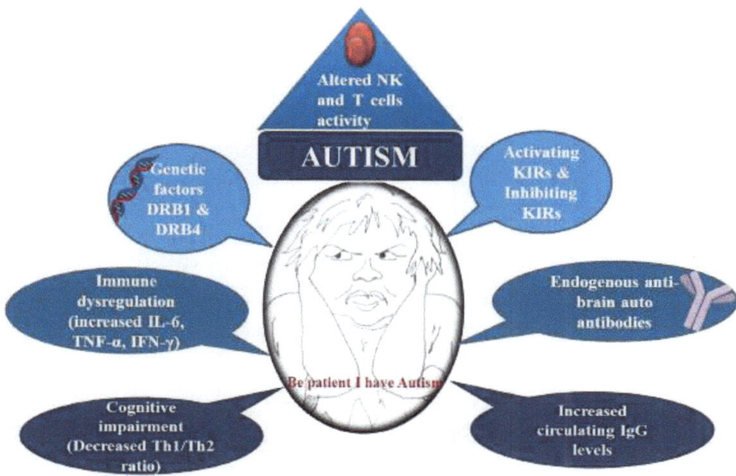

Fig. (8). Novel ethiopathogenetic of autism spectrum disorders.

Both rare mutations with a high effect and common variants with a moderate effect play a role in risk. Early detection is critical for early intervention, and multidisciplinary and developmental assessments are required. Many people with autism spectrum disorder have a better outlook than 50 years ago; more people can read, speak, and live in the community rather than in institutions. Some will be essentially free of symptoms by maturity. Genetics and neurology have discovered intriguing risk patterns, but they aren't really useful. Comprehensive and focused early behavioural therapy can improve social communication while also reducing anxiety and aggression. Drugs can help with comorbid symptoms, but they have no direct effect on social communication. It's crucial to create a welcoming environment that values and respects individual diversity. According to recent studies, 1 in every 68 children in the United States has ASD (Center for Disease Control and Prevention. The intensity of symptoms varies depending on the severity of illness development in each patient with behavioural problems. As a result, these children's families bear various emotional, social, and financial responsibilities [184].

Clinical features-ASD differs from other neurological illnesses in terms of its clinical manifestations. Repetitive and Restricted Activities (RRBs), stereotyped motor behaviours, and limited social interaction are characteristics of children with ASD. Food restriction, problematic eating patterns, and GI issues were among the medical illnesses linked to ASDs that were easily noted. Indeed, children with ASD are fussy eaters, and the majority of them dislike specific food colours, textures, odours, or other qualities. This has a direct negative impact on diet quality, nutritional deficit, and gut flora composition. In addition, most ASD patients have GI issues, which are influenced by specific eating habits that might exacerbate ASD symptoms [185].

Pathomechanism of ASD

Genetic Causes

Several factors have been identified in the development of autistic disorders, although the specific pathomechanism of ASD is unknown. The genetic origin has long been implicated as a solid evidence-based aetiology in cases of several co-occurring or linked disorders with ASD, such as tuberous sclerosis, fragile X syndrome, Rett syndrome, and others. Autism is more common among siblings of autistic offspring than in the general population, and twin studies have also revealed a significant role in inheritance [186]. There is a wide range of phenotypes, however, ASD patients with more genetic homogeneity had less phenotypic heterogeneity. The etiopathogenesis of ASD is still unknown, but it has been indicated that pathogenic variables in ASD have a strong genetic foundation, with more than 100 (60%) of genes and genomic regions associated with ASD pathogenesis. The hallmarks of both *de novo* mutations and deletions in genes linked to ASD, such as SH3, multiple ankyrin repeat domain 3 (SHANK3), contacting associated protein-like 2 (CNTNAP2), and chromodomain helicase DNA binding protein 8 (CHD8), have been emphasised by accumulating data [187]. Aside from genetic mutation, numerous environmental factors have been linked to an increased risk of ASD, including air pollution, pesticide exposure, dietary factors, stress, maternal illnesses, medicines, inflammatory disorders, and antibiotic use during pregnancy. Furthermore, maternal prenatal and perinatal folate, iron, and polyunsaturated fatty acid (PUFA) intake have been identified as dietary risk factors. Furthermore, Obrenovich *et al.* discovered that metal ion homeostasis is disrupted in children with ASD, resulting in the deposition of numerous divalent cations. This was established as a complicated autosomal dominant disorder characterised by ASD, known as Timothy syndrome [188].

Male to Female Ratio in ASD

The rationale for the 4/1 male to female ratio in ASD isn't widely known, but it's critical [189]. Recent research has linked sex-specific effects of Y-linked genes, balanced and skewed X-inactivation, escaping X-inactivation, and the parent-of-origin allelic gene, among other epigenetic phenomena, to the aetiology of ASD and heterogeneity in gene regulation at the allelic level as well as total gene expression. These sex differences could be related to genetic and hormonal variances that arise early in life due to how people respond to and interact with numerous environmental factors like nutrition, stress, infection, and medicines [190]. The aetiology of ASD has long been linked to internal and exterior environmental variables. Prenatal stress may be caused by early maternal immune activation, with boys being more affected due to a vulnerable genotype. Children with autism have brains that expand faster than normal immediately after birth, followed by regular or slower growth throughout infancy. It's unclear whether all autistic children have early overgrowth [191].

Mechanism

Autism, unlike other neurological illnesses like Alzheimer's, Multiple Sclerosis, and Parkinson's, lacks a specific molecular or cellular origin. It appears to be caused by developmental variables that impact many or all brain systems, and it appears to be more concerned with the timing of brain development than with the result. Autism's causation is significantly linked to abnormalities in brain development immediately after conception, according to neuroanatomical research and teratogen linkages. This aberration appears to set off a chain reaction of degenerative brain processes controlled by the environment. It appears to be more noticeable in brain areas linked to increased cognitive specialisation [192].

The Gut Microbiome's Role in ASD's Underlying Mechanisms

Despite various studies looking into the suspected pathology of ASD, the disease's true origins are unknown, and the current consensus is that ASD is caused by a combination of inherited and environmental risk factors [193]. According to growing evidence, environmental risk factors and related comorbidities, in particular, play a role in the fundamental neurobehavioral symptoms of ASD. In addition, the gene-environment interaction functions of the microbiome that dwells inside the body are well-known. With the growing understanding of the microbiome, it is thought that ASD-related changes in the gut microbiota and its metabolites can modulate corresponding immunological and GI problems both directly and indirectly and may play a role in the aetiology of ASD [194].

Role of Gut Microbiota in ASD

Microbiota in ASD

The gut microbiome's role in ASD's underlying mechanisms Despite several research focusing on the suspected pathogenesis of ASD, the actual origins of the disease remain unknown, and the current view suggests that ASD is the result of a combination of hereditary and environmental risk factors. According to growing studies, environmental risk factors and associated comorbidities, in particular, have contributed to critical neurobehavioral symptoms of the disorder [195]. The microbiome that lives in the body is well-known for its gene-environment interaction functions. Gut dysbiosis has been implicated in the aetiology of several diseases. The gut microbiota of ASD patients has been observed to have a lower level of fermenters. The potential of gut bacteria to connect with the brain and regulate behaviour is the microbiota-gut-brain axis. Although faecal microbiota transplantation has been used to treat various GI illnesses, further research and control trials are required before being widely used in the clinic [196].

Other investigations, however, found no difference in the GI microbiota of ASD children with and without GI disturbances. Intestinal mucosa may be vulnerable to infections, inflammation, injuries, improper digestion, immune reactivity, cross-reaction in other tissues, including the brain, and immunological imbalance if the gut microbiota population is out of balance [197]. In this area, more study is being conducted. Dysbiosis of the gut microbiota is linked to a variety of GI, immunological, and neurobehavioral problems. As a result, ASD youngsters may contribute to both GIT and CNS symptoms. There are several pieces of evidence pointing to close ties between the microbiota, stomach, and brain. The CNS, for example, uses peptides to control the composition of the gut microbiome, which affects nutritional availability. In addition, immunological and neurological pathways influence mucin secretion from intestinal epithelial cells, which is known to exert control over microbial populations in the gut. On the other hand, the gut microbiota has been proven to exert control over CNS activity *via* several neuronal, endocrine, and other mechanisms [196].

As previously stated, bacteria and their ligands play an essential role in maintaining cell-cell barrier integrity. As a result, a damaged epithelium barrier has been dubbed "leaky gut," which allows bacteria, toxins, and metabolites to flow through, potentially activating the immune system, and related to various intestinal and systemic illnesses.

Increased circulation of bacteria-derived endotoxins such as LPS and phenols is another consequence of increased intestinal permeability, which further activates inflammatory and immunological responses. On the other hand, Cytokines are essential for normal neurodevelopment, and disturbances can affect this process. Immunologic dysfunction and elevated cytokine levels have also been found in children with ASD [198]. Pro-inflammatory cytokines such as IL-1β, IL-6, IL-8, and IL-12p40 and macrophage migration inhibitory factor (MIF) and platelet-derived growth factor (PDGF) have been related to ASD. In addition, stereotypy, hyperactivity, and communication issues have all been linked to changes in peripheral immune profiles, specifically TGF beta, p-selectin, and MIF levels [199].

The Therapeutic Perspective of ASD by Targeting Gut Microbiota

Techniques based on the gut microbiome may aid in disease progression as well as symptom treatment. Some of the medical approaches addressing gut microbiota are as follows:

i. **Probiotics-** Because probiotics can heal the intestinal mucosa, protect the epithelial barrier through mucin production and tight junction fortification, increase digestive enzymes and antioxidants, and modulate the immune response. Thus, they may be a better alternative to restrictive dietary interventions. Dietary supplementation with one capsule of "Children Dophilus" (3 strains of Lactobacillus, two strains of Bifidobacterium, and 1 strain of Streptococcus) three times a day normalised the Bacteroidetes/Firmicutes ratio, Desulfovibrio spp. abundance, and reduced TNF- levels in the stool of autistic toddlers in a four-month randomised clinical trial [200].

ii. **Prebiotics**- Prebiotics are non-digestible chemicals that aid in the proliferation of beneficial gut bacteria such as Lactobacilli and Bifidobacteria by being digested by the digestive tract. GOS, for example, has been demonstrated to have a bifidogenic effect in both autistic and non-autistic children and lower neuroendocrine stress and boost attentional alertness in healthy volunteers [201].

iii. **Faecal Microbiota Transplant (FMT)**- Researchers have recently become interested in FMT and microbiota transfer therapy (MTT), a modified FMT technique, due to their efficacy in treating recurrent Clostridium difficile infections their potential participation in the treatment of IBD. FMT involves transplanting faecal microbiota from healthy volunteers to patients with gut dysbiosis to re-establish the physiological intestinal microbiota. It has been found to help children with ASD with GI and neurobehavioral problems [202].

Gastrointestinal issues and gut microbiota dysbiosis are common in children with neurodevelopmental disorders, including ASD. On the other hand, humans have co-evolved with various bacteria that live on exterior and internal surfaces [203]. The microbiome, a collection of microorganisms and their genetic material, aids the host in various important aspects of life while also responsible for some illnesses. According to a significant amount of preclinical research, the gut microbiota plays a vital role in the bidirectional gut-brain axis that communicates between the gut and the central nervous system [204]. Furthermore, mounting evidence suggests that the gut microbiome plays a role in ASD pathophysiology.

CONCLUSION

The benefits of probiotics therapy in reducing comorbid symptoms associated with ASD have been demonstrated in human and animal studies, suggesting that addressing the gut microbiota as a potential treatment for ASD patients could be beneficial. However, the exact role of the gut microbiome in the development of ASD is uncertain, demanding more research into mechanisms including immunological, neurological, and endocrine pathways in microbiome-brain communications. In addition, the disputed microbiological abnormalities seen in ASD patients need to be thoroughly studied. Furthermore, when adding probiotics into human clinics, caution should be maintained. Finally, the effects of a wide range of symbiotic species and their metabolites on hosts remain unclear, and they may act as carriers for medicinal chemicals.

REFERENCES

[1] Ogunrinola GA, Oyewale JO, Oshamika OO, Olasehinde GI. The human microbiome and its impacts on health. Int J Microbiol 2020; 2020: 1-7. [http://dx.doi.org/10.1155/2020/8045646] [PMID: 32612660]

[2] Anal AK, Koirala S, Shrestha S. Gut Microbiome and Their Possible Roles in Combating Mycotoxins. Mycotoxins in Food and Beverages. 213-35. [http://dx.doi.org/10.1201/9781003176046-9]

[3] Tang WHW, Li DY, Hazen SL. Dietary metabolism, the gut microbiome, and heart failure. Nat Rev Cardiol 2019; 16(3): 137-54. [http://dx.doi.org/10.1038/s41569-018-0108-7] [PMID: 30410105]

[4] Montalban-Arques A, De Schryver P, Bossier P, *et al.* Selective manipulation of the gut microbiota improves immune status in vertebrates. Front Immunol 2015; 6: 512. [http://dx.doi.org/10.3389/fimmu.2015.00512] [PMID: 26500650]

[5] Kolmeder CA, de Vos WM. Metaproteomics of our microbiome — Developing insight in function and activity in man and model systems. J Proteomics 2014; 97: 3-16. [http://dx.doi.org/10.1016/j.jprot.2013.05.018] [PMID: 23707234]

[6] Ursell LK, Metcalf JL, Parfrey LW, Knight R. Defining the human microbiome. Nutr Rev 2012; 70 (Suppl. 1): S38-44. [http://dx.doi.org/10.1111/j.1753-4887.2012.00493.x] [PMID: 22861806]

[7] Al Omran Y, Aziz Q. The brain-gut axis in health and disease. Microbial endocrinology: the microbiota-gut-brain axis in health and disease 2014; 135-53. [http://dx.doi.org/10.1007/978-1-4939-0897-4_6]

[8] Breit S, Kupferberg A, Rogler G, Hasler G. Vagus nerve as modulator of the brain–gut axis in psychiatric and inflammatory disorders. Front Psychiatry 2018; 9: 44. [http://dx.doi.org/10.3389/fpsyt.2018.00044] [PMID: 29593576]

[9] Foster JA, McVey Neufeld KA. Gut–brain axis: how the microbiome influences anxiety and depression. Trends Neurosci 2013; 36(5): 305-12. [http://dx.doi.org/10.1016/j.tins.2013.01.005] [PMID: 23384445]

[10] Zheng D, Liwinski T, Elinav E. Interaction between microbiota and immunity in health and disease. Cell Res 2020; 30(6): 492-506.
[http://dx.doi.org/10.1038/s41422-020-0332-7] [PMID: 32433595]

[11] La Rosa F, Clerici M, Ratto D, *et al.* The gut-brain axis in Alzheimer's disease and omega-3. A critical overview of clinical trials. Nutrients 2018; 10(9): 1267.
[http://dx.doi.org/10.3390/nu10091267] [PMID: 30205543]

[12] Mu Q, Kirby J, Reilly CM, Luo XM. Leaky gut as a danger signal for autoimmune diseases. Front Immunol 2017; 8: 598.
[http://dx.doi.org/10.3389/fimmu.2017.00598] [PMID: 28588585]

[13] Nagpal R, Mainali R, Ahmadi S, *et al.* Gut microbiome and aging: Physiological and mechanistic insights. Nutr Healthy Aging 2018; 4(4): 267-85.
[http://dx.doi.org/10.3233/NHA-170030] [PMID: 29951588]

[14] Kim M, Benayoun BA. The microbiome: An emerging key player in aging and longevity. Transl Med Aging 2020; 4: 103-16.
[http://dx.doi.org/10.1016/j.tma.2020.07.004] [PMID: 32832742]

[15] Yang LL, Millischer V, Rodin S, MacFabe DF, Villaescusa JC, Lavebratt C. Enteric short-chain fatty acids promote proliferation of human neural progenitor cells. J Neurochem 2020; 154(6): 635-46.
[http://dx.doi.org/10.1111/jnc.14928] [PMID: 31784978]

[16] Ramos S, Martín MÁ. Impact of diet on gut microbiota. Curr Opin Food Sci 2020.

[17] Bauer KC, Huus KE, Finlay BB. Microbes and the mind: emerging hallmarks of the gut microbiota-brain axis. Cell Microbiol 2016; 18(5): 632-44.
[http://dx.doi.org/10.1111/cmi.12585] [PMID: 26918908]

[18] Furness JB, Callaghan BP, Rivera LR, Cho HJ. The enteric nervous system and gastrointestinal innervation: integrated local and central control. Microbial endocrinology: The microbiota-gut-brain axis in health and disease 2014; 39-71.
[http://dx.doi.org/10.1007/978-1-4939-0897-4_3]

[19] Lyte M. Microbial endocrinology in the microbiome-gut-brain axis: how bacterial production and utilization of neurochemicals influence behavior. PLoS Pathog 2013; 9(11): e1003726.
[http://dx.doi.org/10.1371/journal.ppat.1003726] [PMID: 24244158]

[20] Wang SZ, Yu YJ, Adeli K. Role of gut microbiota in neuroendocrine regulation of carbohydrate and lipid metabolism *via* the microbiota-gut-brain-liver axis. Microorganisms 2020; 8(4): 527.
[http://dx.doi.org/10.3390/microorganisms8040527] [PMID: 32272588]

[21] Macfabe DF. Short-chain fatty acid fermentation products of the gut microbiome: implications in autism spectrum disorders. Microb Ecol Health Dis 2012; 23(1): 19260.
[PMID: 23990817]

[22] Amato KR. Incorporating the gut microbiota into models of human and non-human primate ecology and evolution. Am J Phys Anthropol 2016; 159 (Suppl. 61): 196-215.
[http://dx.doi.org/10.1002/ajpa.22908] [PMID: 26808106]

[23] Morais LH, Schreiber HL IV, Mazmanian SK. The gut microbiota–brain axis in behaviour and brain disorders. Nat Rev Microbiol 2021; 19(4): 241-55.
[http://dx.doi.org/10.1038/s41579-020-00460-0] [PMID: 33093662]

[24] Sampson TR, Debelius JW, Thron T, *et al.* Gut microbiota regulate motor deficits and neuroinflammation in a model of Parkinson's disease. Cell 2016; 167(6): 1469-1480.e12.
[http://dx.doi.org/10.1016/j.cell.2016.11.018] [PMID: 27912057]

[25] Eltokhi A, Janmaat IE, Genedi M, Haarman BCM, Sommer IEC. Dysregulation of synaptic pruning as a possible link between intestinal microbiota dysbiosis and neuropsychiatric disorders. J Neurosci Res 2020; 98(7): 1335-69.

[http://dx.doi.org/10.1002/jnr.24616] [PMID: 32239720]

[26] Borre YE, O'Keeffe GW, Clarke G, Stanton C, Dinan TG, Cryan JF. Microbiota and neurodevelopmental windows: implications for brain disorders. Trends Mol Med 2014; 20(9): 509-18. [http://dx.doi.org/10.1016/j.molmed.2014.05.002] [PMID: 24956966]

[27] Kim S, Jazwinski SM. The gut microbiota and healthy aging: a mini-review. Gerontology 2018; 64(6): 513-20. [http://dx.doi.org/10.1159/000490615] [PMID: 30025401]

[28] Rinninella E, Raoul P, Cintoni M, *et al.* What is the healthy gut microbiota composition? A changing ecosystem across age, environment, diet, and diseases. Microorganisms 2019; 7(1): 14. [http://dx.doi.org/10.3390/microorganisms7010014] [PMID: 30634578]

[29] Lebel C, Deoni S. The development of brain white matter microstructure. Neuroimage 2018; 182: 207-18. [http://dx.doi.org/10.1016/j.neuroimage.2017.12.097] [PMID: 29305910]

[30] Leeming ER, Johnson AJ, Spector TD, Le Roy CI. Effect of diet on the gut microbiota: rethinking intervention duration. Nutrients 2019; 11(12): 2862. [http://dx.doi.org/10.3390/nu11122862] [PMID: 31766592]

[31] Chew LJ, Fusar-Poli P, Schmitz T. Oligodendroglial alterations and the role of microglia in white matter injury: relevance to schizophrenia. Dev Neurosci 2013; 35(2-3): 102-29. [http://dx.doi.org/10.1159/000346157] [PMID: 23446060]

[32] Ottman N, Smidt H, de Vos WM, Belzer C. The function of our microbiota: who is out there and what do they do? Front Cell Infect Microbiol 2012; 2: 104. [http://dx.doi.org/10.3389/fcimb.2012.00104] [PMID: 22919693]

[33] Naegele M, Martin R. The good and the bad of neuroinflammation in multiple sclerosis. Handb Clin Neurol 2014; 122: 59-87. [http://dx.doi.org/10.1016/B978-0-444-52001-2.00003-0] [PMID: 24507513]

[34] Skaper SD, Facci L, Zusso M, Giusti P. An inflammation-centric view of neurological disease: beyond the neuron. Front Cell Neurosci 2018; 12: 72. [http://dx.doi.org/10.3389/fncel.2018.00072] [PMID: 29618972]

[35] Correale J, Marrodan M, Ysrraelit M. Mechanisms of neurodegeneration and axonal dysfunction in progressive multiple sclerosis. Biomedicines 2019; 7(1): 14. [http://dx.doi.org/10.3390/biomedicines7010014] [PMID: 30791637]

[36] Di Meo F, Donato S, Di Pardo A, Maglione V, Filosa S, Crispi S. New therapeutic drugs from bioactive natural molecules: the role of gut microbiota metabolism in neurodegenerative diseases. Curr Drug Metab 2018; 19(6): 478-89. [http://dx.doi.org/10.2174/1389200219666180404094147] [PMID: 29623833]

[37] Miller JS, Rodriguez-Saona L, Hackshaw KV. Metabolomics in central sensitivity syndromes. Metabolites 2020; 10(4): 164. [http://dx.doi.org/10.3390/metabo10040164] [PMID: 32344505]

[38] Kaur G, Behl T, Bungau S, *et al.* Dysregulation of the gut-brain axis, dysbiosis and influence of numerous factors on gut microbiota associated parkinson's disease. Curr Neuropharmacol 2020; 19(2): 233-47. [http://dx.doi.org/10.2174/1570159X18666200606233050] [PMID: 32504503]

[39] Baxter MG, Crimins JL. Acetylcholine Receptor Stimulation for Cognitive Enhancement: Better the Devil You Know? Neuron 2018; 98(6): 1064-6. [http://dx.doi.org/10.1016/j.neuron.2018.06.018] [PMID: 29953868]

[40] Duranti S, Ruiz L, Lugli GA, *et al.* Bifidobacterium adolescentis as a key member of the human gut microbiota in the production of GABA. Sci Rep 2020; 10(1): 14112. [http://dx.doi.org/10.1038/s41598-020-70986-z] [PMID: 32839473]

[41] Sanchez S, Rodríguez-Sanoja R, Ramos A, Demain AL. Our microbes not only produce antibiotics, they also overproduce amino acids. J Antibiot (Tokyo) 2018; 71(1): 26-36. [http://dx.doi.org/10.1038/ja.2017.142] [PMID: 29089597]

[42] Yaghoubfar R, Behrouzi A, Ashrafian F, *et al.* Modulation of serotonin signaling/metabolism by Akkermansia muciniphila and its extracellular vesicles through the gut-brain axis in mice. Sci Rep 2020; 10(1): 22119. [http://dx.doi.org/10.1038/s41598-020-79171-8] [PMID: 33335202]

[43] Liu G, Chong HX, Chung FYL, Li Y, Liong MT. Lactobacillus plantarum DR7 modulated bowel movement and gut microbiota associated with dopamine and serotonin pathways in stressed adults. Int J Mol Sci 2020; 21(13): 4608. [http://dx.doi.org/10.3390/ijms21134608] [PMID: 32610495]

[44] Rooks MG, Garrett WS. Gut microbiota, metabolites and host immunity. Nat Rev Immunol 2016; 16(6): 341-52. [http://dx.doi.org/10.1038/nri.2016.42] [PMID: 27231050]

[45] Giau V, Wu S, Jamerlan A, An S, Kim S, Hulme J. Gut microbiota and their neuroinflammatory implications in Alzheimer's disease. Nutrients 2018; 10(11): 1765. [http://dx.doi.org/10.3390/nu10111765] [PMID: 30441866]

[46] Utzeri E, Usai P. Role of non-steroidal anti-inflammatory drugs on intestinal permeability and nonalcoholic fatty liver disease. World J Gastroenterol 2017; 23(22): 3954-63. [http://dx.doi.org/10.3748/wjg.v23.i22.3954] [PMID: 28652650]

[47] Wang L, Hu C, Shao L. The antimicrobial activity of nanoparticles: present situation and prospects for the future. Int J Nanomedicine 2017; 12: 1227-49. [http://dx.doi.org/10.2147/IJN.S121956] [PMID: 28243086]

[48] Deane R, Bell RD, Sagare A, Zlokovic BV. Clearance of amyloid-β peptide across the blood-brain barrier: implication for therapies in Alzheimer's disease. CNS & Neurological Disorders-Drug Targets (Formerly Current Drug Targets-CNS & Neurological Disorders) 2009; 8(1): 16-30. [http://dx.doi.org/10.2174/187152709787601867]

[49] Tognini P. Gut microbiota: a potential regulator of neurodevelopment. Front Cell Neurosci 2017; 11: 25. [http://dx.doi.org/10.3389/fncel.2017.00025] [PMID: 28223922]

[50] Wang P, Li D, Ke W, Liang D, Hu X, Chen F. Resveratrol-induced gut microbiota reduces obesity in high-fat diet-fed mice. Int J Obes 2020; 44(1): 213-25. [http://dx.doi.org/10.1038/s41366-019-0332-1] [PMID: 30718820]

[51] Goyal D, Ali SA, Singh RK. Emerging role of gut microbiota in modulation of neuroinflammation and neurodegeneration with emphasis on Alzheimer's disease. Prog Neuropsychopharmacol Biol Psychiatry 2021; 106: 110112. [http://dx.doi.org/10.1016/j.pnpbp.2020.110112] [PMID: 32949638]

[52] Zhang YJ, Li S, Gan RY, Zhou T, Xu DP, Li HB. Impacts of gut bacteria on human health and diseases. Int J Mol Sci 2015; 16(12): 7493-519. [http://dx.doi.org/10.3390/ijms16047493] [PMID: 25849657]

[53] Zhan X, Stamova B, Sharp FR. Lipopolysaccharide associates with amyloid plaques, neurons and oligodendrocytes in Alzheimer's disease brain: a review. Front Aging Neurosci 2018; 10: 42. [http://dx.doi.org/10.3389/fnagi.2018.00042] [PMID: 29520228]

[54] Khan AA, Sirsat AT, Singh H, Cash P. Microbiota and cancer: current understanding and mechanistic implications. Clin Transl Oncol 2021; 1-0. [PMID: 34387847]

[55] Hardy H, Harris J, Lyon E, Beal J, Foey A. Probiotics, prebiotics and immunomodulation of gut mucosal defences: homeostasis and immunopathology. Nutrients 2013; 5(6): 1869-912.

[http://dx.doi.org/10.3390/nu5061869] [PMID: 23760057]

[56] Xiao J, Katsumata N, Bernier F, *et al.* Probiotic bifidobacterium breve in improving cognitive functions of older adults with suspected mild cognitive impairment: a randomized, double-blind, placebo-controlled trial. J Alzheimer's Dis 2020; 1-9. [http://dx.doi.org/10.3233/JAD-200488]

[57] Markowiak P, Śliżewska K. Effects of probiotics, prebiotics, and synbiotics on human health. Nutrients 2017; 9(9): 1021. [http://dx.doi.org/10.3390/nu9091021] [PMID: 28914794]

[58] Shao X, Zhu G. Associations among monoamine neurotransmitter pathways, personality traits, and major depressive disorder. Front Psychiatry 2020; 11: 381. [http://dx.doi.org/10.3389/fpsyt.2020.00381] [PMID: 32477180]

[59] Otte C, Gold SM, Penninx BW, *et al.* Major depressive disorder. Nat Rev Dis Primers 2016; 2(1): 16065. [http://dx.doi.org/10.1038/nrdp.2016.65] [PMID: 27629598]

[60] Hidaka BH. Depression as a disease of modernity: Explanations for increasing prevalence. J Affect Disord 2012; 140(3): 205-14. [http://dx.doi.org/10.1016/j.jad.2011.12.036] [PMID: 22244375]

[61] Pedrelli P, Nyer M, Yeung A, Zulauf C, Wilens T. College students: mental health problems and treatment considerations. Acad Psychiatry 2015; 39(5): 503-11. [http://dx.doi.org/10.1007/s40596-014-0205-9] [PMID: 25142250]

[62] Burcusa SL, Iacono WG. Risk for recurrence in depression. Clin Psychol Rev 2007; 27(8): 959-85. [http://dx.doi.org/10.1016/j.cpr.2007.02.005] [PMID: 17448579]

[63] Brigitta B. Pathophysiology of depression and mechanisms of treatment. Dialogues Clin Neurosci 2002; 4(1): 7-20. [http://dx.doi.org/10.31887/DCNS.2002.4.1/bbondy] [PMID: 22033824]

[64] Jacques A, Chaaya N, Beecher K, Ali SA, Belmer A, Bartlett S. The impact of sugar consumption on stress driven, emotional and addictive behaviors. Neurosci Biobehav Rev 2019; 103: 178-99. [http://dx.doi.org/10.1016/j.neubiorev.2019.05.021] [PMID: 31125634]

[65] Castro-Marrero J, Sáez-Francàs N, Santillo D, Alegre J. Treatment and management of chronic fatigue syndrome/myalgic encephalomyelitis: all roads lead to Rome. Br J Pharmacol 2017; 174(5): 345-69. [http://dx.doi.org/10.1111/bph.13702] [PMID: 28052319]

[66] Smith BA, Georgiopoulos AM, Quittner AL. Maintaining mental health and function for the long run in cystic fibrosis. Pediatr Pulmonol 2016; 51(S44): S71-8. [http://dx.doi.org/10.1002/ppul.23522] [PMID: 27662107]

[67] Krishnan V, Nestler EJ. The molecular neurobiology of depression. Nature 2008; 455(7215): 894-902. [http://dx.doi.org/10.1038/nature07455] [PMID: 18923511]

[68] Walker AK, Kavelaars A, Heijnen CJ, Dantzer R. Neuroinflammation and comorbidity of pain and depression. Pharmacol Rev 2014; 66(1): 80-101. [http://dx.doi.org/10.1124/pr.113.008144] [PMID: 24335193]

[69] Nemeroff CB, Weinberger D, Rutter M, *et al.* DSM-5: a collection of psychiatrist views on the changes, controversies, and future directions. BMC Med 2013; 11(1): 202. [http://dx.doi.org/10.1186/1741-7015-11-202] [PMID: 24229007]

[70] Krystal AD. Psychiatric disorders and sleep. Neurol Clin 2012; 30(4): 1389-413. [http://dx.doi.org/10.1016/j.ncl.2012.08.018] [PMID: 23099143]

[71] Juruena MF, Calil HM, Fleck MP, Del Porto JA. Melancholia in Latin American studies: a distinct mood disorder for the ICD-11. Br J Psychiatry 2011; 33 (Suppl. 1): S37-58. [http://dx.doi.org/10.1590/S1516-44462011000500005] [PMID: 21845334]

[72] Hollon SD, Jacobson V. Cognitive approaches InHandbook of clinical behavior therapy with adults. Boston, MA: Springer 1985; pp. 169-99. [http://dx.doi.org/10.1007/978-1-4613-2427-0_7]

[73] Diener E, Seligman MEP. Beyond Money. Psychol Sci Public Interest 2004; 5(1): 1-31. [http://dx.doi.org/10.1111/j.0963-7214.2004.00501001.x] [PMID: 26158992]

[74] Shih RA, Belmonte PL, Zandi PP. A review of the evidence from family, twin and adoption studies for a genetic contribution to adult psychiatric disorders. Int Rev Psychiatry 2004; 16(4): 260-83. [http://dx.doi.org/10.1080/09540260400014401] [PMID: 16194760]

[75] van Heyningen V, Yeyati PL. Mechanisms of non-Mendelian inheritance in genetic disease. Hum Mol Genet 2004; 13(Spec No 2) (Suppl. 2): R225-33. [http://dx.doi.org/10.1093/hmg/ddh254] [PMID: 15358729]

[76] Hayashi T, Hou Y, Glasser MF, *et al.* The nonhuman primate neuroimaging and neuroanatomy project. Neuroimage 2021; 229: 117726. [http://dx.doi.org/10.1016/j.neuroimage.2021.117726] [PMID: 33484849]

[77] Silva AR, Grosso C, Delerue-Matos C, Rocha JM. Comprehensive review on the interaction between natural compounds and brain receptors: Benefits and toxicity. Eur J Med Chem 2019; 174: 87-115. [http://dx.doi.org/10.1016/j.ejmech.2019.04.028] [PMID: 31029947]

[78] Hussain LS, Reddy V, Maani CV. Physiology, Noradrenergic Synapse. StatPearls. 2020.

[79] Rebecchi MJ, Pentyala SN. Anaesthetic actions on other targets: protein kinase C and guanine nucleotide-binding proteins. Br J Anaesth 2002; 89(1): 62-78. [http://dx.doi.org/10.1093/bja/aef160] [PMID: 12173242]

[80] Abou-haila A, Tulsiani DRP. Signal transduction pathways that regulate sperm capacitation and the acrosome reaction. Arch Biochem Biophys 2009; 485(1): 72-81. [http://dx.doi.org/10.1016/j.abb.2009.02.003] [PMID: 19217882]

[81] Pandey UB, Nichols CD. Human disease models in Drosophila melanogaster and the role of the fly in therapeutic drug discovery. Pharmacol Rev 2011; 63(2): 411-36. [http://dx.doi.org/10.1124/pr.110.003293] [PMID: 21415126]

[82] Filatova EV, Shadrina MI, Slominsky PA. Major depression: one brain, one disease, one set of intertwined processes. Cells 2021; 10(6): 1283. [http://dx.doi.org/10.3390/cells10061283] [PMID: 34064233]

[83] Ikram H, Haleem DJ. Repeated treatment with reserpine as a progressive animal model of depression. Pak J Pharm Sci 2017; 30(3): 897-902. [PMID: 28653936]

[84] Goel N, Workman JL, Lee TT, Innala L, Viau V. Sex differences in the HPA axis. Compr Physiol 2014; 4(3): 1121-55. [http://dx.doi.org/10.1002/cphy.c130054] [PMID: 24944032]

[85] Limbana T, Khan F, Eskander N. Gut Microbiome and Depression: How Microbes Affect the Way We Think. Cureus 2020; 12(8): e9966. [http://dx.doi.org/10.7759/cureus.9966] [PMID: 32983670]

[86] Holmes SE. Neuroinflammation in Major Depressive Disorder and Schizophrenia: A PET Study. United Kingdom: The University of Manchester 2016.

[87] García-Cabrerizo R, Carbia C, O'Riordan KJ, Schellekens H, Cryan JF. Microbiota-gut-brain axis as a regulator of reward processes. J Neurochem 2021; 157(5): 1495-524. [http://dx.doi.org/10.1111/jnc.15284] [PMID: 33368280]

[88] Moulton CD, Pavlidis P, Norton C, *et al.* Depressive symptoms in inflammatory bowel disease: an extraintestinal manifestation of inflammation? Clin Exp Immunol 2019; 197(3): 308-18. [http://dx.doi.org/10.1111/cei.13276] [PMID: 30762873]

[89] Sarkar A, Yoo JY, Valeria Ozorio Dutra S, Morgan KH, Groer M. The association between early-life gut microbiota and long-term health and diseases. J Clin Med 2021; 10(3): 459. [http://dx.doi.org/10.3390/jcm10030459] [PMID: 33504109]

[90] Martin CR, Osadchiy V, Kalani A, Mayer EA. The brain-gut-microbiome axis. Cell Mol Gastroenterol Hepatol 2018; 6(2): 133-48. [http://dx.doi.org/10.1016/j.jcmgh.2018.04.003] [PMID: 30023410]

[91] Bonaz B, Sinniger V, Pellissier S. Anti-inflammatory properties of the vagus nerve: potential therapeutic implications of vagus nerve stimulation. J Physiol 2016; 594(20): 5781-90. [http://dx.doi.org/10.1113/JP271539] [PMID: 27059884]

[92] Sarkar A, Lehto SM, Harty S, Dinan TG, Cryan JF, Burnet PWJ. Psychobiotics and the manipulation of bacteria–gut–brain signals. Trends Neurosci 2016; 39(11): 763-81. [http://dx.doi.org/10.1016/j.tins.2016.09.002] [PMID: 27793434]

[93] Carabotti M, Scirocco A, Maselli MA, Severi C. The gut-brain axis: interactions between enteric microbiota, central and enteric nervous systems. Ann Gastroenterol 2015; 28(2): 203-9. [PMID: 25830558]

[94] Sochocka M, Donskow-Łysoniewska K, Diniz BS, Kurpas D, Brzozowska E, Leszek J. The gut microbiome alterations and inflammation-driven pathogenesis of Alzheimer's disease—a critical review. Mol Neurobiol 2019; 56(3): 1841-51. [http://dx.doi.org/10.1007/s12035-018-1188-4] [PMID: 29936690]

[95] Kho ZY, Lal SK. The human gut microbiome–a potential controller of wellness and disease. Front Microbiol 2018; 9: 1835. [http://dx.doi.org/10.3389/fmicb.2018.01835] [PMID: 30154767]

[96] Kaplan BJ, Rucklidge JJ, Romijn A, McLeod K. The emerging field of nutritional mental health: Inflammation, the microbiome, oxidative stress, and mitochondrial function. Clin Psychol Sci 2015; 3(6): 964-80. [http://dx.doi.org/10.1177/2167702614555413]

[97] Cheung SG, Goldenthal AR, Uhlemann AC, Mann JJ, Miller JM, Sublette ME. Systematic review of gut microbiota and major depression. Front Psychiatry 2019; 10: 34. [http://dx.doi.org/10.3389/fpsyt.2019.00034] [PMID: 30804820]

[98] Barandouzi ZA, Starkweather AR, Henderson WA, Gyamfi A, Cong XS. Altered composition of gut microbiota in depression: a systematic review. Front Psychiatry 2020; 11: 541. [http://dx.doi.org/10.3389/fpsyt.2020.00541] [PMID: 32587537]

[99] Arzani M, Jahromi SR, Ghorbani Z, *et al.* Gut-brain Axis and migraine headache: a comprehensive review. J Headache Pain 2020; 21(1): 15. [http://dx.doi.org/10.1186/s10194-020-1078-9] [PMID: 32054443]

[100] Huang F, Wu X. Brain neurotransmitter modulation by gut microbiota in anxiety and depression. Front Cell Dev Biol 2021; 9: 649103. [http://dx.doi.org/10.3389/fcell.2021.649103] [PMID: 33777957]

[101] Mayer EA. Gut feelings: the emerging biology of gut–brain communication. Nat Rev Neurosci 2011; 12(8): 453-66. [http://dx.doi.org/10.1038/nrn3071] [PMID: 21750565]

[102] Ma Q, Xing C, Long W, Wang HY, Liu Q, Wang RF. Impact of microbiota on central nervous system and neurological diseases: the gut-brain axis. J Neuroinflammation 2019; 16(1): 53. [http://dx.doi.org/10.1186/s12974-019-1434-3] [PMID: 30823925]

[103] Sternini C, Anselmi L, Rozengurt E. Enteroendocrine cells: a site of 'taste' in gastrointestinal chemosensing. Curr Opin Endocrinol Diabetes Obes 2008; 15(1): 73-8. [http://dx.doi.org/10.1097/MED.0b013e3282f43a73] [PMID: 18185066]

[104] Meijerink J. The Intestinal Fatty Acid-Enteroendocrine Interplay, Emerging Roles for Olfactory Signaling and Serotonin Conjugates. Molecules 2021; 26(5): 1416.
[http://dx.doi.org/10.3390/molecules26051416] [PMID: 33807994]

[105] Fricker LD. Neuropeptides and other bioactive peptides: from discovery to function. Colloquium series on neuropeptides 2012 Jun 30 (Vol 1, No 2, pp 1-122).
[http://dx.doi.org/10.4199/C00058ED1V01Y201205NPE003]

[106] Fung C, Vanden Berghe P. Functional circuits and signal processing in the enteric nervous system. Cell Mol Life Sci 2020; 77(22): 4505-22.
[http://dx.doi.org/10.1007/s00018-020-03543-6] [PMID: 32424438]

[107] Lach G, Schellekens H, Dinan TG, Cryan JF. Anxiety, depression, and the microbiome: a role for gut peptides. Neurotherapeutics 2018; 15(1): 36-59.
[http://dx.doi.org/10.1007/s13311-017-0585-0] [PMID: 29134359]

[108] Zeng N, Athmann C, Kang T, *et al.* PACAP type I receptor activation regulates ECL cells and gastric acid secretion. J Clin Invest 1999; 104(10): 1383-91.
[http://dx.doi.org/10.1172/JCI7537] [PMID: 10562300]

[109] Parker HE, Reimann F, Gribble FM. Molecular mechanisms underlying nutrient-stimulated incretin secretion. Expert Rev Mol Med 2010; 12(Jan): e1.
[http://dx.doi.org/10.1017/S146239940900132X] [PMID: 20047700]

[110] Ringseis R, Gessner DK, Eder K. The Gut–Liver Axis in the Control of Energy Metabolism and Food Intake in Animals. Annu Rev Anim Biosci 2020; 8(1): 295-319.
[http://dx.doi.org/10.1146/annurev-animal-021419-083852] [PMID: 31689373]

[111] Liu L, Zhu G. Gut–brain axis and mood disorder. Front Psychiatry 2018; 9: 223.
[http://dx.doi.org/10.3389/fpsyt.2018.00223] [PMID: 29896129]

[112] Hurwitz B. Urban observation and sentiment in James Parkinson's essay on the shaking palsy (1817). Lit Med 2014; 32(1): 74-104.
[http://dx.doi.org/10.1353/lm.2014.0002] [PMID: 25055707]

[113] Obeso JA, Stamelou M, Goetz CG, *et al.* Past, present, and future of Parkinson's disease: A special essay on the 200th Anniversary of the Shaking Palsy. Mov Disord 2017; 32(9): 1264-310.
[http://dx.doi.org/10.1002/mds.27115] [PMID: 28887905]

[114] Hughes RC. Parkinson's disease and its management. BMJ 1994; 308(6923): 281.
[http://dx.doi.org/10.1136/bmj.308.6923.281]

[115] Henry E, Lai EC. Hypokinesia and movement challenges in Parkinson's disease. Neurology Care Line 1(1): 1-2.

[116] Park JH, Kang YJ, Horak FB. What is wrong with balance in Parkinson's disease? J Mov Disord 2015; 8(3): 109-14.
[http://dx.doi.org/10.14802/jmd.15018] [PMID: 26413237]

[117] Schneider SA, Obeso JA. Clinical and pathological features of Parkinson's disease. Curr Top Behav Neurosci 2015; 22: 205-20.
[http://dx.doi.org/10.1007/7854_2014_317]

[118] Onaolapo OJ, Odeniyi AO, Onaolapo AY. Parkinson's Disease: Is there a Role for Dietary and Herbal Supplements? CNS & Neurological Disorders-Drug Targets (Formerly Current Drug Targets-CNS & Neurological Disorders) 2021.

[119] Martinez-Ramirez D, Okun MS, Jaffee MS. Parkinson's disease psychosis: therapy tips and the importance of communication between neurologists and psychiatrists. Neurodegener Dis Manag 2016; 6(4): 319-30.
[http://dx.doi.org/10.2217/nmt-2016-0009] [PMID: 27408981]

[120] Dias V, Junn E, Mouradian MM. The role of oxidative stress in Parkinson's disease. J Parkinsons Dis

2013; 3(4): 461-91.
[http://dx.doi.org/10.3233/JPD-130230] [PMID: 24252804]

[121] Zhao Z, Ning J, Bao X, *et al.* Fecal microbiota transplantation protects rotenone-induced Parkinson's disease mice *via* suppressing inflammation mediated by the lipopolysaccharide-TLR4 signaling pathway through the microbiota-gut-brain axis. Microbiome 2021; 9(1): 226.
[http://dx.doi.org/10.1186/s40168-021-01107-9] [PMID: 34784980]

[122] Crispi S, Filosa S, Di Meo F. Polyphenols-gut microbiota interplay and brain neuromodulation. Neural Regen Res 2018; 13(12): 2055-9.
[http://dx.doi.org/10.4103/1673-5374.241429] [PMID: 30323120]

[123] Meade RM, Fairlie DP, Mason JM. Alpha-synuclein structure and Parkinson's disease – lessons and emerging principles. Mol Neurodegener 2019; 14(1): 29.
[http://dx.doi.org/10.1186/s13024-019-0329-1] [PMID: 31331359]

[124] Bartels AL, Leenders KL. Parkinson's disease: The syndrome, the pathogenesis and pathophysiology. Cortex 2009; 45(8): 915-21.
[http://dx.doi.org/10.1016/j.cortex.2008.11.010] [PMID: 19095226]

[125] Travagli RA, Browning KN, Camilleri M. Parkinson disease and the gut: new insights into pathogenesis and clinical relevance. Nat Rev Gastroenterol Hepatol 2020; 17(11): 673-85.
[http://dx.doi.org/10.1038/s41575-020-0339-z] [PMID: 32737460]

[126] Rogers GB, Keating DJ, Young RL, Wong M-L, Licinio J, Wesselingh S. From gut dysbiosis to altered brain function and mental illness: mechanisms and pathways. Mol Psychiatry 2016; 21(6): 738-48.
[http://dx.doi.org/10.1038/mp.2016.50] [PMID: 27090305]

[127] Shishov VA, Kirovskaia TA, Kudrin VS, Oleskin AV. Amine neuromediators, their precursors, and oxidation products in the culture of Escherichia coli K-12. Prikl Biokhim Mikrobiol 2009; 45(5): 550-4.
[PMID: 19845286]

[128] Pokusaeva K, Johnson C, Luk B, *et al.* GABA-producing *Bifidobacterium dentium* modulates visceral sensitivity in the intestine. Neurogastroenterol Motil 2017; 29(1): e12904.
[http://dx.doi.org/10.1111/nmo.12904] [PMID: 27458085]

[129] Özoğul F. Production of biogenic amines by Morganella morganii, Klebsiella pneumoniae and Hafnia alvei using a rapid HPLC method. Eur Food Res Technol 2004; 219(5): 465-9.
[http://dx.doi.org/10.1007/s00217-004-0988-0]

[130] Liu J, Xu F, Nie Z, Shao L. Gut Microbiota Approach—A New Strategy to Treat Parkinson's Disease. Front Cell Infect Microbiol 2020; 10: 570658.
[http://dx.doi.org/10.3389/fcimb.2020.570658] [PMID: 33194809]

[131] Herold R, Schroten H, Schwerk C. Virulence factors of meningitis-causing bacteria: enabling brain entry across the blood–brain barrier. Int J Mol Sci 2019; 20(21): 5393.
[http://dx.doi.org/10.3390/ijms20215393] [PMID: 31671896]

[132] Javed I, Cui X, Wang X, *et al.* Implications of the human gut–brain and gut–cancer axes for future nanomedicine. ACS Nano 2020; 14(11): 14391-416.
[http://dx.doi.org/10.1021/acsnano.0c07258] [PMID: 33138351]

[133] Menozzi E, Macnaughtan J, Schapira AHV. The gut-brain axis and Parkinson disease: clinical and pathogenetic relevance. Ann Med 2021; 53(1): 611-25.
[http://dx.doi.org/10.1080/07853890.2021.1890330] [PMID: 33860738]

[134] Yang D, Zhao D, Ali Shah SZ, *et al.* The role of the gut microbiota in the pathogenesis of Parkinson's disease. Front Neurol 2019; 10: 1155.
[http://dx.doi.org/10.3389/fneur.2019.01155] [PMID: 31781020]

[135] Mogensen TH. Pathogen recognition and inflammatory signaling in innate immune defenses. Clin

Microbiol Rev 2009; 22(2): 240-73.
[http://dx.doi.org/10.1128/CMR.00046-08] [PMID: 19366914]

[136] Haddad F, Sawalha M, Khawaja Y, Najjar A, Karaman R. Dopamine and levodopa prodrugs for the treatment of Parkinson's disease. Molecules 2017; 23(1): 40.
[http://dx.doi.org/10.3390/molecules23010040] [PMID: 29295587]

[137] Yu XX, Fernandez HH. Dopamine agonist withdrawal syndrome: A comprehensive review. J Neurol Sci 2017; 374: 53-5.
[http://dx.doi.org/10.1016/j.jns.2016.12.070] [PMID: 28104232]

[138] Kulisevsky J, Pagonabarraga J. Tolerability and safety of ropinirole *versus* other dopamine agonists and levodopa in the treatment of Parkinson's disease: meta-analysis of randomized controlled trials. Drug Saf 2010; 33(2): 147-61.
[http://dx.doi.org/10.2165/11319860-000000000-00000] [PMID: 20082541]

[139] Naoi M, Maruyama W, Shamoto-Nagai M. Neuroprotective Function of Rasagiline and Selegiline, Inhibitors of Type B Monoamine Oxidase, and Role of Monoamine Oxidases in Synucleinopathies. Int J Mol Sci 2022; 23(19): 11059.
[http://dx.doi.org/10.3390/ijms231911059] [PMID: 36232361]

[140] Katsaiti I, Nixon J. Are there benefits in adding catechol-O methyltransferase inhibitors in the pharmacotherapy of Parkinson's disease patients? A systematic review. J Parkinsons Dis 2018; 8(2): 217-31.
[http://dx.doi.org/10.3233/JPD-171225] [PMID: 29614697]

[141] Kim A, Kim YE, Yun JY, *et al.* Amantadine and the risk of dyskinesia in patients with early Parkinson's disease: an open-label, pragmatic trial. J Mov Disord 2018; 11(2): 65-71.
[http://dx.doi.org/10.14802/jmd.18005] [PMID: 29860788]

[142] Brocks DR. Anticholinergic drugs used in Parkinson's disease: An overlooked class of drugs from a pharmacokinetic perspective. J Pharm Pharm Sci 1999; 2(2): 39-46.
[PMID: 10952768]

[143] Roy D, Delcenserie V. The protective role of probiotics in disturbed enteric microbiota InProbiotic Bacteria and Enteric Infections. Dordrecht: Springer 2011; pp. 221-61.
[http://dx.doi.org/10.1007/978-94-007-0386-5_11]

[144] Davani-Davari D, Negahdaripour M, Karimzadeh I, *et al.* Prebiotics: Definition, types, sources, mechanisms, and clinical applications. Foods 2019; 8(3): 92.
[http://dx.doi.org/10.3390/foods8030092] [PMID: 30857316]

[145] Bereded NK, Abebe GB, Fanta SW, *et al.* The impact of sampling season and catching site (wild and aquaculture) on gut microbiota composition and diversity of nile tilapia (oreochromis niloticus). Biology (Basel) 2021; 10(3): 180.
[http://dx.doi.org/10.3390/biology10030180] [PMID: 33804538]

[146] Agim ZS, Cannon JR. Dietary factors in the etiology of Parkinson's disease. BioMed Research International 2015; 2015: 672838.
[http://dx.doi.org/10.1155/2015/672838]

[147] Robins-Browne RM. Traditional enteropathogenic Escherichia coli of infantile diarrhea. Clin Infect Dis 1987; 9(1): 28-53.
[http://dx.doi.org/10.1093/clinids/9.1.28] [PMID: 3547577]

[148] Rudick RA, Miller D, Clough JD, Gragg LA, Farmer RG. Quality of life in multiple sclerosis. Comparison with inflammatory bowel disease and rheumatoid arthritis. Arch Neurol 1992; 49(12): 1237-42.
[http://dx.doi.org/10.1001/archneur.1992.00530360035014] [PMID: 1449401]

[149] Dobson R, Giovannoni G. Multiple sclerosis-a review. Eur J Neurol 2019; 26(1): 27-40.
[http://dx.doi.org/10.1111/ene.13819] [PMID: 30300457]

[150] Patsopoulos NA. Genetics of multiple sclerosis: An overview and new directions. Cold Spring Harb Perspect Med 2018; 8(7): a028951.
[http://dx.doi.org/10.1101/cshperspect.a028951] [PMID: 29440325]

[151] Holberg C, Finlayson M. Factors influencing the use of energy conservation strategies by persons with multiple sclerosis. Am J Occup Ther 2007; 61(1): 96-107.
[http://dx.doi.org/10.5014/ajot.61.1.96] [PMID: 17302111]

[152] Yamout B, Alroughani R, Al-Jumah M, *et al.* Consensus guidelines for the diagnosis and treatment of multiple sclerosis. Curr Med Res Opin 2013; 29(6): 611-21.
[http://dx.doi.org/10.1185/03007995.2013.787979] [PMID: 23514115]

[153] Rolak LA, Fleming JO. The differential diagnosis of multiple sclerosis. Neurologist 2007; 13(2): 57-72.
[http://dx.doi.org/10.1097/01.nrl.0000254705.39956.34] [PMID: 17351525]

[154] Pelidou SH, Giannopoulos S, Tzavidi S, Lagos G, Kyritsis AP. Multiple sclerosis presented as clinically isolated syndrome: the need for early diagnosis and treatment. Ther Clin Risk Manag 2008; 4(3): 627-30.
[PMID: 18827858]

[155] Love S. Demyelinating diseases. J Clin Pathol 2006; 59(11): 1151-9.
[http://dx.doi.org/10.1136/jcp.2005.031195] [PMID: 17071802]

[156] Stadelmann C, Wegner C, Brück W. Inflammation, demyelination, and degeneration — Recent insights from MS pathology. Biochim Biophys Acta Mol Basis Dis 2011; 1812(2): 275-82.
[http://dx.doi.org/10.1016/j.bbadis.2010.07.007] [PMID: 20637864]

[157] Lassmann H. Multiple sclerosis pathology. Cold Spring Harb Perspect Med 2018; 8(3): a028936.
[http://dx.doi.org/10.1101/cshperspect.a028936] [PMID: 29358320]

[158] Mayo L, Quintana FJ, Weiner HL. The innate immune system in demyelinating disease. Immunol Rev 2012; 248(1): 170-87.
[http://dx.doi.org/10.1111/j.1600-065X.2012.01135.x] [PMID: 22725961]

[159] Engelhardt B. T cell migration into the central nervous system during health and disease: Different molecular keys allow access to different central nervous system compartments. Clin Exp Neuroimmunol 2010; 1(2): 79-93.
[http://dx.doi.org/10.1111/j.1759-1961.2010.009.x]

[160] Höftberger R, Lassmann H. Inflammatory demyelinating diseases of the central nervous system. Handb Clin Neurol 2018; 145: 263-83.
[http://dx.doi.org/10.1016/B978-0-12-802395-2.00019-5] [PMID: 28987175]

[161] Schepici G, Silvestro S, Bramanti P, Mazzon E. The gut microbiota in multiple sclerosis: an overview of clinical trials. Cell Transplant 2019; 28(12): 1507-27.
[http://dx.doi.org/10.1177/0963689719873890] [PMID: 31512505]

[162] Cryan JF, O'Riordan KJ, Cowan CSM, *et al.* The microbiota-gut-brain axis. Physiol Rev 2019; 99(4): 1877-2013.
[http://dx.doi.org/10.1152/physrev.00018.2018] [PMID: 31460832]

[163] Maglione A, Zuccalà M, Tosi M, Clerico M, Rolla S. Host Genetics and Gut Microbiome: Perspectives for Multiple Sclerosis. Genes (Basel) 2021; 12(8): 1181.
[http://dx.doi.org/10.3390/genes12081181] [PMID: 34440354]

[164] Annalisa N, Alessio T, Claudette TD, Erald V, Antonino DL, Nicola DD. Gut microbioma population: an indicator really sensible to any change in age, diet, metabolic syndrome, and life-style. Mediators Inflamm 2014; 2014: 1-11.
[http://dx.doi.org/10.1155/2014/901308] [PMID: 24999296]

[165] Lavelle A, Sokol H. Gut microbiota-derived metabolites as key actors in inflammatory bowel disease.

Nat Rev Gastroenterol Hepatol 2020; 17(4): 223-37.
[http://dx.doi.org/10.1038/s41575-019-0258-z] [PMID: 32076145]

[166] Sintzel MB, Rametta M, Reder AT. Vitamin D and multiple sclerosis: a comprehensive review. Neurol Ther 2018; 7(1): 59-85.
[http://dx.doi.org/10.1007/s40120-017-0086-4] [PMID: 29243029]

[167] Alharbi FM. Update in vitamin D and multiple sclerosis. Neurosciences 2015; 20(4): 329-35.
[http://dx.doi.org/10.17712/nsj.2015.4.20150357] [PMID: 26492110]

[168] Di Rosa M, Malaguarnera M, Nicoletti F, Malaguarnera L. Vitamin D3: a helpful immuno-modulator. Immunology 2011; 134(2): 123-39.
[http://dx.doi.org/10.1111/j.1365-2567.2011.03482.x] [PMID: 21896008]

[169] Forbes JD, Van Domselaar G, Bernstein CN. The gut microbiota in immune-mediated inflammatory diseases. Front Microbiol 2016; 7: 1081.
[http://dx.doi.org/10.3389/fmicb.2016.01081] [PMID: 27462309]

[170] Hasan N, Yang H. Factors affecting the composition of the gut microbiota, and its modulation. PeerJ 2019; 7: e7502.
[http://dx.doi.org/10.7717/peerj.7502] [PMID: 31440436]

[171] Dinan K, Dinan TG. Gut Microbes and Neuropathology: Is There a Causal Nexus? Pathogens 2022; 11(7): 796.
[http://dx.doi.org/10.3390/pathogens11070796] [PMID: 35890040]

[172] Cristofori F, Dargenio VN, Dargenio C, Miniello VL, Barone M, Francavilla R. Anti-inflammatory and immunomodulatory effects of probiotics in gut inflammation: a door to the body. Front Immunol 2021; 12: 578386.
[http://dx.doi.org/10.3389/fimmu.2021.578386] [PMID: 33717063]

[173] Brahmer JR, Abu-Sbeih H, Ascierto PA, *et al.* Society for Immunotherapy of Cancer (SITC) clinical practice guideline on immune checkpoint inhibitor-related adverse events. J Immunother Cancer 2021; 9(6): e002435.
[http://dx.doi.org/10.1136/jitc-2021-002435] [PMID: 34172516]

[174] Berkovich R, Agius MA. Mechanisms of action of ACTH in the management of relapsing forms of multiple sclerosis. Ther Adv Neurol Disord 2014; 7(2): 83-96.
[http://dx.doi.org/10.1177/1756285613518599] [PMID: 24587825]

[175] Weber F, Janovskaja J, Polak T, Poser S, Rieckmann P. Effect of interferon on human myelin basic protein-specific T-cell lines: Comparison of IFN -1a and IFN -1b. Neurology 1999; 52(5): 1069-71.
[http://dx.doi.org/10.1212/WNL.52.5.1069] [PMID: 10102432]

[176] Dhib-Jalbut S. Mechanisms of action of interferons and glatiramer acetate in multiple sclerosis. Neurology 2002; 58(8, Supplement 4) (Suppl. 4): S3-9.
[http://dx.doi.org/10.1212/WNL.58.8_suppl_4.S3] [PMID: 11971121]

[177] Polman CH, O'Connor PW, Havrdova E, *et al.* A randomized, placebo-controlled trial of natalizumab for relapsing multiple sclerosis. N Engl J Med 2006; 354(9): 899-910.
[http://dx.doi.org/10.1056/NEJMoa044397] [PMID: 16510744]

[178] Fox EJ. Management of worsening multiple sclerosis with mitoxantrone: A review. Clin Ther 2006; 28(4): 461-74.
[http://dx.doi.org/10.1016/j.clinthera.2006.04.013] [PMID: 16750460]

[179] Oh J, O'Connor PW. An update of teriflunomide for treatment of multiple sclerosis. Ther Clin Risk Manag 2013; 9: 177-90.
[PMID: 23761970]

[180] Linker RA, Lee DH, Ryan S, *et al.* Fumaric acid esters exert neuroprotective effects in neuroinflammation *via* activation of the Nrf2 antioxidant pathway. Brain 2011; 134(3): 678-92.
[http://dx.doi.org/10.1093/brain/awq386] [PMID: 21354971]

[181] Kohl HM, Castillo AR, Ochoa-Repáraz J. The Microbiome as a Therapeutic Target for Multiple Sclerosis: Can Genetically Engineered Probiotics Treat the Disease? Diseases 2020; 8(3): 33. [http://dx.doi.org/10.3390/diseases8030033] [PMID: 32872621]

[182] Kim KO, Gluck M. Fecal microbiota transplantation: an update on clinical practice. Clin Endosc 2019; 52(2): 137-43. [http://dx.doi.org/10.5946/ce.2019.009] [PMID: 30909689]

[183] Singh RK, Chang HW, Yan D, *et al.* Influence of diet on the gut microbiome and implications for human health. J Transl Med 2017; 15(1): 73. [http://dx.doi.org/10.1186/s12967-017-1175-y] [PMID: 28388917]

[184] Peng M, Tabashsum Z, Anderson M, *et al.* Effectiveness of probiotics, prebiotics, and prebiotic-like components in common functional foods. Compr Rev Food Sci Food Saf 2020; 19(4): 1908-33. [http://dx.doi.org/10.1111/1541-4337.12565] [PMID: 33337097]

[185] Hodges H, Fealko C, Soares N. Autism spectrum disorder: definition, epidemiology, causes, and clinical evaluation. Transl Pediatr 2020; 9(S1) (Suppl. 1): S55-65. [http://dx.doi.org/10.21037/tp.2019.09.09] [PMID: 32206584]

[186] Samsam M, Ahangari R, Naser SA. Pathophysiology of autism spectrum disorders: Revisiting gastrointestinal involvement and immune imbalance. World J Gastroenterol 2014; 20(29): 9942-51. [http://dx.doi.org/10.3748/wjg.v20.i29.9942] [PMID: 25110424]

[187] De Rubeis S, Buxbaum JD. Genetics and genomics of autism spectrum disorder: embracing complexity. Hum Mol Genet 2015; 24(R1): R24-31. [http://dx.doi.org/10.1093/hmg/ddv273] [PMID: 26188008]

[188] Karimi P, Kamali E, Mousavi SM, Karahmadi M. Environmental factors influencing the risk of autism. 2017; 22. [http://dx.doi.org/10.4103/1735-1995.200272]

[189] Werling DM, Geschwind DH. Sex differences in autism spectrum disorders. Curr Opin Neurol 2013; 26(2): 146-53. [http://dx.doi.org/10.1097/WCO.0b013e32835ee548] [PMID: 23406909]

[190] Migeon B. Females are mosaics: X inactivation and sex differences in disease. Oxford university press 2007.

[191] Beversdorf DQ, Stevens HE, Margolis KG, Van de Water J. Prenatal stress and maternal immune dysregulation in autism spectrum disorders: Potential points for intervention. Curr Pharm Des 2020; 25(41): 4331-43. [http://dx.doi.org/10.2174/1381612825666191119093335] [PMID: 31742491]

[192] Xiong J, Chen S, Pang N, *et al.* Neurological diseases with autism spectrum disorder: role of ASD risk genes. Front Neurosci 2019; 13: 349. [http://dx.doi.org/10.3389/fnins.2019.00349] [PMID: 31031587]

[193] Chaste P, Leboyer M. Autism risk factors: genes, environment, and gene-environment interactions. Dialogues Clin Neurosci 2012; 14(3): 281-92. [http://dx.doi.org/10.31887/DCNS.2012.14.3/pchaste] [PMID: 23226953]

[194] Fattorusso A, Di Genova L, Dell'Isola GB, Mencaroni E, Esposito S. Autism spectrum disorders and the gut microbiota. Nutrients 2019; 11(3): 521. [http://dx.doi.org/10.3390/nu11030521]

[195] Roussin L, Prince N, Perez-Pardo P, Kraneveld AD, Rabot S, Naudon L. Role of the gut microbiota in the pathophysiology of autism spectrum disorder: clinical and preclinical evidence. Microorganisms 2020; 8(9): 1369. [http://dx.doi.org/10.3390/microorganisms8091369] [PMID: 32906656]

[196] Li Q, Han Y, Dy ABC, Hagerman RJ. The gut microbiota and autism spectrum disorders. Front Cell

Neurosci 2017; 11: 120.
[http://dx.doi.org/10.3389/fncel.2017.00120] [PMID: 28503135]

[197] Adams JB, Johansen LJ, Powell LD, Quig D, Rubin RA. Gastrointestinal flora and gastrointestinal status in children with autism – comparisons to typical children and correlation with autism severity. BMC Gastroenterol 2011; 11(1): 22.
[http://dx.doi.org/10.1186/1471-230X-11-22] [PMID: 21410934]

[198] Belkaid Y, Hand TW. Role of the microbiota in immunity and inflammation. Cell 2014; 157(1): 121-41.
[http://dx.doi.org/10.1016/j.cell.2014.03.011] [PMID: 24679531]

[199] Szachta P, Skonieczna-Żydecka K, Adler G, Karakua-Juchnowicz H, Madlani H, Ignyś I. Immune related factors in pathogenesis of autism spectrum disorders. Eur Rev Med Pharmacol Sci 2016; 20(14): 3060-72.
[PMID: 27460736]

[200] Santocchi E, Guiducci L, Prosperi M, *et al.* Effects of probiotic supplementation on gastrointestinal, sensory and core symptoms in autism spectrum disorders: a randomized controlled trial. Front Psychiatry 2020; 11: 550593.
[http://dx.doi.org/10.3389/fpsyt.2020.550593] [PMID: 33101079]

[201] Ng Q, Loke W, Venkatanarayanan N, Lim D, Soh A, Yeo W. A systematic review of the role of prebiotics and probiotics in autism spectrum disorders. Medicina (Kaunas) 2019; 55(5): 129.
[http://dx.doi.org/10.3390/medicina55050129] [PMID: 31083360]

[202] Kang DW, Adams JB, Coleman DM, *et al.* Long-term benefit of Microbiota Transfer Therapy on autism symptoms and gut microbiota. Sci Rep 2019; 9(1): 5821.
[http://dx.doi.org/10.1038/s41598-019-42183-0] [PMID: 30967657]

[203] Srikantha P, Mohajeri MH. The possible role of the microbiota-gut-brain-axis in autism spectrum disorder. Int J Mol Sci 2019; 20(9): 2115.
[http://dx.doi.org/10.3390/ijms20092115] [PMID: 31035684]

[204] Butt RL, Volkoff H. Gut microbiota and energy homeostasis in fish. Front Endocrinol (Lausanne) 2019; 10: 9.
[http://dx.doi.org/10.3389/fendo.2019.00009] [PMID: 30733706]

CHAPTER 4

The Role of Age in Pediatric Tumors of the Central Nervous System

Nesibe S. Kutahyalioglu[1,*] and **Dylan V. Scarton**[2]

[1] *Karabuk University, Faculty of Health Science, Karabuk, Turkey*

[2] *George Mason University, Interdisciplinary Program in Neuroscience, Fairfax, Virginia, USA*

Abstract: Pediatric tumors of the central nervous system (CNS) are the second most common type of solid childhood cancer. As such, they have a major effect on the rates of morbidity and mortality in children. CNS tumors originate from abnormal cells in the brain and/or spinal cord, which can be classified as either benign or malignant. They can be further subdivided into different categories based on several principal aspects, such as tumor location, histopathology, and developmental age. Among these various characteristics, age is one of the most consequential determinants for CNS tumors. Specific groups between 0 and 21 years of age, for instance, have radically divergent landscapes in terms of their tumor incidence and unique biology. Depending on the age of the child, key case features may differ like the clinical evaluation, medical diagnosis and prognosis, recommended therapy and treatment courses, anticipated responses and tolerability to treatment, and management of side effects. Effective teamwork is another crucial component for the successful management of pediatric CNS tumors. In patient-and-family-centered care, ensuring a detailed education of the children and their families, as well as their involvement in the decision-making process where appropriate, is imperative. To determine the best available options for the patient, multidisciplinary medical teams will often deliberate over all of the possible procedures. The holistic care provided by these inter-professional collaborations for this vulnerable population will depend on the age of the child, in addition to the level of patient and family participation. Evidence shows that support and counseling of the patient and their family during the entire treatment process can have a significant impact on outcomes. This chapter will review the essential diagnostic and prognostic considerations of childhood CNS tumors, with special emphasis placed on favorable therapies and treatments, including in-depth discussions around the multi-faceted responses to treatment and the management of its side effects. In particular, this content will highlight the critical role that age, and interdisciplinary healthcare teams play in comprehensive disease management.

[*] **Corresponding author Nesibe S. Kutahyalioglu:** Karabuk University, Faculty of Health Science, Karabuk, Turkey; E-mail: nesibekutahyalioglu@karabuk.edu.tr

Zareen Amtul (Ed.)

Keywords: Age, Brain, Central Nervous System, CNS, Disease Management, Multidisciplinary, Pediatric, Quality of Life, Spinal Cord, Tumors, Team, Treatment.

INTRODUCTION

2020 was a harrowing year in human health history. For Andrew Kaczynski and Rachel Louise Ensign, it was no different. Six months after then-President Donald Trump declared a state of emergency over COVID-19 in the United States (U.S.), Andrew and Rachel received news that no parent would wish upon another—their six-month-old daughter, Francesca "Beans" Kaczynski, had just been diagnosed with a rare and aggressive brain cancer known as atypical teratoid rhabdoid tumor (ATRT). This type of tumor often manifests in the cerebellum or brain stem, which are largely responsible for coordinating movement and controlling basic body functions, respectively. Symptoms include headaches, nausea, and vomiting, as well as balance loss and an abnormally large head size in infants. Prior to Beans' supervising physician developing a novel treatment protocol at the Dana-Farber Cancer Institute in Boston, Massachusetts, ATRT was effectively a death sentence for kids. Now, the survival rate for children with ATRT over the age of three can be as high as 70%; however, if the child is under one, then their chance of survival plummets to less than 10%. How can merely two years of life account for more than a seven-fold difference in surviving such a pernicious disease? This tragic discrepancy is due in part to treatment protocols not recommending radiation therapy for babies before they reach 12 months. With limited options in the wake of this grim prognosis, Beans passed away just three months later on Christmas Eve [1].

Although this case is fortunately far and few between, pediatric tumors of the central nervous system (CNS) continue to affect families around the nation and across the world every day. While an average of only 60 cases of ATRT are identified in the U.S. each year [2], 500 more children are diagnosed with medulloblastoma [3]. Approximately one in every four American children with a brain tumor has medulloblastoma, making it the most common type among adolescence [4]. Overall, depending on the country, the annual incidence of pediatric brain tumors ranges from 1.15 to 5.14 cases per 100,000 children [5]. These sobering rates translate to tens of thousands of children developing CNS tumors worldwide, wherein the causes are mostly unknown a number do not survive. The survivors encounter immense challenges related to their physical and psychological health, in addition to various social and intellectual obstacles. Nevertheless, children with brain tumors generally have a better prognosis than adults with similar conditions [6]. For all forms of brain tumors, approximately 3 out of 4 children survive for at least 5 years after their diagnosis [7]. Clinical outcomes for these vulnerable patients have improved tremendously in recent decades, with the implementation of advanced screening measures and the development of targeted treatment strategies [8]. The field of pediatric oncology continues to grow and progress in what is a very active area of research.

This chapter will explore the essential diagnostic and prognostic considerations of childhood CNS tumors, with special emphasis placed on favorable therapies and treatments, including in-depth discussions around the multi-faceted responses to treatment and the management of its side effects. In particular, this content will highlight the critical role that age and multidisciplinary healthcare teams play in comprehensive disease management. The key learning objectives are to:

1. Discuss the procedures for performing a clinical evaluation and rendering medical diagnoses/prognoses of CNS tumors, according to the fundamental features of the case.

2. Examine the suggested, evidence-based treatment and therapy options for CNS tumors in children, depending on the medical diagnosis and prognosis.

3. Investigate the patient response to treatment and management of side effects, given relevant circumstances like developmental age and tumor presentation.

4. Review the importance of enhancing holistic care approaches among multidisciplinary healthcare teams by improving the quality of care for vulnerable pediatric populations affected by childhood CNS tumors.

The ultimate goal of this chapter is to provide a comprehensive summary on how CNS tumors are managed in children of different ages while also emphasizing the role of multidisciplinary team care, from diagnosis to post-treatment. Clinically relevant descriptions are given for the most common tumor types in addition to their overall epidemiology, pathological presentation, clinical diagnosis, and treatment regimens. Tumor classifications are separated by their severity scale, from low- to high-grade, with some genetic data included as well. Although the diagnostic criteria and treatment modalities may be similar for these two broad groups of tumors, their management and outcomes are quite different. Furthermore, despite a degree of standardization in treatment strategies among many practitioners and institutions, a variety of emerging and experimental techniques are also employed on a case-by-case basis. General management principles are described in the context of standard therapy, but other approaches may be considered equally valid. This complete presentation should better inform the reader to the different management options for pediatric CNS tumors.

Definition of Pediatric Tumors in the CNS

Pediatric tumors of the CNS encompass many diseases, where normal brain and/or spinal cord cells undergo genetic alterations and result in a growing mass of abnormal cells. Primary brain tumors involve a growth that starts in the brain, rather than spreading to the brain from another part of the body. Although the exact cause of primary brain tumors remains unknown, some tumors have germ-line mutations and thus hereditary origins [9 - 11]. Two inherited risk factors that have been well-established are ionizing radiation, such as high-dose cranial irradiation, and specific genetic syndromes, like neurofibromatosis type 1, Li-Fraumeni syndrome, Turcot syndrome, Lynch syndrome, and constitutional mismatch repair deficiency. These tumors are of acute concern in pediatric patients because of their long-life expectancies. The majority of CNS tumors, however, result from somatic mutations and are not hereditary [9, 12].

These tumors may be either malignant and cancerous or benign and non-cancerous. Benign brain tumors may grow and press on nearby areas of the brain. They rarely spread into other brain tissue. Malignant brain tumors may be either low-grade or high-grade. High-grade tumors are likely to grow quickly and spread into other brain tissue. Low-grade tumors tend to grow and spread more slowly than high-grade tumors. When a tumor grows into or presses on an area of the brain, it may stop that part of the brain from working the way it should. Both benign and malignant brain tumors can cause signs or symptoms, need treatment, and can recur [13].

Brain tumors are classified by the presentation age, histological type, and extent of respectability or ability to be removed by surgery. Congenital tumors present antenatally in the first 60 days of life, while tumors that appear within the first year are tumors of infancy. Otherwise, tumors that develop in older children before the age of 15 are considered pediatric cases. The most common types of congenital brain tumors include teratoma, choroid plexus papilloma, desmoplastic infantile tumors, glioblastoma multiforme, and medulloblastoma [14]. Fig. (**1**) below shows some of the major neuroanatomical landmarks and primary locations of several common brain tumors [15].

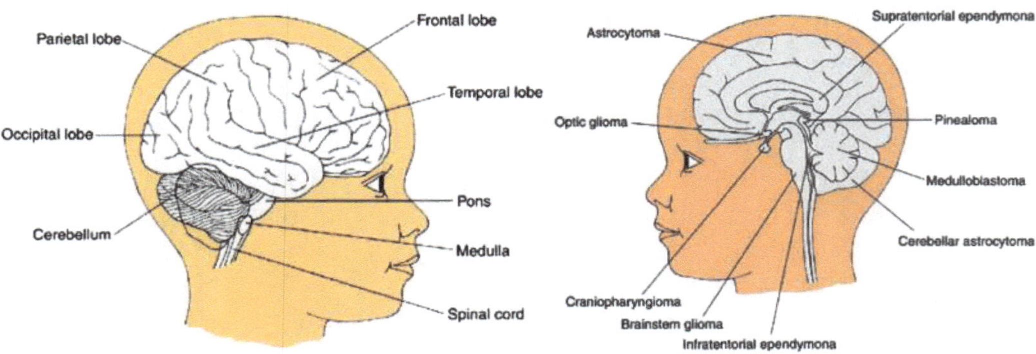

Fig. (1). Side-by-side schematics showing the major landmarks of basic neuroanatomy (*left*) and the primary location of prevalent brain tumors (*right*) among children [15].

The histopathology of brain tumors varies greatly among the different subtypes. Teratomas, for instance, arise from all three germ layers that comprise the CNS—ectoderm, mesoderm, and endoderm. They can be either mature or immature and can behave malignantly. This malignant variation, however, is the least common among congenital brain tumors [14, 16, 17]. Gliomas, on the other hand, are derived directly from glial precursor cells of the CNS. Within this category, astrocytic tumors constitute a large group of neoplasms that display tremendous heterogeneity in terms of pathology, histology, and severity according the World Health Organization (WHO) grading scale. Gliomas may be classified from low-grade (*i.e.*, WHO grade I-II) to high-grade (*i.e.*, WHO grade III-IV). Examples of low-grade tumors include pilocytic astrocytoma, desmoplastic infantile astrocytoma, and ganglioglioma, all of which are WHO grade I. On the opposite end of the spectrum would be glioblastoma multiforme, which is a high-grade tumor at WHO grade IV. The congenital form of this severe brain tumor is extremely rare and exhibits clinical characteristics that are different and have a better prognosis compared to its pediatric variant. Among the embryonal brain tumors, medulloblastomas are by far the most common. Recent molecular studies have demonstrated that they make up a group of unique neoplasms that have distinct cellular origins and thus behave discretely from one another. Since 2016, the revised WHO classification lists four subgroups of medulloblastoma: wingless, sonic hedgehog, group 3, and group 4 [18, 19]. Table **1** below outlines this grading scale with typical characteristics of different tumor types [20].

Table 1. International classification system of low- and high-grade brain tumors from the WHO with sample characteristics of different tumor types [20].

<table>
<tr><th colspan="2">Severity Grade</th><th>Typical Characteristics</th><th>Tumor Types</th></tr>
<tr><td rowspan="2">Low-Grade</td><td>WHO Grade I</td><td>Least malignant (benign)
Possibly curable *via* surgery alone
Non-infiltrative
Long-term survival
Slow growing</td><td>Pilocytic astrocytoma
Craniopharyngioma
Gangliocytoma
Ganglioglioma</td></tr>
<tr><td>WHO Grade II</td><td>Relatively slow growing
Somewhat infiltrative
May recur as higher grade</td><td>"Diffuse" Astrocytoma
Pineocytoma
Pure oligodendroglioma</td></tr>
<tr><td rowspan="2">High-Grade</td><td>WHO Grade III</td><td>Malignant
Infiltrative
Tend to recur as higher grade</td><td>Anaplastic astrocytoma
Anaplastic ependymoma
Anaplastic oligodendroglioma</td></tr>
<tr><td>WHO Grade IV</td><td>Most malignant
Rapid growth, aggressive
Widely infiltrative
Rapid recurrence
Necrosis prone</td><td>Glioblastoma multiforme
Pineoblastoma
Medulloblastoma
Ependymoblastoma</td></tr>
</table>

[a]Data adapted from Metro Health HMO Limited [20].

Overview of the Current State of the Field

Second only to leukemia as a cause of pediatric malignancy, pediatric brain tumors are the most common type of solid childhood cancer. Among all childhood cancers, brain tumors are the leading cause of death. CNS tumors constitute about 20% of all cancers in patients under the age of 15 [21]. In children younger than 3 years of age, supratentorial tumors are more common than infratentorial tumors, whereas in children between 4 and 10 years of age, posterior fossa tumors are the most prevalent type. Following puberty, the tumor incidence in each of these youth groups occurs at an equal frequency [22]. Among congenital brain tumors, which are diagnosed antenatally or within the first two months of life, teratoma is the most common type of congenital brain tumor. The incidence of this germ cell tumor is equally distributed between male and female patients [23, 24]. After teratomas and gliomas, choroid plexus papillomas are the third most common brain tumor in the pediatric population. These masses can arise from anywhere that the choroid plexus is present, which often occurs in the lateral ventricles of infants [25]. Medulloblastoma is the most common posterior fossa tumor among children. They arise from the vermis of the cerebellum and can cause obstructive hydrocephalus and leptomeningeal seeding along the spinal cord [26].

The international incidence of childhood CNS tumors differs among different countries. Their prognoses and survival rates depend on multivariate factors, from histological subtype to anatomical location. Incidence per 100,000 children varies from 1.15 to 5.14 cases, where the highest rates are reported in the United States [5]. The overall incidence of congenital brain tumor varies ranges between 0.3 to 2.9 cases per 100,000 live births worldwide [27]. Within the congenital brain tumors, the most common type is teratoma (26.6% to 48%), followed by astrocytoma (7.4% to 28.8%), choroid plexus papilloma (3.7% to 13.2%), embryonal tumor (3% to 13%), craniopharyngioma (5.6% to 6.8%), and ependymoma (4.4%) [28].

CNS tumors in children persist as a tremendous challenge for clinical teams. Their diverse biological behaviors, in the unique context of the developing nervous system, require flexible and tailored treatment plans. Over the last few decades, great strides have been made in our understanding of the molecular and genetic basis of human malignancy. This progress is particularly true for pediatric tumors like teratoma, glioma, and medulloblastoma. The challenge for clinicians is leveraging this wealth of new biologic information in a directed and rational manner to select effective and less toxic therapeutic agents [29].

Key Evidence-Based Factors Impacting Patients Outcomes

Pediatric brain tumors have differential prognoses depending on the age at presentation, histological type, and extent of resection. The pathology of congenital brain tumors manifests in different ways than similar tumors of older children. Infants with high-grade gliomas, for instance, have a better prognosis than older children with the same tumor type [30, 31]. Significant advances in imaging techniques, as well as molecular biology and genetics, have enabled clinicians to better diagnose pediatric brain tumors earlier in the course of disease, thereby allowing them to apply more targeted treatment strategies [29].

Children with brain and spinal cord tumors have unique needs from other children with cancer. Some of the symptoms that initially bring children with CNS tumors to medical attention — seizures, visual changes, cognitive and behavioral changes, weakness of arms or legs, and hormonal changes, to name a few — require highly trained specialists to help manage these problems while treatment for the tumor is ongoing. It is therefore crucial that an interconnected, multidisciplinary team of experienced medical professionals evaluate these children.

CLINICAL ASPECTS

Second to leukemia, pediatric tumors of the CNS are the most common type of solid childhood cancer. For this reason, they have significant repercussions on both the morbidity and mortality rates in children [21]. CNS tumors originate from abnormal cells in the brain and/or spinal cord, which can be classified as either benign or malignant. They can be further subdivided into distinct categories based on several principal aspects, such as developmental age (*e.g.*, antepartum, infancy, adolescence), histopathology (*e.g.*, teratoma, astrocytoma, embryonal), and tumor location (*e.g.*, supratentorial, infratentorial, intradural) [9, 14]. The most common symptoms of pediatric CNS cancers are the result of increased intracranial pressure and manifest as headache, nausea, and vomiting. Other symptoms that may present include seizure, hemi- and monoparesis, cranial nerve deficits, ataxia, hemisensory loss, dysphasia and aphasia, and memory impairment. Infants and young children may experience an enlarged head circumference and miss critical developmental milestones [32 - 34].

The pathogenesis of brain tumors is believed to be the result of combinatorial environmental and genetic factors. Direct casual relationships have not been identified, despite significant advancements in medical knowledge. Family history may be involved in the development of CNS tumors, as many studies report an association between brain tumors and siblings. Moreover, parental age at birth could be implicated. Some studies have shown children of women over the age of 40 are at an increased risk for brain cancer, especially astrocytoma and ependymoma [14]. Additionally, cancer predisposition syndromes, such as DICER1, Li-Fraumeni, and other non-neoplastic neurocutaneous syndromes like Sturge-Weber syndrome and hereditary telangiectasia, have all shown CNS manifestations and associations with brain tumors. In other words, these syndromes have been established as predisposing specific types of brain tumors. Neurofibromatosis type 1, for example, could be associated with low-grade gliomas, while tuberous sclerosis may be associated with subependymal giant cell astrocytoma, and Von-Hippel Lindau has been associated with hemangioblastomas [9 - 11]. Conclusive associations remain a challenge in the case of interpreting clinical presentations. For instance, differentiating between myelin vacuolation spots resulting from non-neoplastic neurofibromatosis *versus* those caused by low-grade glioma in response to neurofibromatosis type 1 may be difficult. Likewise, cortical and subcortical tubers in tuberous sclerosis may mimic gliomas if the T2-weighted images appear as hyperintense signals. Some studies have implied a link between infectious exposure during childhood and brain cancer onset, but this connection is still debatable. Table **2** below describes some of the CNS manifestations that occur in these familial syndromes and which chromosomes and genes are implicated in their hereditary nature [35].

Table 2. List of common familial or hereditary syndromes, including their corresponding chromosomal-gene locations, and their associated manifestations in the CNS [35].

Hereditary Syndrome	CNS Manifestations	Chromosome	Gene
Neurofibromatosis type 1	Optic pathway gliomas Astrocytoma Malignant peripheral nerve sheath tumors Neurofibromas	17q11	*NF1*
Neurofibromatosis type 2	Vestibular schwannomas Meningiomas Spinal cord ependymoma Spinal cord astrocytoma Hamartomas	22q12	*NF2*
von Hippel-Lindau	Hemangioblastoma	3p25-26	*VHL*
Tuberous sclerosis	Sub-ependymal giant cell astrocytoma Cortical tubers	9q34	*TSC1*
		16q13	*TSC2*
Li-Fraumeni	Astrocytoma Primitive neuroectodermal tumor	17q13	*TP53*
Cowden	Dysplastic gangliocytoma of the cerebellum (Lhermitte-Duclos disease)	10q23	*PTEN*
Turcot	Medulloblastoma	5q21	*APC*
	Glioblastoma	3p21	*hMLH1*
		7p22	*hPSM2*
Nevoid basal cell carcinoma	Medulloblastoma	9q31	*PTCH*

[a]Data adapted from Kleihues *et al.*, 2000 [35].

Environmental risks like high-dose radiation have also been linked to brain malignancies as well, which has been established in children who receive radiation treatment for leukemia [36].

The following section will discuss common procedures for performing clinical screenings and evaluations of pediatric CNS tumors and rendering their medical diagnoses, according to the patient case's fundamental features. Additionally, it will review different treatment considerations and cover how clinicians manage the side effects, with a look towards the long-term outcomes and their impact on quality of life.

Diagnosis and Screening

Childhood brain and spinal cord tumors are detected by examining the brain and spinal cord through a battery of medical tests and procedures. They may include a physical exam, wherein the body is inspected for general signs of health, including checking for signs of disease like lumps or anything else abnormal. These results are then documented in the patient's health history. Classic symptoms of CNS tumors that an evaluating physician would be searching for are often caused by increased intracranial pressure, which include headache, nausea, and vomiting, as well as irritability, lethargy, changes in behavior, and gait and balance disorders. Other common symptoms are related to tumor location, including vision loss and seizures. A history of the patient's health habits and past illnesses and treatments will also be taken. Once all of this information is collected, a neurological exam may be conducted in which a series of questions and tests are used to check for normal brain, spinal cord, and nerve function. This exam assesses the patient's mental status, physical coordination and gait, and how well the muscles, senses, and reflexes are performing. After these basic medical checks, additional tests in the form of imaging may be required to more conclusively diagnose some childhood brain and spinal cord tumors [37].

If medical imaging is required, then magnetic resonance imaging (MRI) is the current gold standard and test of choice among clinicians. MRI uses a magnet, radio waves, and computer to make a series of detailed pictures of the brain and spinal cord. For greater anatomical clarity, a substance called gadolinium can be injected intravenously to illuminate the targeted cancer cells in a procedure known as nuclear MRI (NMRI). With this non-invasive yet informative technique, high-resolution visuals can detect parenchymal tumors, as well as tumors within the posterior fossa, subarachnoid spaces, and arachnoid and pia mater meningeal layers. This method is especially valuable when a biopsy cannot be safely performed because of where the tumor has formed in the brain or spinal cord. Similarly, a computerized tomography (CT) scan may be done, but it is less sensitive and less specific. Serum tumor marker tests may complement these imaging techniques. These procedures involve testing a sample of blood to measure the amounts of target substances released into the blood by organs, tissues, or tumor cells in the body. Certain substances are linked to specific types of cancer when found in increased levels in the blood, which are called tumor markers [38, 39].

Congenital brain tumors may be discovered prenatally *via* ultrasound during the second or third trimester. Fetal MRI is becoming more common to visually confirm the findings originally detected by ultrasound. Postnatally, ultrasound may reveal an intracranial mass, hydrocephalus, macrocrania, and/or polyhydramnios, among other characteristic neurological signs [22]. Clinical presentations in newborns and older children, on the other hand, depend on the specific tumor location. Supratentorial tumors may present with limb weakness, body convulsions, and altered consciousness, whereas infratentorial tumors commonly present with signs of heightened intracranial pressure and bodily imbalance. As previously mentioned, the former variant is more common than the latter [40].

Teratomas are often visualized antenatally by ultrasound during the second and third trimesters, where they appear as a mixed echogenic and cystic intracranial mass [41]. MRI results should show heterogeneous signal intensity on T1/T2 weighting, and the presence of fat signal intensity on T1 would be pathognomonic. Similarly, CT scans should reveal a mixed-density lesion with frequent coarse calcification, and the presence of fat would suggest an identical diagnosis. Glioblastoma multiforme, as a high-grade astrocytoma, generally manifests in a large and ill-defined intracranial mass that occupies most of one cerebral hemisphere or spreads through the corpus callosum into the other hemisphere. Ultrasonography defines it as an echogenic and heterogeneous mass that typically reveals intralesional hemorrhage and necrosis. MRI diffusion restriction enables visualization of the high-grade nature of the lesion, including increased intracranial pressure [42]. CT scans of choroid plexus papilloma usually present it as an enhancing intraventricular mass, due to considerable hydrocephalus, and MRI images appear iso-hypo intense on T1 and iso-hyperintense on T2 [43]. Medulloblastomas typically develop along the midline of the posterior fossa and grow into the fourth ventricle. They appear hyperdense on CT scans and display variable intensity on T1/T2 MRI with diffusion restriction and varying contrast enhancement [44]. For all tumor types, scanning the whole neural axis is imperative to identify cerebrospinal fluid seeding and spinal drop metastasis [14, 45].

Differential diagnoses come in many forms, such as the extracranial neoplasms of congenital CNS infections like TORCH (toxoplasmosis, rubella, cytomegalovirus, and herpes simplex) mimicking metastatic disease [46]. Likewise, for intracranial hemorrhage, different developmental stages of hematoma may mimic tumors upon imaging. In particular, CT and MRI of the subacute and chronic stages may reveal central clot retraction and peripheral enhancement with/without surrounding edema. MRI sequences for blood products like susceptibility weighted imaging, diffusion weighted imaging, and magnetic resonance spectroscopy could help to clarify follow-up images. Dedicated CT or MRI angiography can be performed if underlying vascular malformation or vascular lesion is suspected [47]. Additionally, ischemic arterial or venous stroke may mimic tumors in imaging by showing mass effects and/or irregular contrast enhancements. Advanced clinical history and medical knowledge of arterial and venous imaging are essential to ensure an accurate diagnosis. In this case, advanced MRI sequences like diffusion weighted imaging and perfusion MRI can aid in resolving the ambiguity. Generally, if the edema shows restricted diffusion then it is caused by the tumor [14].

Once the presence of a brain tumor is confirmed by medical imaging, a biopsy is often necessary in most cases to verify the diagnosis and determine tumor type and grade. The international WHO classification system, last updated in 2021, provides relevant clinical information to further categorize CNS tumors by shared properties. This staging system, which considers tumor spread or metastasis, allows risk assessments to be conducted that then inform prognosis and treatment options. These assessments are based on the patient age and the amount of residual tumor, molecular findings, and evidence of disease spread. Residual tumors can be identified by performing an MRI of the entire spine, a lumbar puncture for cerebrospinal fluid cytology, and a postoperative MRI [38, 39]. For reference, Table **3** shows the results of an 18-year retrospective study that computed the average age (in years) of pediatric tumor diagnosis from over 250 children that were admitted to Mofid Pediatric Hospital in Tehran, Iran. Of note is the wide variance in mean age, from 0.2 in the case of atypical meningioma to 12 in the case of (typical) meningioma [48].

Regardless of these sophisticated imaging techniques, the majority of pediatric brain tumors are diagnosed and removed in surgery. If a patient has a suspected tumor, then a biopsy may be performed to remove a tissue sample for analysis. For primary tumors already growing in the brain, a craniotomy may be conducted wherein part of the skull is removed and a needle, sometimes guided by a computer robot, is used to remove the desired tissue sample. A pathologist can then view the tissue sample under a microscope to check for the presence of cancer cells, including the type and grade of brain tumor. This grading evaluation is based on how abnormal the cancer cells appear under the microscope and how quickly the tumor is likely to metastasize. If the tumor is determined to be cancerous, then the neurosurgeon may remove as many tumors as safely possible during the same surgery [33]. The diagnostic and prognostic details of several common pediatric tumors originating from the posterior fossa are listed below in Table **4** [49].

Table 3. Average of pediatric tumor diagnosis from an 18-year retrospective study of over 250 children that were admitted to Mofid Pediatric Hospital in Tehran, Iran [48].

Tumor Category	Age Mean ± SD in Years
Ependymoma	4.3 ± 4
Ependymoblastoma	2 ± 2.2
Choroid plexus papilloma	0.6 ± 0.3
Astrocytomas	6.2 ± 3.4
Pilocytic astrocytoma	5.6 ± 3.5
Glioblastoma multiforme	6.6 ± 4.8

(Table 3) cont.....

Oligodendroglioma	8.0 ± 0
Unspecified gliomas	4.4 ± 3.7
Astroblastoma	1.0 ±0
Mixed neuronal-glial tumors	7.6 ± 4.7
Medulloblastoma	5.4 ± 3.5
Primitive neuroectodermal tumors	4.2 ± 3.5
Atypical teratoid/rhabdoid tumor	1.5 ±0.8
Craniopharyngioma	9.0 ± 2.6
Pineal parenchymal tumors	12 ± 2.1
Meningioma	12 ± 2.2
Angioblastic meningioma	6 ± 0
Atypical meningioma	0.2 ± 0.1
Lymphoma	3.8 ± 2.8
Metastasis	4.6 ± 1.3

[a]Data adapted from Golkashani *et al.*, 2015 [48].

Treatment Considerations and Management of Side Effects

The signs and symptoms of pediatric CNS tumors will vary among children. They depend on the age and development of the child, where the tumor forms in the brain or spinal cord, and the size of the tumor and how fast it grows. Common signs and symptoms include morning headache or a headache that goes away after vomiting, frequent nausea and vomiting, problems with vision, hearing, and speech, loss of balance and trouble walking, unusual sleepiness or change in activity level, unusual changes in personality or behavior, seizures, and increased in the head size, especially in infants. Additional indications may also include back or extremity pain, changes in bowel habits or trouble urinating, and weakness in the legs. Critically, some children with CNS tumors are unable to meet important growth and development milestones like sitting up, crawling, walking, and certain aspects of speech production [33].

Table 4. Diagnostic and prognostic summary of typical pediatric tumors originating from the posterior fossa, as well as details regarding clinical presentation and relative incidence [49].

Tumor Type	Relative Incidence	Clinical Presentation	Diagnosis	Prognosis
Medulloblastoma	35-40%	2-3 months of: Headaches Vomiting Truncal ataxia	Heterogeneously or homogenously enhancing fourth ventricular mass May be disseminated	65-85% survival Dependent on stage/type Poorer in infants (20-70%)

(Table 4) cont.....

Cerebellar astrocytoma	35-40%	3-6 months of: Limb ataxia Secondary headaches Vomiting	Cerebellar hemisphere mass, usually with cystic and solid (mural nodule) components	90-100% survival in totally resected pilocytic type
Brain stem glioma	10-15%	1-4 months of: Double vision Unsteadiness Weakness Cranial nerve deficits Facial weakness Swallowing deficits	Diffusely expanded, minimally or partially enhancing mass in 80% 20% more focal tectal or cervicomedullary lesion	> 90% mortality in diffuse tumors Better in localized
Ependymoma	10-15%	2-5 months of: Unsteadiness Headaches Double vision Facial asymmetry	Usually enhancing, fourth ventricular mass with cerebellopontine predilection	> 75% survival in totally resected lesions
Atypical teratoid/rhabdoid	> 5%	2-3 months of: Headaches Vomiting Truncal ataxia Facial weakness Strabismus	Heterogeneously or homogenously enhancing fourth ventricular mass May be disseminated Laterally extended	≤ 10-20% survival in infants

[a]Data adapted from Packer *et al.*, 2008 [49].

The treatment approaches that clinicians decide apply will depend on the disease and tumor diagnosis, grade, stage, and risk assessment. In general, after any initial surgical procedure, radiation therapy, chemotherapy, or both may be required. The main treatment for congenital brain tumors is gross total resection. For this reason, surgical outcomes are largely dependent on the growth extent of the tumor and the general condition of the patient. Adjuvant therapies are not commonly used in newborns due to inherent risks, while craniospinal radiotherapy results in significant developmental retardation and is avoided in patients below the age of 3. Although radiation can be an effective option when treating adult patients, it is often necessary to avoid or reduce radiation in young children [14, 50, 51]. Table **5** below provides a brief summary of some standard treatment options for CNS embryonal tumors that present in childhood [52].

The National Comprehensive Cancer Network (NCCN) recently issued new guidelines recommending evidence-based treatment options for children with brain cancers across treatment disciplines, including oncologists/neuro-oncologists, radiation oncologists, pathologists, and pediatric neurosurgeons. The guidelines recommend that all patients with high-grade gliomas should receive care from a multidisciplinary team with experience managing CNS tumors. Treatment for pediatric CNS tumors often includes a combination of surgery, radiation therapy, and chemotherapy. The NCCN also recommends evaluating the patient for genetic counseling and/or evaluation for cancer predisposition. Treating children with cancer is quite different from adults. Clinicians exercise extreme caution to not impact physical and cognitive development and to protect against long-term adverse effects [53].

Table 5. Brief summary of some standard treatment options for CNS embryonal tumors that present in childhood [52].

Clinical Diagnosis	Childhood Age	Standard Treatment Options
Medulloblastoma	≤ 3 years	Surgery
		Adjuvant chemotherapy
	> 3 years (average- and high-risk)	Surgery
		Adjuvant Therapy: Radiation therapy Chemotherapy
CNS primitive neuroectodermal tumor	≤ 3 years	Surgery
		Adjuvant chemotherapy
	> 3 years	Surgery
		Adjuvant Therapy: Radiation therapy Chemotherapy
Medulloepithelioma and Ependymoblastoma	≤ 3 years	Surgery
		Adjuvant chemotherapy
	> 3 years	Surgery
		Adjuvant Therapy: Radiation therapy Chemotherapy
Pineoblastoma	≤ 3 years	Biopsy (for diagnosis)
		Chemotherapy
	> 3 years	Surgery
		Adjuvant Therapy: Radiation therapy Chemotherapy

[a]Data adapted from Childhood Central Nervous System Embryonal Tumors Treatment (PDQ®), 2016 [52].

Lastly, entry into a clinical trial, if available, may be considered for all children with a CNS tumor. Optimal treatment requires a multidisciplinary team of pediatric oncologists, pediatric neuro-oncologists, pediatric neurosurgeons, neuropathologists, neuroradiologists, and radiation oncologists who have experience treating CNS tumors in children [54]. Radiation therapy for CNS tumors is technically demanding, so children should be sent to collaborative medical centers that have experience in this area, if possible. These suggestions for evidence-based treatment and therapy options of CNS tumors in children, according to the medical diagnosis, should profound effects on the patient's prognosis.

Long-Term Outcomes and Quality of Life

The chance of recovery for childhood survivors of CNS tumors is affected by a number of certain factors. These factors include the patient's age, what was the type and location of the tumor, whether any cancer cells remain after surgery, and had the tumor just been diagnosed or had it recurred. Complications will also vary along these same lines but are also strongly impacted by the treatment protocol(s). Congenital brain tumors, in particular, have a poor prognosis overall, with a survival of less than 30%. The key factors of long-term health in this case are similar to those for other tumor survivability and complications, namely stage of fetal development, histological features, and treatment-related complications [33].

Disease complications and treatment effects can have a lasting impact on the long-term outcomes and quality of life for young patients. Survivors may experience several tumor- and treatment-related side effects requiring medical attention, which can affect their daily lives. They may experience health deficits like endocrinopathies, obesity, seizures, visual and hearing impairments, neurocognitive impairments, and cerebrovascular disease, even for low-grade gliomas and glioneuronal tumors. These chronic health conditions are prevalent in childhood cancer survivors and affect their well-being [33]. In a recent Childhood Cancer Survivor Study, 62% of the survivors had at least one chronic condition at a mean age of 26.6 years, with survivors of CNS tumors at the second highest risk of severe or life-threatening events. Furthermore, 66% of survivors as young as 5 years old had described chronic health conditions, with the prevalence increasing to 88% in survivors between the ages of 40 and 49 years [55]. Chronic health deficits are highly prevalent and multi-systemic in survivors, and they warrant preemptive counseling and dedicated multidisciplinary management.

AGE-RELATED TRENDS

Among these various characteristics, age is one of the most significant determinants for CNS tumors. Understanding the role of age is essential for children with CNS tumors because it can impact their diagnosis, treatment, and long-term outcomes. For instance, specific groups between 0 and 21 years of age have radically divergent landscapes regarding their tumor incidence and unique biology. Depending on the child's age, the critical case features like clinical evaluation, medical diagnosis, prognosis, recommended therapy, treatment courses, anticipated responses and tolerability to treatment, and management of side effects may differ.

It is difficult to generalize and compare different age groups regarding diagnosis, treatment, and prevalence of specific central nervous system tumors. Since age is not the only predictor, but has a crucial role in disease management. Another limitation is no consistency among studies about different age groups and critical disease features (diagnosis, prognosis, treatment options, *etc.*). This content will highlight age's vital role in comprehensive disease management by investigating the patient response to treatment and management of side effects, given relevant circumstances like developmental age and tumor presentation.

Importance of Understanding the Role of Age in the 0-14 Years Range

Age can affect the type of CNS tumor that a child develops. For example, the type of pediatric brain tumor is age dependent. Between ages 0-4, medulloblastomas and other embryonal tumors (arising from fetal cells in the brain) are the most common. Between ages 5-9, the most common tumor tends to be a pilocytic astrocytoma. Malignant gliomas are most common in ages 10-15.

The most common site for brain tumors in young children tends to be the posterior fossa, also known as the infratentorial space. This is where the cerebellum and brainstem are in the back of the head. As children age, the number of tumors presenting in the posterior fossa decreases, and the number of tumors in the supratentorial space (where the brain hemispheres are located) increases.

Although new developing therapies improve survivors' quality of life, reduce long-term side effects of treatment, and increase patient survival, CNS tumors remain the most common cause of cancer-related deaths in pediatric age groups. Around 25% of all tumors in those 0–14 years of age involve the CNS. The age-standardized rate for 0-14 years of age is 3.56 per 100,000 person-years with an annual average of approximately 330 new cases, according to one study in England from a group of nearly 3,000. Within the malignant group, the astrocytomas in those 0–14 years of age were mainly low grade (WHO grade I and II) and primarily in the supra- and infratentorial brain originating from neuroepithelial tissue [14, 21, 22].

Symptoms and presentation depend on the child's age and the tumor's location. Young children tend to get infratentorial tumors; these lesions often block fluid flow in the brain, causing hydrocephalus – too much fluid in the brain, leading to elevated blood pressure. This causes headaches, vomiting, and balance problems.

Young infants and babies are often present with failure to thrive and vomiting. Often these patients are current after multiple visits to other specialists, including gastroenterologists, for treatment of continued, unexplained vomiting. For babies with a soft spot, the fontanelle — where the skull bones have not fused yet — can present with a pulsatile soft spot due to elevated pressure. Tumors involving the brainstem often present with cranial nerve deficits such as double vision or failure of the eyes to turn together in the same direction, difficulty swallowing, balance, coordination problems, and weakness of one side of the face. Younger children may have trouble communicating their symptoms, while older children may be more likely to report headaches, vision problems, or changes in personality or behavior.

Although age is a crucial factor in determining the long-term prognosis for children with CNS tumors, it can be challenging to decide on the effect of age on prognosis too. Because younger patients typically undergo different treatment techniques than older patients. Research utilizing various chemotherapy regimens has been carried out to postpone or avoid radiation therapy [62]. The mortality range is higher (40%) in children under one year at diagnosis [81]. Also, treatment-related mortality and morbidity increase while the child's age decreases [56, 81]. The 5-year observed survival rate for brain and spinal cord malignancies in children aged 0 to 14 is 72%. This suggests that children with brain or spinal cord malignancies have an average survival rate of 72% or at least 5 years.

Age can influence the treatment options available for children with CNS tumors. Pediatric neuro-oncology faces significant challenges in providing therapies that are both effective and of tolerable toxicity, particularly when treating young patients. Since younger children are more sensitive to the side effects of radiation therapy, it can affect their cognitive and physical development. Immature brain structure is particularly susceptible to the toxicity of current treatment options. Children (under the age of 5) are especially vulnerable to the long-term side effects of cytotoxic medicines/therapies. Therefore, healthcare professionals frequently must choose between intensifying treatment to better control disease while reducing acute toxicity and the danger of long-term side effects. Due to radiation therapy is often entirely skipped or postponed in the treatment of brain tumors in young children, healthcare providers recommend alternative treatments, such as chemotherapy or surgery for children under the age of 5 [56, 80, 81].

Importance of Understanding the Role of Age for 14-21 Years Range

Similarly, the approach to a CNS tumor can vary as children age. As age increases, the mortality and treatment-related morbidity rates decrease, tolerability to treatment improves, side effects become more manageable, and treatment and therapy options get broader [56, 80, 81]—also, the survival rate increase with age. For teenagers aged 15 to 19, the 5-year survival rate is 76%, which is slightly significant compared to younger children (72%) [80].

When we look at ages 15-19, the most common tumors tend to be tumors in the pituitary gland region, lesions like craniopharyngiomas, and germ cell tumors. As the children age, the tumors can present in different fashions depending on their location, size, and aggressiveness. One joint presentation is nausea and vomiting from elevated intracranial pressure — either from hydrocephalus or the generous size of the tumor; balance and coordination problems for lesions located near or on the brainstem and cerebellum; weakness, numbness or some form of neurological deficit related to the location of the tumor. Older children and adolescents tend to be present with worsening headaches and seizures.

In summary, understanding the role of age is crucial for the diagnosis, treatment, and long-term outcomes of children with CNS tumors. By considering age, healthcare providers can tailor their approach to each child's unique needs and optimize their chances for a successful outcome. However, further research is needed for specific age groups to understand the mechanism of age in disease management.

MULTIDISCIPLINARY HEALTHCARE TEAMS

Neonates and children are vulnerable members of society. Other members need to be a voice for infants and young children as they are not competent to express complaints about neurological issues such as headaches, blurred vision, or impaired visual acuity. Delays in recognizing brain tumors at the early age group are frequently caused by the variety of early neurological features, the lack of specificity of the most prevalent signs and symptoms (headaches, vomiting, *etc.*), which may mimic other more common diseases. Any delay in either diagnosis or treatment may be life-threatening and cause long-term sequelae. Children with central nervous system (CNS) tumors have complicated medical issues, which require multidisciplinary specialized care to manage all treatment related aspects in children [68].

Multidisciplinary care has been established as the cornerstone of care for patients diagnosed with cancer. It is described as the collaboration of many specialist professionals participating in cancer care with the overarching objective of enhancing patient care and treatment effectiveness [58, 62]. Due to the need for a variety of specialist professionals to determine the best treatment pathway for children with CNS tumors, multidisciplinary care teams, which include medical, nursing, allied professionals, and diagnostic experts, are established [60]. An efficient team adopts an integrated, synergistic strategy, and the knowledge of each team member supports creating and implementing an efficient care plan.

The management of CNS tumors in children is affected by several factors. Histological diagnosis, disease stage, and the patient's unique characteristics, especially age, may affect treatment decisions and subsequent results. The listed factors play a role in determining the best course of action for pediatric CNS malignancies. However, one of the major obstacles is having lack of routine multidisciplinary care teams [58].

Role of Collaboration

Collaborative medicine plays a crucial role in the diagnosis, treatment, and management of CNS tumors in children. The collaborative team involved in the care of a child with a CNS tumor may include but not limited with neurologists, neurosurgeons, pediatrician, clinical and medical oncologists, radiation oncologists, pathologists, psychologists, specialist nurses, other healthcare professionals, parents, and children [56, 57, 64]. As a team, they coordinate their efforts and work together to develop a comprehensive treatment plan that considers the unique needs of the child and the specific characteristics of the tumor. Multidisciplinary teamwork throughout the care pathway, from diagnosis through follow-up and beyond.

The collaborative approach begins with the accurate diagnosis of the tumor. Once the diagnosis is made, the team works together to determine the best course of treatment, which may include surgery, chemotherapy, radiation therapy, or a combination of these modalities. During treatment, the collaborative team closely monitors the child's progress, assessing the efficacy of treatment and managing any side effects. They also provide support and resources to the child and family, including psychological counseling, social services, and education about the disease and its treatment. After treatment, the collaborative team continues to follow up with the child to monitor for any signs of recurrence or late effects of treatment. This ongoing care is essential for ensuring the child's long-term health and quality of life [59, 63]. Each team member from a different discipline plays the specific job that has been given and specializes in a certain area. Yet, decisions are often taken and endorsed jointly by several or all relevant disciplines [67].

Roles of Multidisciplinary Care Team During Diagnosis

To determine early medical diagnosis and immediate and appropriate treatment, a multidisciplinary team including pediatricians, oncologists, neurologists, and neurosurgeons should work collaboratively. After the clinical examination and thorough analysis, any modest suspicion of a pediatric CNS tumor should be explored by CT scan as soon as possible. The pediatrician generally takes the leading role and when suspected CNS tumors consults it with a specialist (neurologist or neurosurgeon). Then, pathologists become responsible for grading and classifying the tumor. Patients and their parents/caregivers should get assistance with decision-making from a neurologist who has been involved in the care ever since their diagnosis. Also, the neurosurgeon may provide histological diagnosis through biopsy and resection [57, 59].

Another team member is the nurse, who will first assess the patient and family at the time of diagnosis. It is crucial to build trust to involve the patient and family in decision-making, teach them how to control treatment toxicity, recognize new symptoms, and involve them in the decision-making process. This partnership will maximize healthcare resources while ensuring patient adherence and compliance to treatment [63, 67].

Roles of Multidisciplinary Care Team During Treatment

Compared to patients with adult brain tumors, neurological signs, and symptoms in children with brain tumors can vary. Thus, it is recommended to have a pediatric neurologist and/or pediatric neurosurgeon with extensive experience in pediatric neuro-oncology in the team. If it is not possible, neurosurgeons and neuro-oncologists in the team should work collaboratively with pediatricians [57, 63, 65].

Chemotherapy, interval monitoring, and radiation-based treatments all show promise in the treatment of specific populations of patients; however, surgery is a crucial first-line intervention for children with a heavy burden of disease, a mass impact from tumor proliferation, the necessity for a precise pathologic diagnosis, or for those with unbearable presenting symptoms. From a surgical standpoint, the neurosurgeon describes the potential surgical approaches, such as if a tumor resection is feasible or whether a biopsy can be done instead. The neurosurgeon creates an operative plan and manages postoperative complications and risks. Then, s/he needs to take the leading role and discuss with other team members about significant postoperative prognosis, likelihood of morbidity, and quality of life after surgery [57, 65, 67].

The neuro-oncologist emphasizes the size of the resection in accordance with the suspected diagnosis. Several possibilities can be reviewed collectively. The oncologist must consider inclusion in ongoing clinical studies. Planning of chemotherapy treatments, which are individualized to the patient and differ based on tumor histology, location, and goals of care is one of the roles of oncologist. Additionally, neuro-oncologists and pediatricians are in charge of providing palliative care for patients. Thus, they focus on symptom relief, particularly pain relieving. Pediatricians should be involved in decision making for managing medical treatment choices and controls [57, 65, 67].

The neuro-oncologists and radiation oncologists frequently collaborate and co-monitor the progression of therapy based on the type of tumor and its response to treatment. The role of radiation oncologists is designing a radiation therapy plan and whether to use radio-sensitizing agents during therapy. This decision is based on the age of the child, location, type, size of tumor and other factors and given with other team members according to balance between treatment benefits and side effects. Depending on the child's age, chemotherapy may be used as a time-saving method initially, delaying the start of radiation therapy until the child's brain is more developed [57, 63].

The expanded role of endocrinologists is especially true in the treatment of pediatric brain cancers, where radio therapy can negatively impact growth and development, and survivors have longer lifespans. Because long-term survivors are at a greater risk for cardiovascular disease and obesity, weight and lifestyle concerns are a crucial part of primary care management for patients of any age. To handle the common neurologic problems found in patients with brain tumors, multidisciplinary teams caring for those patients should include a neurologist as well. The management of seizures and strokes, the serious issues requiring urgent care, is one area where neurologists are involved in patient care [65].

Patients with brain tumors may have psychiatric alterations during the initial presentation and treatment. Brain tumors can first appear with the complete spectrum of psychiatric symptoms, including depression, manic, auditory, and visual hallucinations, anxiety, and amnesia. Hence, psychiatrists assist patients in managing their symptoms. The psychologist may already be able to support the family and better manage concerns and psychological problems. Children and parents/caregivers need psychological care for not only acute situations, but also long-term effects [63, 65, 67].

Children with CNS tumors will require all-encompassing physiological, psychological, social, and emotional care. To finish the intended course of treatment, patients must get complete support from the moment of diagnosis. The specialized clinical nurse's role, as a crucial member of the multidisciplinary team, is to support patients throughout the diagnostic and therapeutic process. This will involve not only providing nursing interventions (such as symptom, toxicity, and/or wound management), but also operational case management, like planning and coordinating treatment [63, 65].

Roles of Multidisciplinary Care Team After Treatment

All team members should participate in decision making process, implementation of treatment, support, and follow-up after treatments. The neurosurgeon role includes follow up children for necessity of second look surgery. The constant and appropriate monitoring of patients by radiation oncologists and neuro-oncologists is usually ensured, and interval scans are carried out to verify the lack or presence of tumor progression or recurrence and managing side effects. Knowledge about these side effects should be shared with children and families. Endocrinologists will be more involved in monitoring the aftereffects of various therapy modalities as brain tumor medicines advance and the number of long-term survivors rises.

Patients are still at a significant risk of experiencing symptoms after a diagnosis even if no prior psychiatric history is given. The development of the tumor or the course of treatment itself can lead to new symptoms. Some common preoperative and postoperative drugs can lead to mood disorders. Therefore, psychiatrists are responsible to closely follow-up those patients. Continued involvement of psychiatrists beyond the hospitalization is important. Psychotherapy can be helpful for both children and caregivers who frequently deal with a neurocognitive condition that is rapidly changing [65, 66]. Increased neurologic reliance and impairment may lead to more mood swings and stress. In these circumstances, psychotherapy and support groups may be beneficial [65].

The family should be informed of all potential disease effects following a multidisciplinary discussion among specialists. Typically, one of the disciplines leads the first dialogue with the family in which the other experts contribute further information. A psychologist assists in introducing the family to all the relevant disciplines and carefully guiding them through the discussion. The interdisciplinary team can better communicate to the patient's family and all pertinent information. This communication must include children in an age-appropriate way [67].

Effective Teamwork and Clinical Management

Multidisciplinary team approach is widely utilized for chronic conditions [69], and it is keystone of care for all cancer patients. It boosted team members' communication, strengthened continuity of coordinated care, and the decision-making process [60].

Parents and children are given devastating news, a ton of information, and little time to comprehend it all and make decisions during the first diagnosis and treatment of pediatric CNS tumors. Also, they need pain and symptom management, which should be available from the time of diagnosis. A variety of medical specialists who can each contribute in a different way to improving results and providing better care and parent and child who can participate in decision making process are included in a multidisciplinary approach to healthcare [59].

For a multidisciplinary team to function well, there should be a flow that includes the sharing of all patient information among team members, the presenting of therapeutic choices, gathering patient feedback on the many discussed techniques, and developing an individual treatment plan with the patient [71]. Any multidisciplinary cancer clinic, including pediatric CNS tumor clinics, can use these recommendations for multidisciplinary clinics as their skeleton [59].

The key themes that emerged in relation to effective teamworking and successful clinical management were the need for support at an organizational level, the importance of good relationships between team members, and a coordinator who has excellent communication and effective leadership skills. The chair guarantees that team members participate equally, which encourages improved decision-making and raises the standard of care through their impact on team effectiveness and the collegial atmosphere [62, 70].

About 100 recommendations for productive multidisciplinary teamwork were included in a study issued by the UK National Cancer Action Team in 2010 based on clinical consensus among more than 2,000 multidisciplinary team members. These suggestions are divided into five categories: "1- team governance, 2- meeting infrastructure, 3- meeting organization and logistics, 4- patient-centered clinical decision-making, and 5- team meetings." [72]. Other countries, like Australia, have employed comparable frameworks to define effective multidisciplinary teamwork [73].

For children with CNS tumors, multidisciplinary team input is fundamental to improved survival and quality of life. Prompt awareness by caregivers, referral from primary care facilities, informative radiographic scanning, maximum neurosurgical resection that is also safe, prompt neuropathologic interpretation, standardized adjuvant therapy regimens with the necessary support services, and coordinated monitoring over the long term are all inevitable for successfully clinical management [58, 59].

There are several benefits of an effective multidisciplinary team works, which results in enhanced clinical decision making, better coordination of patient care, increased overall quality of care, improved use of evidence-based treatment decisions, and better therapy [72] and provides improved survival rates [59 - 61, 63, 65]. A review by Hong and colleagues in 2010 found twelve studies looking at the association between multidisciplinary teams and survival that demonstrated a substantial improvement in survival for patients who participated in multidisciplinary cancer clinics across a range of cancer types. However, due to methodological variability of the research and the inclusion of retrospective and "before and after" types of studies, the interpretation of the findings was constrained.

Compared to other pediatric cancer survivors, brain tumor survivors are more likely to experience long-term physical health consequences [75]. The high survival rate of pediatric patients has significantly increased the demand for efficient physical and psychological rehabilitation. Although multidisciplinary team approach meets that demand, difficulties exist in the practice. There are few published instructions and guidelines on how to create a multidisciplinary pediatric CNS tumor clinic and how it should be run. Additionally, the ancillary staff, which includes social workers, psychologists, certified child life specialists, and back-to-school specialists, is frequently forgotten in the establishment of multidisciplinary team [59]. The team should inform the family of available services, particularly as children's' transition from the acute stage of the disease to remission and re-adaptation to school and society, including social workers, school interventions, neuropsychological testing, community support groups, and counseling [59].

Patient-and-Family Centered Care

The Institute for Patient-and Family-Centered Care and American Academy of Pediatrics list the following as the fundamental principles of Patient-and-Family Centered Care (PFCC): *dignity and respect, information sharing, family participation in care,* and *care in context of family and community*. The concept of *dignity and respect* refers to the working relationship between healthcare providers and families being respectful of diversity, different perspectives, values, beliefs, care preferences, cultural, and linguistic traditions. *Information sharing* signifies that the exchange of information between health care providers and parents should be complete, accurate, useful, meaningful, unbiased, and timely. *Family participation in care* means medically appropriate decisions should best fit the families' needs and values and be made with both parents and healthcare providers' participation. *Care in context of family and community* indicates family members should collaborate with healthcare providers and administrators about education, policy and program development, implementation, and evaluation [76]. PFCC involves two-way communication and requires reciprocal partnership between parents and healthcare providers [77].

Patient and family centered care is an approach to healthcare that prioritizes the needs and preferences of the patient and their family in all aspects of care, including decision-making, treatment planning, and ongoing support. This approach recognizes that patients and their families are experts in their own care, and that involving them in the care process can improve outcomes and patient satisfaction [78]. When it comes to children with CNS tumors, patient and family centered care is especially important. These tumors can have a significant impact on a child's physical, emotional, and cognitive development, as well as on their family dynamics.

The team should initially conduct a thorough assessment of the patient and family at the time of diagnosis. It is critical to build trust to involve the patient and family in the decision-making process, teach them how to control treatment toxicity and recognize new symptoms. This partnership will maximize healthcare resources while ensuring patient adherence and compliance to treatment [64]. It is necessary to create trusting relationships and provide full involvement, as the family may refuse the treatment once they are not informed or allowed in decision making process [58]. To prevent any treatment refusal the ways for providing patient and family centered care can be listed:

1. Involve the child and family in treatment decisions: Children with CNS tumors and their families should be involved in all treatment decisions, including deciding which treatment options to pursue and developing a treatment plan that is tailored to the child's needs and preferences.

2. Address the child's emotional and psychological needs: A diagnosis of CNS tumor can be overwhelming for both the child and their family. Healthcare providers should be sensitive to the child's emotional and psychological needs and provide support and resources to help them cope.

3. Provide clear and understandable information: Healthcare providers should communicate information about the child's condition and treatment in a clear and understandable way, using age-appropriate language and visuals when necessary.

4. Coordinate care across multidisciplinary care approach: Children with CNS tumors often require care from multiple providers across different specialties. Healthcare providers should work together to coordinate care and ensure that the child and family receive comprehensive and integrated care.

5. Provide ongoing support: Children with CNS tumors and their families may need ongoing support even after treatment is complete. Healthcare providers should be available to provide ongoing support and resources to help the child and family manage any long-term effects of the tumor and its treatment [59, 63, 65, 67].

Overall, it is critical to integrate patient and family centered care approach in the multidisciplinary health care approach for children with CNS tumors. By involving the child in an age-appropriate way and family in all aspects of care and providing ongoing support, healthcare providers can help improve outcomes and satisfaction and ensure that the child and family receive the best possible quality of care.

CONCLUSION

Summary of Findings

The most significant factors, including age and multidisciplinary teamwork that have affected the prognosis for children's CNS malignancies are outlined in this book chapter. Going back to the story of Francesca "Beans," a single hospital or organization is unable to improve care for thousands of children with CNS tumors every year; instead, multi-center collaborative consortia must take the lead in this effort. For children with CNS malignancies, there is a tremendous chance to improve outcomes by understanding the current infrastructure, the skills and limitations of administrators, physicians, researchers, and donors. This chapter is a call for stakeholders, policy makers, researchers, healthcare providers, and parents for research resources. We want to highlight the vulnerabilities of the special population which is children with CNS tumors. Current literature demonstrates the need for long-term assistance and rehabilitation techniques that are especially suited for young children.

Future Directions for Research

The lack of a pediatric focus in current practices restricts stakeholders, healthcare professionals, and researchers from using data from existing resources to plan work and services and create policy pertaining to pediatric cancer. Thus, we have identified several areas in need of further research.

Parents and healthcare providers treating CNS tumor survivors are gravely concerned about their clinical outcomes and quality of life, but there is limited research on young children. More longitudinal studies need to examine long-term effects of disease and treatment, quality of life of children with CNS tumors, so that interventions can be targeted symptoms management after treatment.

While a multidisciplinary team strategy is superior in terms of clinical consensus and research data, it may not be able to conduct randomized control trials to demonstrate effectiveness and improve outcomes in the care of CNS cancer in children. Although tumor boards and multidisciplinary cancer teams have become the norm in healthcare, there is little evidence-based medicine about the advantages and results for patients who are involved.

To acquire bigger samples of patients and caregivers in a timely manner, with long-term postintervention follow-up and with a greater ability to generalize findings, large, well-funded, multisite studies that provide perspective from each side of the team (healthcare providers, parents, children *etc.*) are required. It is difficult to conduct intervention studies with cancer patients and the people who care for them; this is why physicians are needed. They can inform potential participants about the studies that are currently being conducted and persuade them to enroll. To ascertain how successfully these research can be applied in practice settings, they must also be integrated into clinical treatment.

Tumor boards have not been uniformly standardized, and different facilities have different definitions of and arrangements for the elements of a successful multidisciplinary team. Future studies need to assess the quality of teamwork in cancer teams and provide a standardized guideline for diagnosis, treatment and after treatment periods of children with CNS tumors.

Implications for Clinical Practice and Patient Care

A multidisciplinary team approach is necessary to provide care effectively for children. Diagnosis, treatment, and after treatment parts of a child with CNS tumor is not an individual initiative and simply a decision by any one member of the care team. A multidisciplinary care team, including all related healthcare providers, parents, and children based on their age, works together to plan and implement a successful treatment and support approach adapted to children's unique circumstances. When this is achieved children and parents may have a valuable experience.

The team should meet regularly for an individual child from diagnosis to after treatment. It helps patients and families understand the information, promotes collaboration between different specialties, and guarantees that decisions are carried out by having all healthcare professionals engaged in the care available at the same time and place.

Furthermore, institutional support for the clinic is necessary for an efficient team to function, including funding, time, and resources made available to the team to maximize meetings and patient access.

REFERENCES

[1] What 'Beans' taught her parents about cancer and the need for research Available at: https://www.statnews.com/2021/03/22/what-beans-taught-her-parents-about-ped-atric-cancer-the-need-for-research-and-patient-advocacy/ (cited on: Apr 2, 2023).

[2] Smoll NR, Drummond KJ. The incidence of medulloblastomas and primitive neurectodermal tumours in adults and children. J Clin Neurosci 2012; 19(11): 1541-4.
[http://dx.doi.org/10.1016/j.jocn.2012.04.009] [PMID: 22981874]

[3] Atypical TRT. Atypical teratoid rhabdoid tumor (ATRT) - ABTA | learn more. 2022. Available at: https://www.abta.org/tumor_types/atypical-teratoid-rhaboid-tumor-atrt/ (cited on: Apr 2, 2023).

[4] Kumar L, Deepa SFA, Moinca I, Suresh P, Naidu KVJR. Medulloblastoma: A common pediatric tumor: Prognostic factors and predictors of outcome. Asian J Neurosurg 2015; 10(1): 50.
[http://dx.doi.org/10.4103/1793-5482.151516] [PMID: 25767583]

[5] Johnson KJ, Cullen J, Barnholtz-Sloan JS, *et al.* Childhood brain tumor epidemiology: A brain tumor epidemiology consortium review. Cancer Epidemiol Biomarkers Prev 2014; 23(12): 2716-36.
[http://dx.doi.org/10.1158/1055-9965.EPI-14-0207] [PMID: 25192704]

[6] Pediatric and childhood cancers | Boston Children's Hospital. Available at: https://www.childrenshospital.org/conditions/pediatric-and-childhood-cancers (cited 2023 Apr 2).

[7] Key statistics for brain and spinal cord tumors in Children. Available at: https://www.cancer.org/cancer/brain-spinal-cord-tumors-children/about/key-statistics.html (cited 2023 Apr 2).

[8] (8) Reflecting on nearly 40 years of progress against pediatric brain tumors. Available at: https://blogs.stjude.org/progress/reflecting-on-nearly-40-years-of-progress-against-p-diatric-brain-tumors.html (cited on: 2023 Apr 2).

[9] Philadelphia tch of. pediatric brain tumors. The Children's Hospital of Philadelphia. 2014. Available at: https://www.chop.edu/conditions-diseases/pediatric-brain-tumors (cited on: 2023 Mar 30).

[10] (10) Spine MB&. Brain tumor diagnosis and treatment options | Cincinnati, OH Mayfield Brain & Spine Available at: https://www.mayfieldclinic.com/pe-braintumor.htm (cited 2023 Apr 2).

[11] Brain Tumors in Children. 2021. Available at: https://www.hopkinsmedicine.org/health/conditions-and-diseases/brain-tumor/pediatric-brain-tumors (cited on: 2023 Mar 30).

[12] Ostrom QT, Fahmideh MA, Cote DJ, *et al.* Risk factors for childhood and adult primary brain tumors. Neuro-oncol 2019; 21(11): 1357-75.
[http://dx.doi.org/10.1093/neuonc/noz123] [PMID: 31301133]

[13] Stanford medicine children's health. Available at: https://www.stanfordchildrens.org/en/topic/default?id=brain-tumors-in-children-90-P02745 (cited on: 2023 Mar 30).

[14] Subramanian S, Ahmad T. Childhood Brain Tumors.StatPearls. Treasure Island, FL: StatPearls Publishing 2023.http://www.ncbi.nlm.nih.gov/books/NBK535415/ Internet

[15] Oncology brain tumors | Children's Wisconsin. Available at: https://childrenswi.org/medical-care/macc-fund-center/conditions/oncology/brain-tumors

[16] Goyal N, Kakkar A, Singh PK, *et al.* Intracranial teratomas in children: A clinicopathological study. Childs Nerv Syst 2013; 29(11): 2035-42.
[http://dx.doi.org/10.1007/s00381-013-2091-y] [PMID: 23568500]

[17] Raisanen JM, Davis RL. Congenital brain tumors. Pathology 1993; 2(1): 103-16.
[PMID: 9420933]

[18] Louis DN, Perry A, Reifenberger G, *et al.* The 2016 world health organization classification of tumors of the central nervous system: A summary. Acta Neuropathol 2016; 131(6): 803-20.
[http://dx.doi.org/10.1007/s00401-016-1545-1] [PMID: 27157931]

[19] Arslanca SB, Söylemez F, Koç A. Congenital glioblastoma multiforme presented with intracranial bleeding: A case report. J Obstet Gynaecol 2019; 39(3): 427-8.
[http://dx.doi.org/10.1080/01443615.2018.1475470] [PMID: 30226408]

[20] Brain Tumors | MetroHealth HMO. Available at: https://www.metrohealthhmo.com/brain-tumors/ (cited on: 2023 Apr 2).

[21] Cancer in Children and adolescents - NCI 2021. Available at: https://www.cancer.gov/types/childhood-cancers/child-adolescent-cancers-fact-sheet (cited 2023 Apr 2).

[22] Isaacs H Jr. II. Perinatal brain tumors: A review of 250 cases. Pediatr Neurol 2002; 27(5): 333-42. [http://dx.doi.org/10.1016/S0887-8994(02)00459-9] [PMID: 12504200]

[23] Parkes SE, Muir KR, Southern L, Cameron AH, Darbyshire PJ, Stevens MCG. Neonatal tumours: A thirty-year population-based study. Med Pediatr Oncol 1994; 22(5): 309-17. [http://dx.doi.org/10.1002/mpo.2950220503] [PMID: 8127254]

[24] Goyal N, Kakkar A, Singh PK, *et al.* Intracranial teratomas in children: A clinicopathological study. Childs Nerv Syst 2013; 29(11): 2035-42. [http://dx.doi.org/10.1007/s00381-013-2091-y] [PMID: 23568500]

[25] Severino M, Schwartz ES, Thurnher MM, Rydland J, Nikas I, Rossi A. Congenital tumors of the central nervous system. Neuroradiology 2010; 52(6): 531-48. [http://dx.doi.org/10.1007/s00234-010-0699-0] [PMID: 20428859]

[26] Packer RJ, Cogen P, Vezina G, Rorke LB. Medulloblastoma: Clinical and biologic aspects. Neuro-oncol 1999; 1(3): 232-50. [http://dx.doi.org/10.1215/15228517-1-3-232] [PMID: 11550316]

[27] Carstensen H, Juhler M, Bøgeskov L, Laursen H. A report of nine newborns with congenital brain tumours. Childs Nerv Syst 2006; 22(11): 1427-31. [http://dx.doi.org/10.1007/s00381-006-0115-6] [PMID: 16804715]

[28] Alamo L, Beck-Popovic M, Gudinchet F, Meuli R. Congenital tumors: Imaging when life just begins. Insights Imaging 2011; 2(3): 297-308. [http://dx.doi.org/10.1007/s13244-011-0073-8] [PMID: 22347954]

[29] Ostrom QT, de Blank PM, Kruchko C, Petersen CM, Liao P, Finlay JL. Alex's lemonade stand foundation infant and childhood primary brain and central nervous system tumors diagnosed in the United States in 2007-2011. Neuro Oncol 2015; (10)(Suppl. 10): x1-x36. [http://dx.doi.org/10.1093/neuonc/nou327] [PMID: 25542864]

[30] Larouche V, Huang A, Bartels U, Bouffet E. Tumors of the central nervous system in the first year of life. Pediatr Blood Cancer 2007; 49(S7) (Suppl.): 1074-82. [http://dx.doi.org/10.1002/pbc.21351] [PMID: 17943961]

[31] El-Ayadi M, Ansari M, Sturm D, *et al.* High-grade glioma in very young children: A rare and particular patient population. Oncotarget 2017; 8(38): 64564-78. [http://dx.doi.org/10.18632/oncotarget.18478] [PMID: 28969094]

[32] Pediatric brain tumors - Symptoms and causes. Mayo Clinic. Available at: https://www.mayoclinic.org/diseases-conditions/pediatric-bra-n-tumor/symptoms-causes/syc-20361694 (cited on: 2023 Mar 30).

[33] Childhood Brain and Spinal Cord Tumors Treatment Overview.. 2022. Available at: https://www.cancer.gov/types/brain/patient/child-brain-treatment-pdq (cited on: 2023 Mar 30).

[34] Brain tumor center. Available at: https://cincycancercare.org/child/brain-tumor/ (cited on: 2023 Mar 30).

[35] Kleihues P, Cavenee WK. World Health Organization classification of tumors: Pathology and genetics of tumors of the nervous system. Lyon: IARC Press 2000.

[36] Banerjee J, Pääkkö E, Harila M, *et al.* Radiation-induced meningiomas: A shadow in the success story of childhood leukemia. Neuro-oncol 2009; 11(5): 543-9.

[http://dx.doi.org/10.1215/15228517-2008-122] [PMID: 19179425]

[37] PDQ Pediatric Treatment Editorial Board. 2002. Available at: http://www.ncbi.nlm.nih.gov/books/NBK65913/ (cited on: 2023 Apr 3).

[38] How We Diagnose Brain Tumors - Dana-Farber Cancer Institute Boston, MA.. Available at: https://www.dana-farber.org/brain-tumors/diagnosis/ (cited on: 2023 Apr 3).

[39] Pediatric brain tumors-diagnosis and treatment. Available at: https://www.mayoclinic.org/diseases-conditions/pediatric-brain-tumor/diagnosis-treatment/drc-20361706 (cited on: 2023 Apr 3).

[40] Ndubuisi C, Ohaegbulam S, Ejembi G. Paediatric brain tumours managed in Enugu, Southeast Nigeria: Review of one centre experience. Niger Postgrad Med J 2018; 25(3): 186-90. [http://dx.doi.org/10.4103/npmj.npmj_132_18] [PMID: 30264771]

[41] Sherer D, Onyeije C. Prenatal ultrasonographic diagnosis of fetal intracranial tumors: A review. Am J Perinatol 1998; 15(5): 319-28. [http://dx.doi.org/10.1055/s-2007-993951] [PMID: 9643639]

[42] Lee DY, Kim YM, Yoo SJ, *et al.* Congenital glioblastoma diagnosed by fetal sonography. Childs Nerv Syst 1999; 15(4): 197-201. [http://dx.doi.org/10.1007/s003810050369] [PMID: 10361971]

[43] Choroid plexus papilloma: magnetic resonance, computed tomography, and angiographic observations. Available at: https://pubmed.ncbi.nlm.nih.gov/3563861/ (cited on: 2023 Apr 3).

[44] Dangouloff-Ros V, Varlet P, Levy R, *et al.* Imaging features of medulloblastoma: Conventional imaging, diffusion-weighted imaging, perfusion-weighted imaging, and spectroscopy: From general features to subtypes and characteristics. Neurochirurgie 2021; 67(1): 6-13. [http://dx.doi.org/10.1016/j.neuchi.2017.10.003] [PMID: 30170827]

[45] Resende LL, Alves CAPF. Imaging of brain tumors in children: The basics—a narrative review. Transl Pediatr 2021; 10(4): 1138-68. [http://dx.doi.org/10.21037/tp-20-285] [PMID: 34012860]

[46] Brain abscess by citrobacter diversus in infancy: Case report. Available at: https://pubmed.ncbi.nlm.nih.gov/10973119/ (cited on: 2023 Apr 3)

[47] Hakimi R, Garg A. Imaging of Hemorrhagic Stroke. Continuum 2016; 22(5, Neuroimaging): 1424-50. [http://dx.doi.org/10.1212/CON.0000000000000377] [PMID: 27740983]

[48] Green S, Vuong VD, Khanna PC, Crawford JR. Characterization of pediatric brain tumors using pre-diagnostic neuroimaging. Front Oncol 2022; 12: 977814. [http://dx.doi.org/10.3389/fonc.2022.977814] [PMID: 36324580]

[49] Packer RJ, MacDonald T, Vezina G. Central nervous system tumors. Pediatr Clin North Am 2008; 55(1): 121-145, xi. [http://dx.doi.org/10.1016/j.pcl.2007.10.010] [PMID: 18242318]

[50] Brain Tumor - Types of Treatment.. 2012. Available at: https://www.cancer.net/cancer-types/brai--tumor/types-treatment (cited on: 2023 Apr 3).

[51] Hanif F, Muzaffar K, Perveen K, Malhi SM, Simjee ShU. Glioblastoma multiforme: A review of its epidemiology and pathogenesis through clinical presentation and treatment. Asian Pac J Cancer Prev 2017; 18(1): 3-9. [http://dx.doi.org/10.22034/APJCP.2017.18.1.3] [PMID: 28239999]

[52] Childhood central nervous system embryonal tumors treatment. 2016. Available at: https://www.semanticscholar.org/paper/Childhood-Central-Nervous-Sys-em-Embryonal-Tumors/13b4a950f439944ad95a1d1ad1ae502862f4c1ba (cited on: 2023 Apr 3).

[53] Treatment by cancer type. Available at: https://www.nccn.org/guidelines/category_1 (cited on: 2023 Apr 3).

[54] Gajjar A, Mahajan A, Abdelbaki M. 2023.Pediatric central nervous system cancers: version 1. Available at: https://bit.ly/3O96LLY (Accessed on: July 13, 2022).

[55] Liu APY, Hastings C, Wu S, *et al.* Treatment burden and long-term health deficits of patients with low-grade gliomas or glioneuronal tumors diagnosed during the first year of life. Cancer 2019; 125(7): 1163-75.
[http://dx.doi.org/10.1002/cncr.31918] [PMID: 30620400]

[56] Abdullah S, Qaddoumi I, Bouffet E. Advances in the management of pediatric central nervous system tumors. Ann N Y Acad Sci 2008; 1138(1): 22-31.
[http://dx.doi.org/10.1196/annals.1414.005] [PMID: 18837879]

[57] Hammond GD, Bleyer WA, Hartmann JR, Hays DM, Jenkin RDT. The team approach to the management of pediatric cancer. Cancer 1978; 41(1): 29-35.
[http://dx.doi.org/10.1002/1097-0142(197801)41:1<29::AID-CNCR2820410107>3.0.CO;2-D] [PMID: 626937]

[58] Pak-Yin Liu A, Moreira DC, Sun C, *et al.* Challenges and opportunities for managing pediatric central nervous system tumors in China. Pediatr Investig 2020; 4(3): 211-7.
[http://dx.doi.org/10.1002/ped4.12212] [PMID: 33150316]

[59] Abdel-Baki MS, Hanzlik E, Kieran MW. Multidisciplinary pediatric brain tumor clinics: The key to successful treatment? CNS Oncol 2015; 4(3): 147-55.
[http://dx.doi.org/10.2217/cns.15.1] [PMID: 25923018]

[60] Foo JC, Jawin V, Yap TY, *et al.* Conduct of neuro-oncology multidisciplinary team meetings and closing the "gaps" in the clinical management of childhood central nervous system tumors in a middle-income country. Childs Nerv Syst 2021; 37(5): 1573-80.
[http://dx.doi.org/10.1007/s00381-021-05080-4] [PMID: 33580355]

[61] Taylor C, Shewbridge A, Harris J, Green JS. Benefits of multidisciplinary teamwork in the management of breast cancer. Breast Cancer 2013; 5: 79-85.
[http://dx.doi.org/10.2147/BCTT.S35581] [PMID: 24648761]

[62] Lamb BW, Taylor C, Lamb JN, *et al.* Facilitators and barriers to teamworking and patient centeredness in multidisciplinary cancer teams: Findings of a national study. Ann Surg Oncol 2013; 20(5): 1408-16.
[http://dx.doi.org/10.1245/s10434-012-2676-9] [PMID: 23086306]

[63] Taberna M, Gil Moncayo F, Jané-Salas E, *et al.* The multidisciplinary team (MDT) Approach and quality of care. Front Oncol 2020; 10: 85.
[http://dx.doi.org/10.3389/fonc.2020.00085] [PMID: 32266126]

[64] The department of health. Manual for cancer services. London: The Department of Health 2004.

[65] Huang T, Mueller S, Rutkowski MJ, *et al.* Multidisciplinary care of patients with brain tumors. Surg Oncol Clin N Am 2013; 22(2): 161-78.
[http://dx.doi.org/10.1016/j.soc.2012.12.011] [PMID: 23453330]

[66] Northhouse LL, Katapodi MC, Song L, *et al.* Interventions with family caregivers of cancer patients: meta-analysis of randomized trials. CA Cancer J Clin 2010; 60(5): 317-39.
[http://dx.doi.org/10.3322/caac.20081] [PMID: 20709946]

[67] Northouse LL, Katapodi MC, Song L, Zhang L, Mood DW. Interventions with family caregivers of cancer patients: Meta-analysis of randomized trials. CA Cancer J Clin 2010; 60(5): n/a.
[http://dx.doi.org/10.3322/caac.20081] [PMID: 20709946]

[68] Lutz K, Jünger ST, Messing-Jünger M. Essential management of pediatric brain tumors. Children 2022; 9(4): 498.
[http://dx.doi.org/10.3390/children9040498] [PMID: 35455542]

[69] Udaka YT, Packer RJ. Pediatric Brain Tumors. Neurol Clin 2018; 36(3): 533-56.

[http://dx.doi.org/10.1016/j.ncl.2018.04.009] [PMID: 30072070]

[70] Cioffi J, Wilkes L, Cummings J, Warne B, Harrison K. Multidisciplinary teams caring for clients with chronic conditions: Experiences of community nurses and allied health professionals. Contemp Nurse 2010; 36(1-2): 61-70.
[http://dx.doi.org/10.5172/conu.2010.36.1-2.061] [PMID: 21254823]

[71] Soukup T, Lamb B, Arora S, Darzi A, Sevdalis N, Green J. Successful strategies in implementing a multidisciplinary team working in the care of patients with cancer: An overview and synthesis of the available literature. J Multidiscip Healthc 2018; 11: 49-61.
[http://dx.doi.org/10.2147/JMDH.S117945] [PMID: 29403284]

[72] Sitarz R, Kocemba K, Maciejewski R, Polkowski W. Effective cancer treatment by multidisciplinary teams. Pol Przegl Chir 2012; 84(7): 371-6.
[http://dx.doi.org/10.2478/v10035-012-0063-7] [PMID: 22935461]

[73] National Cancer Action TeamThe characteristics of an effective MDT. London: NCAT 2010.http://www.ncin.
org.uk/cancer_type_and_topic_specific_work/multidisciplinary_teams/MDT_development.aspx

[74] National Breast and Ovarian Cancer Centre 4. Canberra: Department of Health and Ageing 2005.

[75] Hong NJL, Wright FC, Gagliardi AR, Paszat LF. Examining the potential relationship between multidisciplinary cancer care and patient survival: An international literature review. J Surg Oncol 2010; 102(2): 125-34.
[http://dx.doi.org/10.1002/jso.21589] [PMID: 20648582]

[76] Macartney G, VanDenKerkhof E, Harrison MB, Stacey D. Symptom experience and quality of life in pediatric brain tumor survivors: A cross-sectional study. J Pain Symptom Manage 2014; 48(5): 957-67.
[http://dx.doi.org/10.1016/j.jpainsymman.2013.12.243] [PMID: 24704799]

[77] Institute for Patient- and Family-Centered Care. Transforming healthcare through partnerships. Available at: http://www.ipfcc.org/about/pfcc.html (Accessed on: 15 March 2023).

[78] Maree C, Downes F. Trends in family-centered care in neonatal intensive care. J Perinat Neonatal Nurs 2016; 30(3): 265-9.
[http://dx.doi.org/10.1097/JPN.0000000000000202] [PMID: 27465463]

[79] Franck LS, O'Brien K. The evolution of family-centered care: from supporting parent-delivered interventions to a model of family integrated care. Birth Defects Res 2019; 111(15): 1044-59.
[http://dx.doi.org/10.1002/bdr2.1521] [PMID: 31115181]

[80] Metzger S, Weiser A, Gerber NU, *et al.* Central nervous system tumors in children under 5 years of age: A report on treatment burden, survival and long-term outcomes. J Neurooncol 2022; 157(2): 307-17.
[http://dx.doi.org/10.1007/s11060-022-03963-3] [PMID: 35147892]

[81] Renzi S, Michaeli O, Ramaswamy V, *et al.* Causes of death in pediatric neuro-oncology: The sickkids experience from 2000 to 2017. J Neurooncol 2020; 149(1): 181-9.
[http://dx.doi.org/10.1007/s11060-020-03590-w] [PMID: 32803658]

CHAPTER 5

Drug Repurposing in CNS and Clinical Trials: Recent Achievements and Perspectives Focusing on Epilepsy and Related Comorbidities

Gabriela Machado Parreira[1], **Antonio Carlos Pinheiro de Oliveira**[2], **Leonardo de Oliveira Guarnieri**[3,*,#] and **Rafael Pinto Vieira**[1,*,#]

[1] *Departamento de Bioquímica e Imunologia, Instituto de Ciências Biológicas, Universidade Federal de Minas Gerais, Brazil*

[2] *Departamento de Farmacologia, Instituto de Ciências Biológicas, Universidade Federal de Minas Gerais, Brazil*

[3] *Departamento de Fisiologia e Biofísica, Instituto de Ciências Biológicas, Universidade Federal de Minas Gerais, Brazil*

Abstract: Central Nervous System (CNS) disorders are a massive burden on the global health system, including a broad range of clinical conditions, such as epilepsies, depression, dementia, multiple sclerosis, and Parkinson's disease. Permanent efforts are being made to find early, non-invasive, and effective diagnostic methods, as well as efficient and safe drug-based treatments for CNS conditions. Nevertheless, many patients displaying these clinical conditions still face the lack of an effective pharmacotherapy to cure the diseases or at least to properly control the progression of symptoms. Currently, epilepsies present an estimated prevalence of 0.5%–1% worldwide, and around 30% of the patients remain refractory to the available drug treatment. The comorbidities that affect epileptic patients, such as cognitive impairment and depression, are major public health challenges. This scenario highlights the urgent need for approving new therapeutic tools for CNS diseases. A successful development process of a new compound presenting therapeutic potential can range up to 20 years and cost hundreds of millions of US dollars, from the initial characterization of the *in vitro* chemical and biological properties until clinical trials. Additionally, drug development has a low success rate in the case of CNS conditions. In this context, drug repurposing (or drug repositioning, DR) is an alternative way to reduce the cost and accelerate the process of a drug-based treatment approach since it identifies a novel clinical application for an existing compound already approved for a distinct indication. In the present chapter, we aim to describe recent outcomes of DR

[*] **Corresponding authors Rafael Pinto Vieira and Leonardo de Oliveira Guarnieri:** Departamento de Bioquímica e Imunologia, Instituto de Ciências Biológicas, Universidade Federal de Minas Gerais and Departamento de Fisiologia e Biofísica, Instituto de Ciências Biológicas, Universidade Federal de Minas Gerais, Brazil; E-mails: leoguarnieri@gmail.com, vieirarp@icb.ufmg.br

[#] These authors equally contributed to this work.

Zareen Amtul (Ed.)

aiming at CNS pathological conditions, especially discussing the recent clinical trials and their impacts on future endeavors in the search for the management of epilepsies and related comorbidities.

Keywords: Central Nervous System, Clinical Trial, CNS Diseases, Cognitive Impairment, Drug Repurposing, Drug Repositioning, Epilepsy, Depression, Pharmacotherapy.

INTRODUCTION

The Drug Repurposing Context

Historically, conventional approaches in relation to investing cost and time in a drug discovery campaign would normally involve millions of dollars and decades as average requirements to achieve the target outcomes and introduce final products into the market. This scenario has not substantially changed in recent years, and apparently, it has no expectations to be far from this pattern in the future. Even considering computational campaigns and further modern strategies to improve the time spent in the first *in silico*, *in vitro* and other preclinical steps, the time required to assure the efficacy and safety of new pharmaceutical entities by clinical studies is mandatory. Furthermore, recent estimation of these data reinforces that the average cost required to bring a new approved compound to clinical use is extremely high, being almost 1 billion dollars. In the specific case of drugs targeting the Central Nervous System (CNS), the mean estimates range from around 500 million to more than 1,8 billion dollars of expenditures. Simultaneously, the percentage of final successful compounds in this high-cost context is considerably low, with clinical trial success rates ranging from only 15% to around 51% [1].

The above-mentioned items that contribute to this scenario are acceptably referred to as top facts that pave the way to the continuing search for less expensive alternatives targeting approved drugs, side-by-side with the urgent pharmacotherapy needs in public health systems. In this context, drug repurposing (or drug repositioning, DR) emerges as one of the most promising pathways.

The main idea of DR is focused on using already approved drugs for new therapeutic applications. The history of DR is relatively recent in therapy, and one of the most representative cases is thalidomide [2]. The drug, which was synthesized in the early 1950s by the company Ciba in Switzerland, has been extensively used for years as an antiemetic agent for pregnant women worldwide [3]. However, the first reports on thalidomide-related congenital malformations were raised in the 1960s [4, 5], and the use of the drug was systematically banned.

Simultaneously, researchers have started investigations on alternative applications for the compound, culminating in surprising and successful applications on lepra skin eruptions [6 - 9] and multiple myeloma, also including thalidomide derivatives in this last case [10 - 13]. Besides thalidomide, other drugs have experienced DR pathways, including the recent and successful administration of systemic dexamethasone in the treatment of patients with COVID-19 [14, 15].

In this chapter, we intend to discuss promising approaches, especially clinical ones, targeting new applications of already approved drugs in the context of CNS disorders. The new applications described here are centered on solving drug-resistant epilepsy and two of its most relevant comorbidities, cognitive impairment and depression.

Drug Repurposing Aiming at Cognitive Improvement and Anticonvulsant Activity in Epilepsy

Epilepsies are chronic conditions mainly characterized by the presence of recurrent seizures. The manifestation of these pathological conditions might occur in different forms, including tonic-clonic convulsions. In terms of biochemical and pathophysiological alterations, epilepsies display modifications in the electrical function of the brain tissue, which might be caused by an imbalance between inhibition and excitation events, resulting in abnormal and excessive neuronal activity [16].

Epileptic seizures present different causes. This etiology scenario ranges from genetic to non-heritable pathological alterations, which might be acquired from other medical conditions. All these possible causes might lead to different types of seizures [17]. In general, the diagnosis pathway of an epileptic patient in clinical practice follows two stages. In this process, the focus of the first stage is based on classifying the seizure type or syndrome, followed by the second stage, which is characterized by the search for the disease's cause. Classifications of epilepsies have been extensively dedicated to the characterization of the first stage. However, the cause of the pathological condition is considered of paramount importance in defining the proper pharmacotherapy, as well as the management of the clinical condition during the patient's life [16].

According to recent data, around 70 million people in the world are diagnosed as epileptic patients, and these numbers suggest a current incidence rate of 61.4 in 100,000 people [18, 19]. Antiepileptic drugs (AEDs) are the first choice in the pharmacotherapy of epilepsies, and approximately 70% of patients properly respond to the treatment by using a monotherapy, while about 20-30% of total patients require polytherapy. Despite the current availability of drugs, about one-third of all epileptic patients are refractory to the treatment or might become drug resistant at some point in pharmacotherapy [20 - 22]. This clinical scenario has been repeatedly observed and, even considering the increased availability of new AEDs in past decades, the percentage of resistance to antiepileptic pharmacotherapy remains essentially unchanged over time since the beginning of the twentieth century [23].

Currently, there are approximately 25 to 30 AEDs with different molecular targets, converging towards the aim of inhibiting excitatory mechanisms or increasing inhibitory pathways (Fig. **1**). Even being approved and considered relatively safe molecules, these AEDs are linked to one or more side effects, including ataxia, episodes of dizziness and sedation. Besides, epilepsies are associated with different comorbidities, such as learning problems, cognitive impairment, autism spectrum disorders, anxiety, and depression [24, 25], which are not simultaneously treated by the same available AEDs in the majority of the cases. Thus, in the present moment and to the best of our knowledge, the current antiepileptic pharmacotherapy is not efficient in treating epilepsies-related comorbidities. Considering the growing impact of these comorbidities on the patients' routines, the need for further studies aiming at the understanding of the mechanisms related to epilepsies or epileptogenic factors and the search for more effective and accurate AEDs have become considerable pharmacological concerns [16, 26, 27]. In addition to the related comorbidities, it is important to highlight that, although the new AEDs have advantages in terms of tolerability, interactions and teratogenicity, what provides gains in the life of the patients, none of these drugs seems to be more effective than first-generation AEDs, since the efficacy in patients with refractory epilepsy is similar [28, 29].

In fact, it is still unclear what are the mechanisms by which epileptic patients become drug resistant. The most accepted hypotheses combine multiple mechanisms extensively reviewed by Loscher *et al.* [30]. Risk factors for the development of drug resistance were identified in a 30-year single-center longitudinal cohort study, such as the number of seizures occurring before treatment initiation, family history of epilepsy, and recreational drug use [31].

To the best of our knowledge, since there are no robust animal models for the study of drug resistance in epilepsy, the challenge is enormous. Besides, clinical studies do not provide enough data to establish resistance mechanisms for each of the AEDs. For example, evidence indicates that sensitivity to carbamazepine, a sodium channel blocker, is absent in a subset of individuals with mesial temporal lobe epilepsy, although the mechanisms involved in such insensitivity are not known [30, 32]. In addition, $GABA_A$ receptors seem to be altered, which would lead to AED resistance. Studies on the pilocarpine model have shown an increase in the internalization of $GABA_A$ receptors, causing a reduction in the number of available postsynaptic receptors [33, 34]. Another hypothesis regarding the resistance is the possible inadequate penetration of AEDs through the blood-brain barrier due to the increased expression of multidrug efflux transporters [35, 36]. Changes in the blood-brain barrier are also related to the increased levels of COX-2 protein and prostaglandin E_2 in epileptogenic regions [37 - 39].

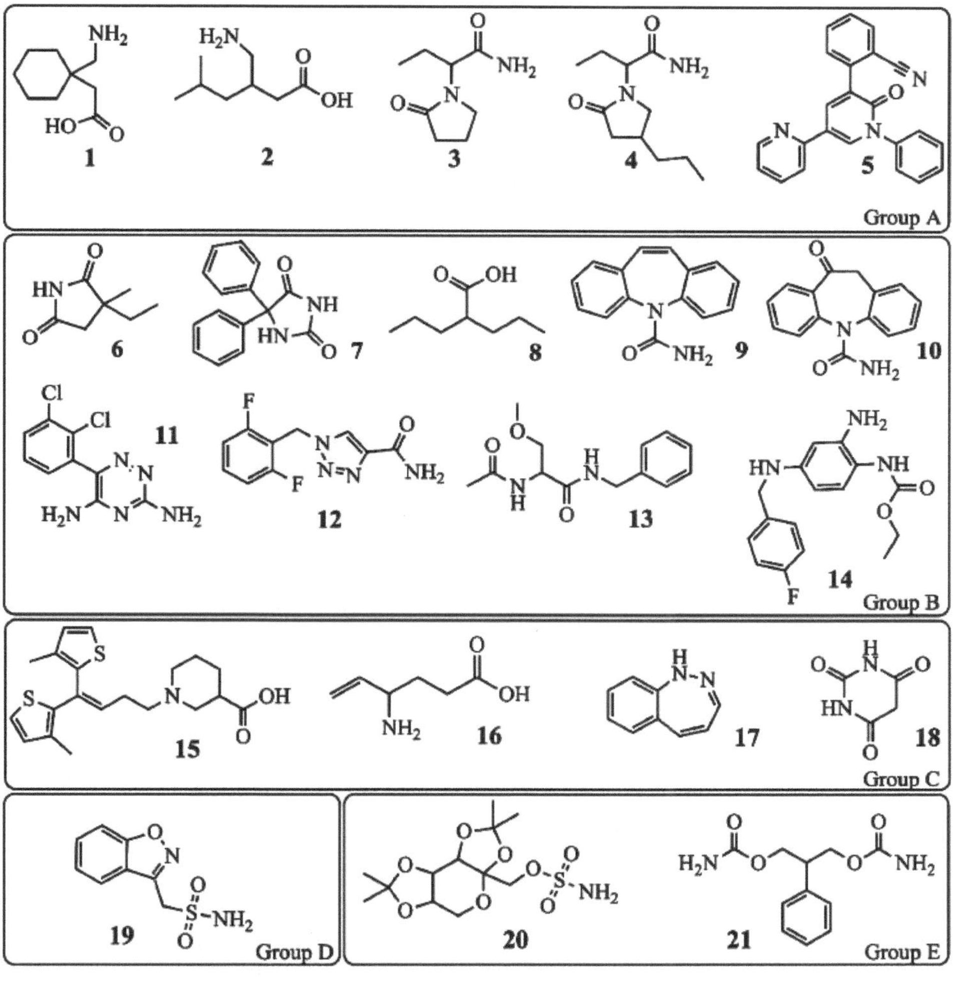

Fig. (1). Chemical structures of the main current AEDs, grouped according to their proposed mechanisms of action. Group A: inhibition of synaptic excitation (mechanism a); Group B: modulation of voltage-gated ion channels (mechanism b); Group C: enhancement of synaptic inhibition (mechanism c); Group D: mechanisms b and c; Group E: mechanisms a, b and c. Compounds: 1 – gabapentin; 2 – pregabalin; 3 – levetiracetam; 4 – brivaracetam; 5 – perampanel; 6 – ethosuximide; 7 – phenytoin; 8 – valproic acid; 9 – carbamazepine; 10 – oxcarbazepine; 11 – lamotrigine; 12 – rufinamide; 13 –lacosamide; 14 – retigabine; 15 – tiagabine; 16 – vigabatrin; 17 – benzodiazepines; 18 – barbiturates; 19 –zonisamide; 20 – topiramate; 21 – felbamate [23].

There are also hypotheses related to the genetic profile of patients, in which changes in the expression of some genes seem to be related to a higher incidence of refractoriness in epileptic patients [40 - 42]. Similarly, it seems that epigenetic factors and microRNAs may also play a role in this phenomenon [43 - 46].

Considering the high prevalence of refractory clinical cases in relation to the current antiepileptic treatment and the need for tools aiming at the current epilepsies-related comorbidities, the DR approach arises as an alternative in order to investigate the possibilities of pharmacotherapy usage among the already clinically approved compounds [47]. Thus, we selected relevant compounds in this context (Fig. **2** and Table **1**), whose clinical studies are in progress or already finished; their available outcomes are also discussed here.

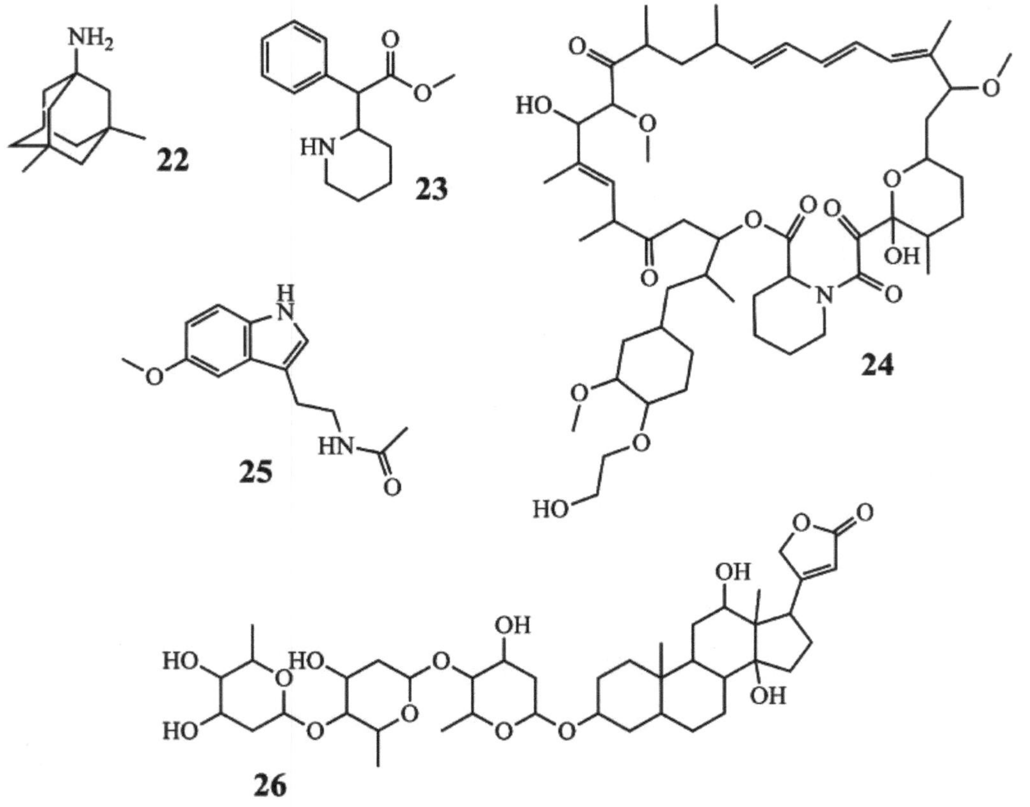

Fig. (2). Chemical structures of selected compounds in the DR context of antiepileptic drugs and cognition-target compounds. 22) memantine; 23) methylphenidate; 24) everolimus; 25) melatonin; 26) digoxin. The structure of natalizumab is not represented here since it is not a small molecule, being a monoclonal antibody with a molecular weight of approximately 149 kilodaltons [48].

Table 1. Selected approved drugs in the DR context of antiepileptic drugs and their related academic and clinical evaluations.

Drug	Initial Target	Evaluation in Epilepsy
Memantine (22)	Alzheimer's disease	[60,62] Clinical trials NCT04417543 and NCT01054599
Methylphenidate (23)	Attention-deficit hyperactivity disorder (ADHD)	[71, 72] Clinical trial NCT04419272
Natalizumab	Multiple sclerosis	[78,79] Clinical trial NCT03283371
Everolimus (24)	Immunosuppressant	[80, 83]
Melatonin (25)	Insomnia	[86 - 89] Clinical trial NCT03590197, NCT00965575, NCT01161108, and NCT02195661.
Digoxin (26)	Congestive heart failure	[92] Clinical trials NCT02172742 and NCT01583036

Memantine

Memantine (3,5-dimethyladamantan-1-amine, MEM, Fig. **2**, compound 22) is a non-competitive NMDA receptor antagonist approved in 2003 for the treatment of moderate to severe Alzheimer's disease (AD) [49]. From that point, no additional small molecule has been incorporated into AD pharmacotherapy. In this challenging scenario, the use of prodrug candidates of MEM aiming to circumvent the compound administration drawbacks in physically compromised patients has been recently explored by research groups, including ours [50].

MEM has been proposed for the therapy of other dementia-related neurodegenerative diseases, such as Parkinson's disease (PD) and Huntington's disease (HD) [51 - 54]. Epilepsies are pathological conditions that frequently display exacerbated activation of NMDA-type receptors [55]. Thus, MEM was proposed as a new DR candidate for the treatment of memory deficits in epilepsy.

Cognitive deficits are frequently found in different types of epilepsy, affecting a plethora of levels, including factors, such as attention, executive function, visual skills, language, and memory [56 - 59]. Among those factors, memory loss seems to be the most distressing scenario for epileptic patients. Unfortunately, there are no current drugs approved for this purpose, aiming at the pharmacotherapy of epilepsy-cognitive disorders.

In this context, recent clinical studies have claimed a promising application of MEM as an adjuvant tool for cognitive impairment in epilepsies. Marimuthu *et al.* conducted a randomized and placebo-controlled clinical trial protocol composed of 55 epileptic patients using antiepileptic drugs, also presenting memory impairment [60]. Patients under MEM 5 mg and 10 mg daily regimen during 16 weeks (8 weeks/5 mg followed by 8 weeks/10 mg) displayed statistically significant improvement in cognition evaluation tools, such as the Mini Mental State Examination (MMSE), in comparison to the placebo group. Simultaneously, no significant increase in seizure frequency was reported. Oustad *et al.* performed a study involving seventy patients with temporal lobe epilepsy, which were divided into two equal groups, 16 weeks of daily treatment [61]. The first group received 10 mg doses of donepezil, an acetylcholinesterase drug that is also used in Alzheimer's disease treatment, and MEM 10 mg was given to the second group. The Montreal Cognitive Assessment (MoCA) test was used as a cognition evaluation tool. The study reports that the MEM group presented better cognition scores than donepezil ones. Leeman-Markowski *et al.* investigated the hypothesis that MEM would be able to improve the performance of patients in different tests of memory in comparison to the placebo-treated group in subjects with focal onset epilepsy [62]. The group also intended to evaluate whether any benefit from the MEM regimen would be specific to memory. The results obtained in the study did not present a significant effect of MEM on cognition, but also highlighted the favorable safety profile of the drug in the epilepsy context. No matter the results, all the studies reported here must be interpreted cautiously, especially due to the number of subjects and the time of the treatments. Therefore, future studies will be necessary to validate or not the use of MEM in the improvement of cognitive function in epileptic patients. Currently, at least two clinical trials are in progress, under the numbers NCT04417543 and NCT01054599, to evaluate the effects of memantine on improving memory and cognitive impairment.

Methylphenidate

Methylphenidate (l-threo-methylphenidate, MPH, Fig. **2**, compound 23) is a FDA-approved psychostimulant whose properties for psychiatric purposes in children have been investigated since the middle of the twentieth century [63, 64]. The structure of the drug is chemically related to amphetamines, consequently leading to a similar mechanism of action, especially focused on the reuptake inhibition of dopamine and norepinephrine. *In vitro* and *in vivo* studies have confirmed this mechanism, which results in the increase of the monoamines concentration in the synaptic cleft [65 - 67]. Due to these effects and their subsequent influence on the ability to increase the attention state of patients, MPH is employed in the pharmacotherapy of attention deficit hyperactivity disorder

(ADHD) and related disorders. Since epileptic patients present deficits in attention, memory, and speed of information processing, the use of drugs, such as MPH, in order to improve patient's cognition and attention scores arises as a promising possibility in the DR context.

Several non-controlled and non-randomized studies suggest the safety and efficacy of MPH use in children with epilepsies, considering a safety profile that is not associated with an increased risk of seizures [68 - 70]. In the case of adults, considering the already known potential of cognitive gains with considerable safety, Adams *et al.* performed a randomized controlled trial including 31 epileptic patients, who presented chronic cognitive impairment and to whom 10 and 20 mg of MPH were administered and compared to placebo [71]. No significant side effects were observed, and a statistically significant performance gain was reported for both MPH doses in comparison to the placebo group. It was demonstrated that the patients when treated with MPH obtained a gain in the performance of tasks. However, further studies are needed since there are limitations, including the small sample group. Afterwards, the same research group performed a one-month trial with 28 adults presenting focal, generalized, or unclassified epilepsies, and administered daily doses of MPH, which ranged from 20 to 40 mg. Cognition improvements in the MPH group were considered moderate to large, with no significant increase in epileptic seizures' frequency [72].

The mechanism of action by which MPH might improve cognition in epileptic patients remains unknown. Cognitive impairment affects approximately 45% of these subjects, and few treatments are proposed to improve their clinical condition. Thus, the availability of a study that aims to use MPH as an adjuvant to improve the deficits is paramount to pave the way for providing significant clinical gains to epileptic patients. In this pathway and since 2020, at least one clinical trial, under the number NCT04419272, is in process to evaluate the effect of MPH in epilepsy-related cognition improvement.

Natalizumab

Natalizumab (NAT) is characterized as a humanized monoclonal antibody with a molecular weight of approximately 149 kilodaltons [48, 73]. Its mechanism of action is based on targeting α4-integrin (CD49d) on the surface of lymphocytes, a process that culminates with the modulation of the inflammation process in the CNS. Based on the effects of these molecular and cellular events, the antibody is a current tool in the pharmacotherapy of multiple sclerosis [74, 75].

Immune and inflammation cascades are recognized as significant pieces in the puzzling pathogenic processes of seizures, especially in the case of refractory epilepsy [76, 77]. Thus, natalizumab may represent an interesting candidate in the pharmacotherapy of this condition.

A case report described by Sotgiu and collaborators describes an adult patient in whom a 12-month treatment with NAT had contributed to significantly ameliorate both multiple sclerosis symptoms and severe refractory epilepsy condition [78]. More recently, French and colleagues have carried out a clinical trial (NCT03283371) based on the administration of 300 mg IV of NAT every 4 weeks/24 weeks, including a comparison to a placebo group [79]. No significant side effects were reported for the drug when compared to the placebo, but no significant reduction in seizure frequency was determined as well. Since changes in doses, administration regimen, and size of participants in NAT and placebo groups must be improved to obtain consistent conclusions about the effects of the drug in seizures, NAT remains an intriguing agent in the context of drug repurposing targeting refractory epileptic patients.

Everolimus

Epilepsy is a condition that is present in approximately 70% of patients with tuberous sclerosis complex. Currently, its treatment is based on the use of conventional antiepileptic drugs, not being specific to the main cause of the disease, the overactivation of mTOR [80].

Everolimus (EVE, Fig. **2**, compound 24) is an mTOR inhibitor derived from rapamycin, and its use was approved in 2004. It has an antiproliferative effect, being a potent immunosuppressant [81]. The mechanism of action targeting cell growth inhibition is based on halting the progression of the cell cycle in the passage from G1 to S phase through the formation of a complex between EVE and FKBP12. Thus, the inhibition of mTOR occurs, consequently stopping the cell cycle in the G1 phase [82]. Therefore, recent studies have investigated its use to decrease the number of seizures in patients with a treatment-resistant tuberous sclerosis complex as a potential therapeutic use based on a DR strategy.

Phase III studies conducted in 2016 observed a significant reduction in the frequency of seizures in patients with treatment-resistant epilepsy and patients with a tuberous sclerosis complex when compared to the placebo group [80, 83]. However, there are some limitations in these studies since the results showed the benefits only in a short period of treatment with EVE as an adjunct to antiepileptic pharmacotherapy. Therefore, a more detailed investigation of the safety, efficacy, and tolerance in long-term treatment is necessary to improve the findings.

Further studies describing the effect of modulators of the mTOR signaling pathway in animal models have been described in the literature. These studies suggest that these modulators have antiepileptogenic effects by modifying different protein signaling and expression pathways, thereby altering different mechanisms involved in epileptogenesis. However, the mechanism of action and clinical effects are still complex [83, 84].

Considering the above-described findings, EVE represents to be promising as a potential treatment for patients with drug-resistant seizures associated with a tuberous sclerosis complex. In addition, the investigation of mTOR inhibitors in patients with other diseases that are resistant to drugs becomes essential to better understand the genetic and molecular causes of epilepsy.

Melatonin

Melatonin (MTn, Fig. **2**, compound 25) is a tryptophan-derived indoleamine whose main biological function is related to the regulation of the circadian cycle. Simultaneously to the knowledge that administering MTn may improve sleep, the prevalence of sleep disturbance in epileptic patients has been extensively described as well [85]. Consequently, many studies aiming to establish the direct or indirect effects of MTn on seizure frequencies of epileptic patients, especially children, have been conducted. Between 2004 and 2015, reports on clinical studies using daily doses of MTn ranging from 5 to 10 mg in isolated or add-on administrations in epileptic children have been described in the literature [86 - 90]. In general, the studies present important but inconclusive discussions about the effects of MTn on epileptic patients since more data on the number and age of subjects, time of administration, and the effect on the daily frequency of epileptic seizures are not sufficient for broader conclusions. Currently, there are four concluded clinical studies at ClinicalTrials.gov (NCT03590197, NCT00965575, NCT01161108, and NCT02195661), and future published results related to these studies must be carefully analyzed to justify further investigations and possible future application of MTn as an anticonvulsant agent.

Digoxin

Digoxin (DIG, Fig. **2**, compound 26) is a cardiac glycoside modulator of Na^+,K^+-ATPase largely used to reduce the morbidity in congestive heart failure. The effect of DIG on astrocytes under *in vitro* conditions has been described previously [91]. Under preclinical conditions in mice, DIG has shown the ability to enhance the anticonvulsant activity of classic anticonvulsants, such as valproate, topiramate, levetiracetam, phenobarbital, and clonazepam [92].

Currently, there are two concluded clinical studies at ClinicalTrials.gov (NCT02172742, NCT01583036) with regards to the effects on the pharmacokinetics of anticonvulsant and digoxin administered simultaneously. The effects of DIG itself on seizures must be investigated in appropriate clinical conditions, emphasizing the importance of conducting the procedures also focused on toxicity since the drug displays a narrow therapeutic window.

DEPRESSION: INTERFACE WITH EPILEPSIES IN A DRUG REPURPOSING CONTEXT

Depression is a common but serious disorder that affects daily life, the ability to carry out common tasks at work and school, sleep, attention, and concentration skills in general, feeding, social interactions, and leisure. It is also one of the most important comorbidities in epileptic patients [27]. Depression might be caused by a combination of genetic, biological, environmental, and psychological factors, being one of the most frequent mental disorders in the population [93]. Worldwide, it is estimated that around 300 million people of all ages suffer from this disorder, contributing to the global relevance of the disease [93 - 95].

Socially, there is a great stigma with regards to depression, frequently associated with personal weakness. This scenario increases the barriers which patients in general must overpass to seek medical and pharmacotherapeutic support. In most cases, the search for medical care is based on complaints that are related to somatic problems or fatigue, challenging the diagnosis of depression to distinguish it from other disorders. Insomnia is a well-known first classic and reliable symptom of depression. Patients who complain of persistent insomnia for more than a year are about three times more likely to be susceptible to depressive processes [96, 97].

Psychosocial stress can play a significant role in the onset of early episodes of depression in patients in general and in epileptic ones. Although patients might be reluctant to admit a depressive process, it is generally assumed that they are under considerable stress and are more willing to accept pharmacotherapy if it is presented to them as a coping strategy. Considering the treatment for depression, there are several available approaches that range from cognitive-behavioral therapy, chronic pharmacotherapy, and the combination of both [98, 99]. Most of the pharmacological regimens prioritize the use of tricyclic drugs (TCA) and selective serotonin reuptake inhibitors (SSRIs) [73]. However, with the current pharmacological approach, between 20% to 40% of patients respond poorly to monotherapy [100], with 50% of patients not responding to a primary antidepressant, only displaying therapy improvements when another is introduced or combined [101]. Additionally, in the case of the depression and epilepsy interface, data have shown that the appropriate management of the depression processes would positively affect epileptic patients by altering the degree and threshold of seizures [27, 102]. And, contrarily, the treatment of seizures would also ameliorate the depressive scenario of the patients [27, 103].

Due to the great refractoriness in the treatment of depression and its relevance in the context of epilepsy, new treatments and pharmacological approaches have been proposed over the years. In this context, DR emerges as an important and strategic approach to finding alternatives to the current drug-resistant patients. In Fig. **3**, Table **2**, and the next pages, we will discuss some of the most relevant pharmacological compounds in the DR context that have been investigated or are under current investigation as alternatives for the treatment of depression, especially as a comorbidity of epilepsy.

Ketamine

Ketamine ((1) 2-(2-Chlorophenyl)-2-(methylamino)-cyclohexanone hydrochloride, KTM, Fig. **3**, compound 27) was first synthesized by Calvin Lee Stevens in 1962, at the Laboratory of Organic Chemistry at Wayne State University, USA, in collaboration with the pharmaceutical company Parke Davis [105]. The objective of obtaining KTM was the idea to provide an anesthetic substitute to phencyclidine (PCP), displaying fewer adverse and psychomimetic effects [106].

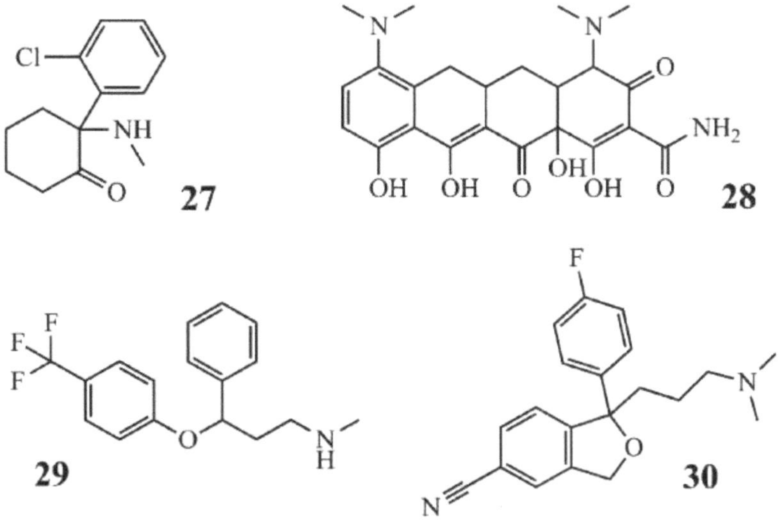

Fig. (3). Chemical structures of selected compounds in the DR context aiming at depressive processes, especially epilepsy-related depression. 27) ketamine; 28) minocycline; 29) fluoxetine; 30) escitalopram. The structure of infliximab is not represented here since it is not a small active compound, being a monoclonal antibody with a molecular weight of approximately 149 kilodaltons [104].

Table 2. DR in depression.

Drug	Initial Target	Evaluation in Depression
Ketamine (27)	General anaesthesia	[121] Clinical trials NCT02544607, NCT03149991 and NCT01700829
Infliximab	Anti-inflammatory – Crohn's disease	[146] Clinical trial NCT00463580.
Minocycline (28)	Tetracycline antibiotic	[172] Clinical trials NCT01659320, NCT01429272, NCT01514422, NCT01403662 NCT02456948, NCT02765100 NCT02362529 and NCT02703363 NCT02263872
Fluoxetine (29)	General antidepressant	[180] Clinical trials NCT02569970, NCT00986310 and NCT02929667
Escitalopram (30)	General antidepressant	[179] Clinical trials NCT01244724 and NCT00595699

Initially, the compound was employed as a veterinary anesthetic after it was patented in Belgium in 1963 [107]. Subsequent human trials have started in 1964, and the resulting outcomes provided conclusions with regard to less hallucinogenic side effects attributed to KTM. The new compound also displayed a shorter effect duration with fewer psychotomic effects in comparison to PCP [108]. KTM was finally approved by the FDA in 1970. From this point, its application as an anesthetic induction has begun in surgical and diagnostic procedures, usually in combination with a muscle relaxant to improve the quality and efficiency of the procedure [109].

With regards to the mechanism of action for KTM, the drug is currently classified as a non-competitive N-methyl-D-aspartate (NMDA) receptor antagonist [110, 111], also presenting additional activities and interactions, including those associated with tachykinin receptors [112], opioid receptors [113, 114], monoaminergic receptors [115 - 118], muscarinic receptors [114], nicotinic receptors [119], and voltage-gated calcium channels [114].

There are currently around 200 studies (more than 30 in phase III and around 40 in phase IV) aiming to assess the antidepressant properties of KTM. One of these studies, identified as NCT02544607, proposed the use of magnetic resonance imaging (MRI) techniques to assess changes before and after the administration of KTM to depressive patients, and to examine the extent of morphological alterations in specific brain regions areas as predictive factors of the antidepressant effects. This work builds on previous work, which suggested that ketamine has the potential to decrease anxiety as a depression symptom [120]. This decrease would be sustained for one month after the KTM administration. In this work, 16 participants between 18 and 65 years old (42±14) were evaluated, being 8 male and 8 female subjects who received KTM 0.5 mg/kg over 40 minutes intravenously.

Pre- and post-infusion diffusion MRI (dMRI) data were available for 13 of the 16 participants. The absence of outcomes from three patients was attributed to events, including scanner malfunction or interruption of scanning before the dMRI sequence acquisition had been completed due to patient discomfort. Data analysis of this study revealed greater pre-infusion fractional anisotropy (FA) in the left cingulate bundle and in the left upper longitudinal fasciculus to be associated with greater improvement in symptoms of depression 24 hours after KTM infusion [121]. In addition, four hours after KTM administration, FA increased rapidly in several white matter (WM) bundles in the brain; this increase was significantly associated with 24-hour symptom improvement in the evaluated patients. KTM administration seems to be associated with rapid changes in WM diffusivity, experiencing rapid changes in its microstructure [121]. Therefore, the study points to the pretreatment effects in WM structure as a potential factor associated with the clinical efficacy of KTM in patients with depression. The observed morphological changes strengthen the idea of the strategy focused on using KTM as a repositioning tool for the treatment of depression. It was reinforced in previous works, such as the clinical trial NCT03149991 (A Study of Brexpiprazole Plus Ketamine in Treatment-Resistant Depression (TRD)) [122], which describes an assessment of 99 outpatients 18–70 years old with treatment-resistant depression (TRD), defined as failure to achieve a response (*e.g.,* less than 50% improvement of depression symptoms) to at least two adequate treatment courses during the current depressive episode. They were randomly assigned to one of five arms in a 1:1:1:1:1 fashion: a single intravenous dose of KTM 0.1 mg/kg (n=18), a single dose of KTM 0.2 mg/kg (n=20), a single dose of KTM 0.5 mg/kg (n=22), a single dose of KTM 1.0 mg/kg (n=20), and a single dose of midazolam 0.045 mg/kg (active placebo) (n=19). In post-hoc paired comparisons controlling for multiple comparisons, standard (0.5 mg/kg) and high (1 mg/kg) doses of intravenous KTM (IV KTM) were superior to active placebo. The results suggest evidence for the antidepressant efficacy of sub-anesthetic doses of 0.5 mg/kg and 1.0 mg/kg of IV KTM and no clear or consistent evidence of clinically significant efficacy for lower doses of IV KTM.

A single-dose administration of KTM was also evaluated, aiming to reduce the suicidal desire in patients with depression in the "Ketamine in the Treatment of Suicidal Depression" (NCT01700829) study conducted by researcher Michael Grunebaum, MD, at the New York State Psychiatric Institute [123]. This randomized clinical trial evaluated the effect of sub-anesthetic intravenous KTM on clinically significant suicidal behavior in major depressive disorder (MDD).

Adults (N = 80) with current MDD and a Suicidal Ideation Scale (SSI) score ≥4, of whom 54% (N = 43) were submitted to the antidepressant medication, randomized to KTM or midazolam infusion. The patients received IV KTM 0.5 mg/kg or midazolam 0.02 mg/kg in 100 mL of normal saline infused within 40 minutes. According to the analyses of the outcomes, greater reductions in general mood disturbance, depression, and fatigue, assessed with the Profile of Mood States (POMS), were described on day 1 after KTM administration when compared to midazolam. Among those subjects with suicidal ideation, there was greater improvement in the KTM group on the "desire and suicidal ideation" subscale (reduction of SSI score was 4.96 points greater after KTM compared with midazolam), which correlated with depression, hopelessness, and previous suicide attempt in a previous study [124].

Despite the promising results, further studies on the effects of KTM in depressive scenarios must be conducted to reinforce its possible application in a repositioning campaign, especially considering randomized studies with greater numbers of subjects. In the context of depression as a comorbidity of epileptic patients, there are studies suggesting the use of KTM as an anticonvulsant agent, especially in the case of refractory status epilepticus (RSE) [125 - 130]. Similarly to the above-reported outcomes, further clinical trials must also be conducted to validate this KTM application.

Apparently, KTM and MEM have similar pharmacological mechanisms of action, raising concerns about the strong divergence with regards to the therapeutic potential of both drugs. It was observed that KTM, in addition to the therapeutic effects already reported, causes memory deficits, reproduces several symptoms of schizophrenia, displays an addiction potential, and may induce the formation of vacuoles in neurons at moderate concentrations and cell death at higher concentrations [131, 132]. MEM is well-tolerated, although cases of psychotic effects have already been reported in clinical studies [52, 133]. Although both drugs present neuroprotective potential at low doses and neurotoxicity at high doses, MEM exhibits much weaker neurotoxicity [52, 134, 135]. KTM and MEM are NMDAR blockers, preferentially binding the opened channels and showing faster inhibition kinetics at higher agonist concentrations. The KTM IC50 value is about half that of the MEM one [136 - 138].

The reasons for the differences mentioned above might be explained based on pharmacokinetic profiles. MEM is eliminated in a slower fashion when compared to KTM, with the human serum elimination half-life being 60-80 h for MEM [133] and 2.5 hours for KTM [139] after oral administration. Another point that has been addressed is the difference with regards to the effects of both substances on NMDA receptors' transitions between blocked states.

Changes in the blocked states of a receptor can have different effects on the inhibitory properties of an antagonist [140]. The antagonists bind the receptor pore, preventing the permeation of ions as well as promoting receptor closure. Once they are trapped in the receptor, these blockers are only turned off when the channel opens again due to a new ligand stimulation [141, 142]. MEM inhibition is known to increase in accordance with agonist concentration [135, 136, 143], an effect that might be related to partial inhibition of channel closure [144]. A higher percentage of blocked channels has also been discussed to preferentially entrap KTM than MEM [137], suggesting that the two drugs may have distinct effects on channel blocking. These findings reinforce the divergent effects on NMDARs' modifications by the inhibition of MEM and KTM [145].

Infliximab

The TNF-alpha antagonist, Infliximab (IFM), is administered by an intravenous-based therapy approved by the FDA for the treatment of inflammatory conditions, such as Crohn's disease and rheumatoid arthritis. A research group led by Andrew H Miller, Emory University, at the National Institute of Mental Health (NIMH), conducted a study to evaluate if IFM would be more effective than placebo in acutely reducing the symptoms of depression in patients who have presented elevation in proinflammatory markers and have not responded to, or been unable to tolerate, at least two previous treatments in the current depressive episode. According to the results, no significant difference in Hamilton Scale for Depression (HAM-D) scores was observed between the treatment groups. However, considerable interactions were observed between factors, such as treatment, time, and log baseline high-sensitivity C-reactive protein (hs-CRP) concentration, suggesting that the use of infliximab would at least improve depressive symptoms in patients presenting high concentration of inflammatory biomarkers [146]. Further studies might be necessary to confirm the results obtained by the group and its possible application in epilepsy-related depression.

Minocycline

Tetracyclines are broad-spectrum bacteriostatic antibiotics that are active against a wide range of bacteria, as well as against other microorganisms, including the genera *Rickettsia*, *Chlamydia*, *Plasmodium* and *Mycoplasma*. The mechanism of action behind the antibiotic properties of tetracyclines is primarily related to their ability to bind the bacterial 30S ribosomal subunit and inhibit the protein synthesis of these organisms [147].

Minocycline (7-dimethylamino-6-dimethyl-6-deoxytetracycline, MINO, Fig. **3**, compound 28) is a second-generation and semi-synthetic tetracycline derivative that has been used for decades [148]. It was also approved by the US Regulatory Agency Medicines and Health Products for the treatment of acne vulgaris and by the Food and US Drug Administration (FDA) for the treatment of sexually transmitted diseases and rheumatoid arthritis [149, 150].

In comparison to first-generation tetracyclines, MINO displays a better pharmacokinetic profile when administered orally, being rapidly absorbed, even in older subjects. It also has a longer half-life and suitable tissue penetration [151 - 153]. Since it is a highly lipophilic compound, MINO can easily cross the blood-brain barrier [154], thus providing its accumulation in the cerebrospinal fluid (CSF) and in the CNS [155 - 157], also enabling its use in the treatment of many CNS pathological conditions [148, 158].

Although the antibiotic properties of tetracyclines were described in the late 1940s, several studies have recently focused on their non-antibiotic effects. It has been reported that tetracyclines can exert a variety of biological actions that are independent of their antimicrobial activity, including anti-inflammatory and anti-apoptotic activities, as well as their inhibitory effects on tumor proteolysis, angiogenesis, and metastasis [159 - 161]. MINO was also more effective than first-generation tetracyclines in terms of neuroprotection, an effect that has been confirmed in experimental models of a plethora of brain conditions, including neurodegenerative diseases [156, 162 - 167].

In the early 2010s, the immune system was identified as a new target in the treatment of depression [168, 169]. Several proof-of-concept randomized controlled clinical trials (RCTs) have been conducted to assess the antidepressant effects of anti-inflammatory agents. Recent meta-analyses have found that mechanically diverse anti-inflammatory agents are potentially effective and well-tolerated as novel treatments for bipolar [169] and unipolar depression [170, 171].

The first case report of MINO used as a treatment for depression was published in 1996 [172], and since then, there has been significant interest and off-label prescription of MINO for depression. Several open clinical trials and RCTs have been conducted to assess the antidepressant effects of the drug targeting depression [171, 173 - 176].

From ClinicalTrials.gov, at least nine studies (NCT01659320, NCT01429272, NCT01514422, NCT01403662, NCT02456948, NCT02765100, NCT02362529, NCT02703363, NCT02263872) display recruitment status as "completed" aiming the evaluation of MINO in different depression scenarios. None of these trials specifically target epilepsy-related depression, but future studies must be conducted with this specific objective. Also, preclinical and preliminary clinical trials' outcomes suggest the promising anticonvulsant activity of MINO [177, 178], but its practical application in epileptic patients requires deeper investigations due to the diversity and complexity of the epilepsy context.

Fluoxetine and Escitalopram

Both fluoxetine (FLX, compound 29, Fig. **2**) and escitalopram (ETP, compound 30, Fig. **2**) are antidepressants that have been used in clinics for decades, being part of the first line of depressants used in epileptic patients [179]. Despite not being characterized as classic examples of DR in epilepsy, these compounds are described here because of their relevance in recent studies showing their activity specifically in depression as a comorbidity of epilepsy, such as escitalopram in concluded studies for major depression in epilepsy (clinical trial: NCT01244724) and temporal lobe epilepsy (clinical trial: NCT00595699). Their direct effects on seizures have also been investigated. Pericić *et al.* have demonstrated both acute and chronic administration of fluoxetine to increase the seizure threshold in mice [180]. Other studies reinforce these data [181 - 183], but there are further data reporting opposite findings, an increase in seizure activity after fluoxetine treatment in rats and isolated cases in patients with Down syndrome [184, 185]. There are three concluded clinical studies at ClinicalTrials.gov (NCT02569970, NCT00986310, NCT02929667) associated with the evaluation of fluoxetine for breathing mechanisms during seizures and the reduction of the risk of ictal/post-ictal hypoxemia. All these outcomes and initiatives are stimulating, requiring further results to endorse practical applications in therapy.

CONCLUSION

A significant percentage of patients with epilepsy are refractory to the available anti-epileptic drugs. Besides, these patients may also develop neuropsychiatric disorders, which are often not treated. Thus, it is paramount to develop new treatments for these conditions.

DR is an important strategy that is being used in the search for new pharmacotherapies. In the scenario of DR for epilepsies, a variety of drugs have been tested. However, due to the reduced number of studies and participants in the trials, no drug is repositioned for epilepsies or their comorbidities yet. Due to the complexity of the epilepsy scenario, which involves altered neurotransmission, neuroinflammation, neuronal cell death, impaired plasticity, and others, it is important to evaluate drugs with different mechanisms. Finally, multi-target drugs or combinations of drugs should also be considered in DR studies.

ACKNOWLEDGEMENTS

GMP, LOG and RPV thank to FAPEMIG (APQ-01532-18) and Conselho Nacional de Desenvolvimento Científico e Tecnológico – CNPq (140585/2019-2). ACPdO acknowledges CNPq for the research productivity fellowship (310347/2018-1). This work was financed in part by Coordenação de Aperfeiçoamento de Pessoal de Nível Superior, Brazil (CAPES) Finance Code 001.

REFERENCES

[1] Wouters OJ, McKee M, Luyten J. Estimated research and development investment needed to bring a new medicine to market, 2009-2018. JAMA 2020; 323(9): 844-53. [http://dx.doi.org/10.1001/jama.2020.1166]

[2] Pushpakom S. Introduction and historical overview of drug repurposing opportunities. Drug Repurposing. 2022; pp. 1-13. [http://dx.doi.org/10.1039/9781839163401-00001]

[3] Rehman W, Arfons LM, Lazarus HM. The rise, fall and subsequent triumph of thalidomide: lessons learned in drug development. Ther Adv Hematol 2011; 2(5): 291-308. [http://dx.doi.org/10.1177/2040620711413165]

[4] Mcbride WG. Thalidomide and congenital abnormalites. Lancet 1961; 278(7216): 1358. [http://dx.doi.org/10.1016/S0140-6736(61)90927-8]

[5] Lenz W, Pfeiffer RA, Kosenow W, Hayman DJ. Thalidomide and congenital abnormalites. Lancet 1962; 279(7219): 45-6. [http://dx.doi.org/10.1016/S0140-6736(62)92665-X]

[6] Sheskin J. Further observation with thalidomide in lepra reactions. Lepr Rev 1965; 36(4): 183-7. [http://dx.doi.org/10.5935/0305-7518.19650036]

[7] Sheskin J. Thalidomide in the treatment of lepra reactions. Clin Pharmacol Ther 1965; 6(3): 303-6. [http://dx.doi.org/10.1002/cpt196563303]

[8] Iyer CG, Languillon J, Ramanujam K, Tarabini-Castellani G, De las Aguas JT, Bechelli LM. WHO co-ordinated short-term double-blind trial with thalidomide in the treatment of acute lepra reactions in male lepromatous patients. Bull World Health Organ 1971; 45(6): 719-32.

[9] Pandhi D, Chhabra N. New insights in the pathogenesis of type 1 and type 2 lepra reaction. Indian J Dermatol Venereol Leprol 2013; 79(6): 739-49. [http://dx.doi.org/10.4103/0378-6323.120719]

[10] Sherwood LM, Parris EE, Folkman J. Tumor angiogenesis: therapeutic implications. N Engl J Med 1971; 285(21): 1182-6. [http://dx.doi.org/10.1056/NEJM197111182852108]

[11] D'Amato RJ, Loughnan MS, Flynn E, Folkman J. Thalidomide is an inhibitor of angiogenesis. Proc Natl Acad Sci USA 1994; 91(9): 4082-5. [http://dx.doi.org/10.1073/pnas.91.9.4082]

[12] Kotla V, Goel S, Nischal S, *et al.* Mechanism of action of lenalidomide in hematological malignancies. J Hematol Oncol 2009; 2(1): 36. [http://dx.doi.org/10.1186/1756-8722-2-36]

[13] Singhal S, Mehta J, Desikan R, *et al.* Antitumor activity of thalidomide in refractory multiple myeloma. N Engl J Med 1999; 341(21): 1565-71. [http://dx.doi.org/10.1056/NEJM199911183412102]

[14] Group RC, Horby P, Lim WS, Emberson JR, Mafham M, Bell JL. Dexamethasone in hospitalized patients with COVID-19. N Engl J Med 2021; 384(8): 693-704.
[http://dx.doi.org/10.1056/NEJMoa2021436]

[15] Association between administration of systemic corticosteroids and mortality among critically ill patients with COVID-19: A meta-analysis. JAMA 2020; 324(13): 1330-41.
[http://dx.doi.org/10.1001/jama.2020.17023]

[16] Scheffer IE, Berkovic S, Capovilla G, *et al.* ILAE classification of the epilepsies: Position paper of the ILAE Commission for Classification and Terminology. Epilepsia 2017; 58(4): 512-21.
[http://dx.doi.org/10.1111/epi.13709]

[17] Falco-Walter JJ, Scheffer IE, Fisher RS. The new definition and classification of seizures and epilepsy. Epilepsy Res 2018; 139: 73-9.
[http://dx.doi.org/10.1016/j.eplepsyres.2017.11.015]

[18] Fiest KM, Sauro KM, Wiebe S, *et al.* Prevalence and incidence of epilepsy. Neurology 2017; 88(3): 296-303.
[http://dx.doi.org/10.1212/WNL.0000000000003509]

[19] Beghi E. The Epidemiology of Epilepsy. Neuroepidemiology 2020; 54(2): 185-91.
[http://dx.doi.org/10.1159/000503831]

[20] Patsalos PN, Spencer EP, Berry DJ. Therapeutic Drug Monitoring of Antiepileptic Drugs in Epilepsy: A 2018 Update. Ther Drug Monit 2018; 40(5): 526-48.
[http://dx.doi.org/10.1097/FTD.0000000000000546]

[21] Johannessen Landmark C, Johannessen SI, Patsalos PN. Therapeutic drug monitoring of antiepileptic drugs: current status and future prospects. Expert Opin Drug Metab Toxicol 2020; 16(3): 227-38.
[http://dx.doi.org/10.1080/17425255.2020.1724956]

[22] Löscher W, Klitgaard H, Twyman RE, Schmidt D. New avenues for anti-epileptic drug discovery and development. Nat Rev Drug Discov 2013; 12(10): 757-76.
[http://dx.doi.org/10.1038/nrd4126]

[23] Wang Y, Chen Z. An update for epilepsy research and antiepileptic drug development: Toward precise circuit therapy. Pharmacol Ther 2019; 201: 77-93.
[http://dx.doi.org/10.1016/j.pharmthera.2019.05.010]

[24] Perucca P, Gilliam FG. Adverse effects of antiepileptic drugs. Lancet Neurol 2012; 11(9): 792-802.
[http://dx.doi.org/10.1016/S1474-4422(12)70153-9]

[25] Stephen LJ, Wishart A, Brodie MJ. Psychiatric side effects and antiepileptic drugs: Observations from prospective audits. Epilepsy Behav 2017; 71(Pt A): 73-8.

[26] Leppik IE. Treatment of epilepsy in the elderly. Curr Treat Options Neurol 2008; 10(4): 239-45.
[http://dx.doi.org/10.1007/s11940-008-0026-9]

[27] Keezer MR, Sisodiya SM, Sander JW. Comorbidities of epilepsy: current concepts and future perspectives. Lancet Neurol 2016; 15(1): 106-15.
[http://dx.doi.org/10.1016/S1474-4422(15)00225-2]

[28] Drew L. Gene therapy targets epilepsy. Nature 2018; 564(7735): S10-1.
[http://dx.doi.org/10.1038/d41586-018-07644-y]

[29] Perucca E, Brodie MJ, Kwan P, Tomson T. 30 years of second-generation antiseizure medications: impact and future perspectives. Lancet Neurol 2020; 19(6): 544-56.
[http://dx.doi.org/10.1016/S1474-4422(20)30035-1]

[30] Löscher W, Potschka H, Sisodiya SM, Vezzani A. Drug Resistance in Epilepsy: Clinical Impact, Potential Mechanisms, and New Innovative Treatment Options. Pharmacol Rev 2020; 72(3): 606-38.
[http://dx.doi.org/10.1124/pr.120.019539]

[31] Chen Z, Brodie MJ, Liew D, Kwan P. Treatment Outcomes in Patients With Newly Diagnosed Epilepsy Treated With Established and New Antiepileptic Drugs. JAMA Neurol 2018; 75(3): 279-86. [http://dx.doi.org/10.1001/jamaneurol.2017.3949]

[32] Remy S, Gabriel S, Urban BW, *et al.* A novel mechanism underlying drug resistance in chronic epilepsy. Ann Neurol 2003; 53(4): 469-79. [http://dx.doi.org/10.1002/ana.10473]

[33] Goodkin HP, Yeh JL, Kapur J. Status epilepticus increases the intracellular accumulation of GABAA receptors. J Neurosci 2005; 25(23): 5511-20. [http://dx.doi.org/10.1523/JNEUROSCI.0900-05.2005]

[34] Naylor DE, Liu H, Wasterlain CG. Trafficking of GABA(A) receptors, loss of inhibition, and a mechanism for pharmacoresistance in status epilepticus. J Neurosci 2005; 25(34): 7724-33. [http://dx.doi.org/10.1523/JNEUROSCI.4944-04.2005]

[35] Löscher W, Potschka H. Drug resistance in brain diseases and the role of drug efflux transporters. Nat Rev Neurosci 2005; 6(8): 591-602. [http://dx.doi.org/10.1038/nrn1728]

[36] Tang F, Hartz AMS, Bauer B. Drug-Resistant Epilepsy: Multiple Hypotheses, Few Answers. Front Neurol 2017; 8: 301. [http://dx.doi.org/10.3389/fneur.2017.00301]

[37] Bauer B, Hartz AMS, Pekcec A, Toellner K, Miller DS, Potschka H. Seizure-induced up-regulation of P-glycoprotein at the blood-brain barrier through glutamate and cyclooxygenase-2 signaling. Mol Pharmacol 2008; 73(5): 1444-53. [http://dx.doi.org/10.1124/mol.107.041210]

[38] Zibell G, Unkrüer B, Pekcec A, *et al.* Prevention of seizure-induced up-regulation of endothelial P-glycoprotein by COX-2 inhibition. Neuropharmacology 2009; 56(5): 849-55. [http://dx.doi.org/10.1016/j.neuropharm.2009.01.009]

[39] van Vliet EA, Zibell G, Pekcec A, *et al.* COX-2 inhibition controls P-glycoprotein expression and promotes brain delivery of phenytoin in chronic epileptic rats. Neuropharmacology 2010; 58(2): 404-12. [http://dx.doi.org/10.1016/j.neuropharm.2009.09.012]

[40] Siddiqui A, Kerb R, Weale ME, *et al.* Association of multidrug resistance in epilepsy with a polymorphism in the drug-transporter gene ABCB1. N Engl J Med 2003; 348(15): 1442-8. [http://dx.doi.org/10.1056/NEJMoa021986]

[41] Löscher W, Klotz U, Zimprich F, Schmidt D. The clinical impact of pharmacogenetics on the treatment of epilepsy. Epilepsia 2009; 50(1): 1-23. [http://dx.doi.org/10.1111/j.1528-1167.2008.01716.x]

[42] Orlandi A, Paolino MC, Striano P, Parisi P. Clinical reappraisal of the influence of drug-transporter polymorphisms in epilepsy. Expert Opin Drug Metab Toxicol 2018; 14(5): 505-12. [http://dx.doi.org/10.1080/17425255.2018.1473377]

[43] Kobow K, El-Osta A, Blümcke I. The methylation hypothesis of pharmacoresistance in epilepsy. Epilepsia 2013; 54 (Suppl. 2): 41-7. [http://dx.doi.org/10.1111/epi.12183]

[44] Miller-Delaney SFC, Bryan K, Das S, *et al.* Differential DNA methylation profiles of coding and non-coding genes define hippocampal sclerosis in human temporal lobe epilepsy. Brain 2015; 138(3): 616-31. [http://dx.doi.org/10.1093/brain/awu373]

[45] Kobow K, Blümcke I. Epigenetics in epilepsy. Neurosci Lett 2018; 667: 40-6. [http://dx.doi.org/10.1016/j.neulet.2017.01.012]

[46] Morris G, Reschke CR, Henshall DC. Targeting microRNA-134 for seizure control and disease modification in epilepsy. EBioMedicine 2019; 45: 646-54. [http://dx.doi.org/10.1016/j.ebiom.2019.07.008]

[47] Klein P, Friedman A, Hameed MQ, *et al.* Repurposed molecules for antiepileptogenesis: Missing an opportunity to prevent epilepsy? Epilepsia 2020; 61(3): 359-86. [http://dx.doi.org/10.1111/epi.16450]

[48] Goyon A, Beck A, Colas O, Sandra K, Guillarme D, Fekete S. Evaluation of size exclusion chromatography columns packed with sub-3 µm particles for the analysis of biopharmaceutical proteins. J Chromatogr A 2017; 1498: 80-9. [http://dx.doi.org/10.1016/j.chroma.2016.11.056]

[49] Rossi M, Freschi M, de Camargo Nascente L, *et al.* Sustainable Drug Discovery of Multi-Targe--Directed Ligands for Alzheimer's Disease. J Med Chem 2021; 64(8): 4972-90. [http://dx.doi.org/10.1021/acs.jmedchem.1c00048]

[50] Araujo de Oliveira AP, Romero Colmenares VC, Diniz R, *et al.* Memantine-Derived Schiff Bases as Transdermal Prodrug Candidates. ACS Omega 2022; 7(14): 11678-87. [http://dx.doi.org/10.1021/acsomega.1c06571]

[51] Johnson J, Kotermanski S. Mechanism of action of memantine. Curr Opin Pharmacol 2006; 6(1): 61-7. [http://dx.doi.org/10.1016/j.coph.2005.09.007]

[52] Parsons CG, Danysz W, Quack G. Memantine is a clinically well tolerated N-methyl-d-aspartate (NMDA) receptor antagonist—a review of preclinical data. Neuropharmacology 1999; 38(6): 735-67. [http://dx.doi.org/10.1016/S0028-3908(99)00019-2]

[53] Beister A, Kraus P, Kuhn W, Dose M, Weindl A, Gerlach M. The N-methyl-D-aspartate antagonist memantine retards progression of Huntington's disease. J Neural Transm Suppl 2004; 68(68): 117-22. [http://dx.doi.org/10.1007/978-3-7091-0579-5_14]

[54] Kilpatrick GJ, Tilbrook GS. Memantine. Merz. Curr Opin Investig Drugs 2002; 3(5): 798-806.

[55] Hanada T. Ionotropic Glutamate Receptors in Epilepsy: A Review Focusing on AMPA and NMDA Receptors. Biomolecules 2020; 10(3): 464. [http://dx.doi.org/10.3390/biom10030464]

[56] Kälviäinen R, Äikiä M, Helkala EL, Mervaala E, Riekkinen PJ. Memory and attention in newly diagnosed epileptic seizure disorder. Seizure 1992; 1(4): 255-62. [http://dx.doi.org/10.1016/1059-1311(92)90034-X]

[57] Samuel P. Visual Motor and Executive Functioning in Adult Patients with Primary Generalized Epilepsy: A Pilot Study. J Epilepsy Res 2020; 10(2): 62-8. [http://dx.doi.org/10.14581/jer.20010]

[58] Li N, Li J, Chen Y, *et al.* One-Year Analysis of Risk Factors Associated With Cognitive Impairment in Newly Diagnosed Epilepsy in Adults. Front Neurol 2020; 11: 594164. [http://dx.doi.org/10.3389/fneur.2020.594164]

[59] Alonazi BK, Keller SS, Fallon N, *et al.* Resting-state functional brain networks in adults with a new diagnosis of focal epilepsy. Brain Behav 2019; 9(1): e01168. [http://dx.doi.org/10.1002/brb3.1168]

[60] Marimuthu P, Varadarajan S, Krishnan M, *et al.* Evaluating the efficacy of memantine on improving cognitive functions in epileptic patients receiving anti-epileptic drugs: A double-blind placebo-controlled clinical trial (Phase IIIb pilot study). Ann Indian Acad Neurol 2016; 19(3): 344-50. [http://dx.doi.org/10.4103/0972-2327.179971]

[61] Oustad M, Najafi M, Mehvari J, Rastgoo A, Mortazavi Z, Rahiminejad M. Effect of donepezil and memantine on improvement of cognitive function in patients with temporal lobe epilepsy. J Res Med Sci 2020; 25: 29.

[62] Leeman-Markowski BA, Meador KJ, Moo LR, *et al.* Does memantine improve memory in subjects with focal-onset epilepsy and memory dysfunction? A randomized, double-blind, placebo-controlled trial. Epilepsy Behav 2018; 88: 315-24. [http://dx.doi.org/10.1016/j.yebeh.2018.06.047]

[63] Wenthur CJ. Classics in Chemical Neuroscience: Methylphenidate. ACS Chem Neurosci 2016; 7(8): 1030-40. [http://dx.doi.org/10.1021/acschemneuro.6b00199]

[64] Conners CK, Eisenberg L. The effects of methylphenidate on symptomatology and learning in disturbed children. Am J Psychiatry 1963; 120(5): 458-64. [http://dx.doi.org/10.1176/ajp.120.5.458]

[65] Heal DJ, Cheetham SC, Smith SL. The neuropharmacology of ADHD drugs *in vivo*: Insights on efficacy and safety. Neuropharmacology 2009; 57(7-8): 608-18. [http://dx.doi.org/10.1016/j.neuropharm.2009.08.020]

[66] Volkow ND, Wang GJ, Fowler JS, *et al.* Dopamine transporter occupancies in the human brain induced by therapeutic doses of oral methylphenidate. Am J Psychiatry 1998; 155(10): 1325-31. [http://dx.doi.org/10.1176/ajp.155.10.1325]

[67] Zimmer L. Contribution of clinical neuroimaging to the understanding of the pharmacology of methylphenidate. Trends Pharmacol Sci 2017; 38(7): 608-20. [http://dx.doi.org/10.1016/j.tips.2017.04.001]

[68] Radziuk AL, Kieling RR, Santos K, Rotert R, Bastos F, Palmini AL. Methylphenidate improves the quality of life of children and adolescents with ADHD and difficult-to-treat epilepsies. Epilepsy Behav 2015; 46: 215-20. [http://dx.doi.org/10.1016/j.yebeh.2015.02.019]

[69] Santos K, Palmini A, Radziuk AL, *et al.* The impact of methylphenidate on seizure frequency and severity in children with attention-deficit-hyperactivity disorder and difficult-to-treat epilepsies. Dev Med Child Neurol 2013; 55(7): 654-60. [http://dx.doi.org/10.1111/dmcn.12121]

[70] Gross-Tsur V, Manor O, van der Meere J, Joseph A, Shalev RS. Epilepsy and attention deficit hyperactivity disorder: Is methylphenidate safe and effective? J Pediatr 1997; 130(4): 670-4. [http://dx.doi.org/10.1016/S0022-3476(97)70258-0]

[71] Adams J, Alipio-Jocson V, Inoyama K, *et al.* Methylphenidate, cognition, and epilepsy. Neurology 2017; 88(5): 470-6. [http://dx.doi.org/10.1212/WNL.0000000000003564]

[72] Adams J, Alipio-Jocson V, Inoyama K, *et al.* Methylphenidate, cognition, and epilepsy: A 1-month open-label trial. Epilepsia 2017; 58(12): 2124-32. [http://dx.doi.org/10.1111/epi.13917]

[73] Costagliola G, Depietri G, Michev A, *et al.* Targeting Inflammatory Mediators in Epilepsy: A Systematic Review of Its Molecular Basis and Clinical Applications. Front Neurol 2022; 13: 741244. [http://dx.doi.org/10.3389/fneur.2022.741244]

[74] Rudick R, Polman C, Clifford D, Miller D, Steinman L. Natalizumab. JAMA Neurol 2013; 70(2): 172-82. [http://dx.doi.org/10.1001/jamaneurol.2013.598]

[75] Polman CH, O'Connor PW, Havrdova E, *et al.* A randomized, placebo-controlled trial of natalizumab for relapsing multiple sclerosis. N Engl J Med 2006; 354(9): 899-910. [http://dx.doi.org/10.1056/NEJMoa044397]

[76] Yamanaka G, Morichi S, Takamatsu T, *et al.* Links between Immune Cells from the Periphery and the Brain in the Pathogenesis of Epilepsy: A Narrative Review. Int J Mol Sci 2021; 22(9): 4395. [http://dx.doi.org/10.3390/ijms22094395]

[77] Rayatpour A, Farhangi S, Verdaguer E, *et al.* The Cross Talk between Underlying Mechanisms of Multiple Sclerosis and Epilepsy May Provide New Insights for More Efficient Therapies. Pharmaceuticals (Basel) 2021; 14(10): 1031.
[http://dx.doi.org/10.3390/ph14101031]

[78] Sotgiu S, Murrighile MR, Constantin G. Treatment of refractory epilepsy with natalizumab in a patient with multiple sclerosis. Case report. BMC Neurol 2010; 10(1): 84.
[http://dx.doi.org/10.1186/1471-2377-10-84]

[79] French JA, Cole AJ, Faught E, *et al.* Safety and Efficacy of Natalizumab as Adjunctive Therapy for People With Drug-Resistant Epilepsy. Neurology 2021; 97(18): e1757-67.
[http://dx.doi.org/10.1212/WNL.0000000000012766]

[80] French JA, Lawson JA, Yapici Z, *et al.* Adjunctive everolimus therapy for treatment-resistant focal-onset seizures associated with tuberous sclerosis (EXIST-3): a phase 3, randomised, double-blind, placebo-controlled study. Lancet 2016; 388(10056): 2153-63.
[http://dx.doi.org/10.1016/S0140-6736(16)31419-2]

[81] Kirchner GI, Meier-Wiedenbach I, Manns MP. Clinical pharmacokinetics of everolimus. Clin Pharmacokinet 2004; 43(2): 83-95.
[http://dx.doi.org/10.2165/00003088-200443020-00002]

[82] Lorenz MC, Heitman J. TOR mutations confer rapamycin resistance by preventing interaction with FKBP12-rapamycin. J Biol Chem 1995; 270(46): 27531-7.
[http://dx.doi.org/10.1074/jbc.270.46.27531]

[83] Curatolo P, Franz DN, Lawson JA, *et al.* Adjunctive everolimus for children and adolescents with treatment-refractory seizures associated with tuberous sclerosis complex: post-hoc analysis of the phase 3 EXIST-3 trial. Lancet Child Adolesc Health 2018; 2(7): 495-504.
[http://dx.doi.org/10.1016/S2352-4642(18)30099-3]

[84] Nikolaeva I, Kazdoba TM, Crowell B, D'Arcangelo G. Differential roles for Akt and mTORC1 in the hypertrophy of Pten mutant neurons, a cellular model of brain overgrowth disorders. Neuroscience 2017; 354: 196-207.
[http://dx.doi.org/10.1016/j.neuroscience.2017.04.026]

[85] van Golde EGA, Gutter T, de Weerd AW. Sleep disturbances in people with epilepsy; prevalence, impact and treatment. Sleep Med Rev 2011; 15(6): 357-68.
[http://dx.doi.org/10.1016/j.smrv.2011.01.002]

[86] Goldberg-Stern H, Oren H, Peled N, Garty BZ. Effect of melatonin on seizure frequency in intractable epilepsy: a pilot study. J Child Neurol 2012; 27(12): 1524-8.
[http://dx.doi.org/10.1177/0883073811435916]

[87] Gupta M, Aneja S, Kohli K. Add-on melatonin improves quality of life in epileptic children on valproate monotherapy: a randomized, double-blind, placebo-controlled trial. Epilepsy Behav 2004; 5(3): 316-21.
[http://dx.doi.org/10.1016/j.yebeh.2004.01.012]

[88] Hancock E, O'Callaghan F, Osborne JP. Effect of melatonin dosage on sleep disorder in tuberous sclerosis complex. J Child Neurol 2005; 20(1): 78-80.
[http://dx.doi.org/10.1177/08830738050200011302]

[89] Jain SV, Horn PS, Simakajornboon N, *et al.* Melatonin improves sleep in children with epilepsy: a randomized, double-blind, crossover study. Sleep Med 2015; 16(5): 637-44.
[http://dx.doi.org/10.1016/j.sleep.2015.01.005]

[90] Gruen R, Weeramanthri T, Knight S, Bailie R. Specialist outreach clinics in primary care and rural hospital settings (Cochrane Review). Community Eye Health 2006; 19(58): 31.

[91] Nguyen KTD, Buljan V, Else PL, Pow DV, Balcar VJ. Cardiac glycosides ouabain and digoxin interfere with the regulation of glutamate transporter GLAST in astrocytes cultured from neonatal rat

brain. Neurochem Res 2010; 35(12): 2062-9.
[http://dx.doi.org/10.1007/s11064-010-0274-4]

[92] Tsyvunin V, Shtrygol' S, Shtrygol' D. Digoxin enhances the effect of antiepileptic drugs with different mechanism of action in the pentylenetetrazole-induced seizures in mice. Epilepsy Res 2020; 167: 106465.
[http://dx.doi.org/10.1016/j.eplepsyres.2020.106465]

[93] Depression. World Health Organization. Available from: https://www.who.int/news-room/fac--sheets/detail/depression

[94] Gilbody S, Gask L. Chapter 6 depressive disorders in primary care: A review. Depressive Disorders. 3rd. 2009; pp. 271-318.

[95] Mitchell AJ, Vaze A, Rao S. Clinical diagnosis of depression in primary care: a meta-analysis. Lancet 2009; 374(9690): 609-19.
[http://dx.doi.org/10.1016/S0140-6736(09)60879-5]

[96] Ford DE, Kamerow DB. Epidemiologic study of sleep disturbances and psychiatric disorders. An opportunity for prevention? JAMA 1989; 262(11): 1479-84.
[http://dx.doi.org/10.1001/jama.1989.03430110069030]

[97] Perlis ML, Giles DE, Buysse DJ, Tu X, Kupfer DJ. Self-reported sleep disturbance as a prodromal symptom in recurrent depression. J Affect Disord 1997; 42(2-3): 209-12.
[http://dx.doi.org/10.1016/S0165-0327(96)01411-5]

[98] Cuijpers P, Noma H, Karyotaki E, Cipriani A, Furukawa TA. Effectiveness and Acceptability of Cognitive Behavior Therapy Delivery Formats in Adults With Depression. JAMA Psychiatry 2019; 76(7): 700-7.
[http://dx.doi.org/10.1001/jamapsychiatry.2019.0268]

[99] Amick HR, Gartlehner G, Gaynes BN, *et al.* Comparative benefits and harms of second generation antidepressants and cognitive behavioral therapies in initial treatment of major depressive disorder: systematic review and meta-analysis. BMJ 2015; 351: h6019.
[http://dx.doi.org/10.1136/bmj.h6019]

[100] Shelton RC. Mood-stabilizing drugs in depression. J Clin Psychiatry 1999; 60 (Suppl. 5): 37-40.

[101] Depression in primary care: Detection, diagnosis, and treatment. J Am Acad Nurse Pract 1994; 6(5): 224-38.
[http://dx.doi.org/10.1111/j.1745-7599.1994.tb00946.x]

[102] Alper K, Schwartz KA, Kolts RL, Khan A. Seizure incidence in psychopharmacological clinical trials: an analysis of Food and Drug Administration (FDA) summary basis of approval reports. Biol Psychiatry 2007; 62(4): 345-54.
[http://dx.doi.org/10.1016/j.biopsych.2006.09.023]

[103] Tellez-Zenteno JF, Dhar R, Hernandez-Ronquillo L, Wiebe S. Long-term outcomes in epilepsy surgery: antiepileptic drugs, mortality, cognitive and psychosocial aspects. Brain 2007; 130(2): 334-45.
[http://dx.doi.org/10.1093/brain/awl316]

[104] Klotz U, Teml A, Schwab M. Clinical pharmacokinetics and use of infliximab. Clin Pharmacokinet 2007; 46(8): 645-60.
[http://dx.doi.org/10.2165/00003088-200746080-00002]

[105] Ivani G, Vercellino C, Tonetti F. Ketamine: a new look to an old drug. Minerva Anestesiol 2003; 69(5): 468-71.

[106] Domino EF, Chodoff P, Corssen G. Pharmacologic effects of CI-581, a new dissociative anesthetic, in man. Clin Pharmacol Ther 1965; 6(3): 279-91.
[http://dx.doi.org/10.1002/cpt196563279]

[107] Kohtala S. Ketamine—50 years in use: from anesthesia to rapid antidepressant effects and neurobiological mechanisms. Pharmacol Rep 2021; 73(2): 323-45.
[http://dx.doi.org/10.1007/s43440-021-00232-4]

[108] Denomme N. The Domino Effect: Ed Domino's early studies of Psychoactive Drugs. J Psychoactive Drugs 2018; 50(4): 298-305.
[http://dx.doi.org/10.1080/02791072.2018.1506599]

[109] Restall J, Tully AM, Ward PJ, Kidd AG. Total intravenous anaesthesia for military surgery. A technique using ketamine, midazolam and vecuronium. Anaesthesia 1988; 43(1): 46-9.
[http://dx.doi.org/10.1111/j.1365-2044.1988.tb05424.x]

[110] Harrison NL, Simmonds MA. Quantitative studies on some antagonists of N-methyl D-aspartate in slices of rat cerebral cortex. Br J Pharmacol 1985; 84(2): 381-91.
[http://dx.doi.org/10.1111/j.1476-5381.1985.tb12922.x]

[111] Sinner B, Graf BM. Ketamine. Handb Exp Pharmacol 2008; 182(182): 313-33.
[http://dx.doi.org/10.1007/978-3-540-74806-9_15]

[112] Okamoto T, Minami K, Uezono Y, *et al.* The inhibitory effects of ketamine and pentobarbital on substance p receptors expressed in Xenopus oocytes. Anesth Analg 2003; 97(1): 104-10.
[http://dx.doi.org/10.1213/01.ANE.0000066260.99680.11]

[113] Smith DJ, Pekoe GM, Martin LL, Coalgate B. The interaction of ketamine with the opiate receptor. Life Sci 1980; 26(10): 789-95.
[http://dx.doi.org/10.1016/0024-3205(80)90285-4]

[114] Hustveit O, Maurset A, Øye I. Interaction of the chiral forms of ketamine with opioid, phencyclidine, sigma and muscarinic receptors. Pharmacol Toxicol 1995; 77(6): 355-9.
[http://dx.doi.org/10.1111/j.1600-0773.1995.tb01041.x]

[115] Kapur S, Seeman P. NMDA receptor antagonists ketamine and PCP have direct effects on the dopamine D2 and serotonin 5-HT2 receptors—implications for models of schizophrenia. Mol Psychiatry 2002; 7(8): 837-44.
[http://dx.doi.org/10.1038/sj.mp.4001093]

[116] Salt PJ, Barnes PK, Beswick FJ. Inhibition of neuronal and extraneuronal uptake of noradrenaline by ketamine in the isolated perfused rat heart. Br J Anaesth 1979; 51(9): 835-8.
[http://dx.doi.org/10.1093/bja/51.9.835]

[117] Martin LL, Bouchal RL, Smith DJ. Ketamine inhibits serotonin uptake *in vivo*. Neuropharmacology 1982; 21(2): 113-8.
[http://dx.doi.org/10.1016/0028-3908(82)90149-6]

[118] Appadu BL, Lambert DG. Interaction of i.v. anaesthetic agents with 5-HT3 receptors. Br J Anaesth 1996; 76(2): 271-3.
[http://dx.doi.org/10.1093/bja/76.2.271]

[119] Kohrs R, Durieux ME. Ketamine: teaching an old drug new tricks. Anesth Analg 1998; 87(5): 1186-93.

[120] Acevedo-Diaz EE, Cavanaugh GW, Greenstein D, *et al.* Can 'floating' predict treatment response to ketamine? Data from three randomized trials of individuals with treatment-resistant depression. J Psychiatr Res 2020; 130: 280-5.
[http://dx.doi.org/10.1016/j.jpsychires.2020.06.012]

[121] Sydnor VJ, Lyall AE, Cetin-Karayumak S, *et al.* Studying pre-treatment and ketamine-induced changes in white matter microstructure in the context of ketamine's antidepressant effects. Transl Psychiatry 2020; 10(1): 432.
[http://dx.doi.org/10.1038/s41398-020-01122-8]

[122] Fava M, Freeman MP, Flynn M, *et al.* Double-blind, placebo-controlled, dose-ranging trial of

intravenous ketamine as adjunctive therapy in treatment-resistant depression (TRD). Mol Psychiatry 2020; 25(7): 1592-603.
[http://dx.doi.org/10.1038/s41380-018-0256-5]

[123] Grunebaum MF, Galfalvy HC, Choo TH, *et al.* Ketamine for Rapid Reduction of Suicidal Thoughts in Major Depression: A Midazolam-Controlled Randomized Clinical Trial. Am J Psychiatry 2018; 175(4): 327-35.
[http://dx.doi.org/10.1176/appi.ajp.2017.17060647]

[124] Rudd MD, Berman AL, Joiner TE Jr, *et al.* Warning signs for suicide: theory, research, and clinical applications. Suicide Life Threat Behav 2006; 36(3): 255-62.
[http://dx.doi.org/10.1521/suli.2006.36.3.255]

[125] Gaspard N, Foreman B, Judd LM, *et al.* Intravenous ketamine for the treatment of refractory status epilepticus: A retrospective multicenter study. Epilepsia 2013; 54(8): 1498-503.
[http://dx.doi.org/10.1111/epi.12247]

[126] Rosati A, L'Erario M, Ilvento L, *et al.* Efficacy and safety of ketamine in refractory status epilepticus in children. Neurology 2012; 79(24): 2355-8.
[http://dx.doi.org/10.1212/WNL.0b013e318278b685]

[127] Mewasingh LD, Sékhara T, Aeby A, Christiaens FJC, Dan B. Oral ketamine in paediatric non-convulsive status epilepticus. Seizure 2003; 12(7): 483-9.
[http://dx.doi.org/10.1016/S1059-1311(03)00028-1]

[128] Synowiec AS, Singh DS, Yenugadhati V, Valeriano JP, Schramke CJ, Kelly KM. Ketamine use in the treatment of refractory status epilepticus. Epilepsy Res 2013; 105(1-2): 183-8.
[http://dx.doi.org/10.1016/j.eplepsyres.2013.01.007]

[129] Zeiler FA, Kaufmann AM, Gillman LM, West M, Silvaggio J. Ketamine for medically refractory status epilepticus after elective aneurysm clipping. Neurocrit Care 2013; 19(1): 119-24.
[http://dx.doi.org/10.1007/s12028-013-9858-6]

[130] Esaian D, Joset D, Lazarovits C, Dugan PC, Fridman D. Ketamine continuous infusion for refractory status epilepticus in a patient with anticonvulsant hypersensitivity syndrome. Ann Pharmacother 2013; 47(11): 1569-76.
[http://dx.doi.org/10.1177/1060028013505427]

[131] Krystal JH, D'Souza CD, Petrakis IL, *et al.* NMDA agonists and antagonists as probes of glutamatergic dysfunction and pharmacotherapies in neuropsychiatric disorders. Harv Rev Psychiatry 1999; 7(3): 125-43.
[http://dx.doi.org/10.3109/hrp.7.3.125]

[132] Sharp FR, Tomitaka M, Bernaudin M, Tomitaka S. Psychosis: pathological activation of limbic thalamocortical circuits by psychomimetics and schizophrenia? Trends Neurosci 2001; 24(6): 330-4.
[http://dx.doi.org/10.1016/S0166-2236(00)01817-8]

[133] Sonkusare SK, Kaul CL, Ramarao P. Dementia of Alzheimer's disease and other neurodegenerative disorders—memantine, a new hope. Pharmacol Res 2005; 51(1): 1-17.
[http://dx.doi.org/10.1016/j.phrs.2004.05.005]

[134] Parsons CG, Quack G, Bresink I, *et al.* Comparison of the potency, kinetics and voltage-dependency of a series of uncompetitive NMDA receptor antagonists *in vitro* with anticonvulsive and motor impairment activity *in vivo*. Neuropharmacology 1995; 34(10): 1239-58.
[http://dx.doi.org/10.1016/0028-3908(95)00092-K]

[135] Rogawski MA. Low affinity channel blocking (uncompetitive) NMDA receptor antagonists as therapeutic agents - toward an understanding of their favorable tolerability. Amino Acids 2000; 19(1): 133-49.
[http://dx.doi.org/10.1007/s007260070042]

[136] Parsons CG, Gruner R, Rozental J, Millar J, Lodge D. Patch clamp studies on the kinetics and

selectivity of N-methyl-d-aspartate receptor antagonism by memantine (1-amino-3-5-dimethyladamantan). Neuropharmacology 1993; 32(12): 1337-50. [http://dx.doi.org/10.1016/0028-3908(93)90029-3]

[137] Mealing GA, Lanthorn TH, Murray CL, Small DL, Morley P. Differences in degree of trapping of low-affinity uncompetitive N-methyl-D-aspartic acid receptor antagonists with similar kinetics of block. J Pharmacol Exp Ther 1999; 288(1): 204-10.

[138] Parsons CG, Panchenko VA, Pinchenko VO, Tsyndrenko AY, Krishtal OA. Comparative patch-clamp studies with freshly dissociated rat hippocampal and striatal neurons on the NMDA receptor antagonistic effects of amantadine and memantine. Eur J Neurosci 1996; 8(3): 446-54. [http://dx.doi.org/10.1111/j.1460-9568.1996.tb01228.x]

[139] Grant IS, Nimmo WS, Clements JA. Pharmacokinetics and analgesic effects of i.m. and oral ketamine. Br J Anaesth 1981; 53(8): 805-10. [http://dx.doi.org/10.1093/bja/53.8.805]

[140] Courtney KR. Mechanism of frequency-dependent inhibition of sodium currents in frog myelinated nerve by the lidocaine derivative GEA. J Pharmacol Exp Ther 1975; 195(2): 225-36.

[141] Blanpied TA, Boeckman FA, Aizenman E, Johnson JW. Trapping channel block of NMDA-activated responses by amantadine and memantine. J Neurophysiol 1997; 77(1): 309-23. [http://dx.doi.org/10.1152/jn.1997.77.1.309]

[142] Sobolevsky AI, Yelshansky MV. The trapping block of NMDA receptor channels in acutely isolated rat hippocampal neurones. J Physiol 2000; 526(3): 493-506. [http://dx.doi.org/10.1111/j.1469-7793.2000.t01-2-00493.x]

[143] Aracava Y, Pereira EFR, Maelicke A, Albuquerque EX. Memantine blocks alpha7* nicotinic acetylcholine receptors more potently than n-methyl-D-aspartate receptors in rat hippocampal neurons. J Pharmacol Exp Ther 2005; 312(3): 1195-205. [http://dx.doi.org/10.1124/jpet.104.077172]

[144] Johnson JW, Qian A. Interaction between channel blockers and channel gating of NMDA receptors. Биол мембраны 2002; 19: 17-22.

[145] Yuan H, Erreger K, Dravid SM, Traynelis SF. Conserved structural and functional control of N-methyl-D-aspartate receptor gating by transmembrane domain M3. J Biol Chem 2005; 280(33): 29708-16. [http://dx.doi.org/10.1074/jbc.M414215200]

[146] Raison CL, Rutherford RE, Woolwine BJ, *et al.* A randomized controlled trial of the tumor necrosis factor antagonist infliximab for treatment-resistant depression: the role of baseline inflammatory biomarkers. JAMA Psychiatry 2013; 70(1): 31-41. [http://dx.doi.org/10.1001/2013.jamapsychiatry.4]

[147] Nelson ML. Chemical and biological dynamics of tetracyclines. Adv Dent Res 1998; 12(1): 5-11. [http://dx.doi.org/10.1177/08959374980120011901]

[148] Yong VW, Wells J, Giuliani F, Casha S, Power C, Metz LM. The promise of minocycline in neurology. Lancet Neurol 2004; 3(12): 744-51. [http://dx.doi.org/10.1016/S1474-4422(04)00937-8]

[149] Good ML, Hussey DL. Minocycline: stain devil? Br J Dermatol 2003; 149(2): 237-9. [http://dx.doi.org/10.1046/j.1365-2133.2003.05497.x]

[150] Garrido-Mesa N, Zarzuelo A, Gálvez J. Minocycline: far beyond an antibiotic. Br J Pharmacol 2013; 169(2): 337-52. [http://dx.doi.org/10.1111/bph.12139]

[151] Barza M, Brown RB, Shanks C, Gamble C, Weinstein L. Relation between lipophilicity and pharmacological behavior of minocycline, doxycycline, tetracycline, and oxytetracycline in dogs. Antimicrob Agents Chemother 1975; 8(6): 713-20.

[http://dx.doi.org/10.1128/AAC.8.6.713]

[152] Kramer PA, Chapron DJ, Benson J, Mercik SA. Tetracycline absorption in elderly patients with achlorhydria. Clin Pharmacol Ther 1978; 23(4): 467-72. [http://dx.doi.org/10.1002/cpt1978234467]

[153] Klein NC, Cunha BA. Tetracyclines. Med Clin North Am 1995; 79(4): 789-801. [http://dx.doi.org/10.1016/S0025-7125(16)30039-6]

[154] Brogden RN, Speight TM, Avery GS. Minocycline. Drugs 1975; 9(4): 251-91. [http://dx.doi.org/10.2165/00003495-197509040-00005]

[155] Aronson AL. Pharmacotherapeutics of the newer tetracyclines. J Am Vet Med Assoc 1980; 176(10): 1061-8.

[156] Yrjänheikki J, Tikka T, Keinänen R, Goldsteins G, Chan PH, Koistinaho J. A tetracycline derivative, minocycline, reduces inflammation and protects against focal cerebral ischemia with a wide therapeutic window. Proc Natl Acad Sci USA 1999; 96(23): 13496-500. [http://dx.doi.org/10.1073/pnas.96.23.13496]

[157] Kielian T, Esen N, Liu S, *et al.* Minocycline modulates neuroinflammation independently of its antimicrobial activity in staphylococcus aureus-induced brain abscess. Am J Pathol 2007; 171(4): 1199-214. [http://dx.doi.org/10.2353/ajpath.2007.070231]

[158] Saivin S, Houin G. Clinical pharmacokinetics of doxycycline and minocycline. Clin Pharmacokinet 1988; 15(6): 355-66. [http://dx.doi.org/10.2165/00003088-198815060-00001]

[159] Golub LM, Ramamurthy NS, McNamara TF, Greenwald RA, Rifkin BR. Tetracyclines inhibit connective tissue breakdown: new therapeutic implications for an old family of drugs. Crit Rev Oral Biol Med 1991; 2(3): 297-321. [http://dx.doi.org/10.1177/10454411910020030201]

[160] Golub LM, Suomalainen K, Sorsa T. Host modulation with tetracyclines and their chemically modified analogues. Curr Opin Dent 1992; 2: 80-90.

[161] Sapadin AN, Fleischmajer R. Tetracyclines: Nonantibiotic properties and their clinical implications. J Am Acad Dermatol 2006; 54(2): 258-65. [http://dx.doi.org/10.1016/j.jaad.2005.10.004]

[162] Yrjänheikki J, Keinänen R, Pellikka M, Hökfelt T, Koistinaho J. Tetracyclines inhibit microglial activation and are neuroprotective in global brain ischemia. Proc Natl Acad Sci USA 1998; 95(26): 15769-74. [http://dx.doi.org/10.1073/pnas.95.26.15769]

[163] Sanchez Mejia RO, Ona VO, Li M, Friedlander RM. Minocycline reduces traumatic brain injury-mediated caspase-1 activation, tissue damage, and neurological dysfunction. Neurosurgery 2001; 48(6): 1393-401. [http://dx.doi.org/10.1227/00006123-200106000-00051]

[164] Raghavendra V, Tanga F, DeLeo JA. Inhibition of microglial activation attenuates the development but not existing hypersensitivity in a rat model of neuropathy. J Pharmacol Exp Ther 2003; 306(2): 624-30. [http://dx.doi.org/10.1124/jpet.103.052407]

[165] Mei XP, Xu H, Xie C, *et al.* Post-injury administration of minocycline: An effective treatment for nerve-injury induced neuropathic pain. Neurosci Res 2011; 70(3): 305-12. [http://dx.doi.org/10.1016/j.neures.2011.03.012]

[166] Du Y, Ma Z, Lin S, *et al.* Minocycline prevents nigrostriatal dopaminergic neurodegeneration in the MPTP model of Parkinson's disease. Proc Natl Acad Sci USA 2001; 98(25): 14669-74. [http://dx.doi.org/10.1073/pnas.251341998]

[167] Thomas M, Le W. Minocycline: neuroprotective mechanisms in Parkinson's disease. Curr Pharm Des 2004; 10(6): 679-86.
[http://dx.doi.org/10.2174/1381612043453162]

[168] Rosenblat JD, Cha DS, Mansur RB, McIntyre RS. Inflamed moods: A review of the interactions between inflammation and mood disorders. Prog Neuropsychopharmacol Biol Psychiatry 2014; 53: 23-34.
[http://dx.doi.org/10.1016/j.pnpbp.2014.01.013]

[169] Rosenblat JD, McIntyre RS. Bipolar Disorder and Inflammation. Psychiatr Clin North Am 2016; 39(1): 125-37.
[http://dx.doi.org/10.1016/j.psc.2015.09.006]

[170] Köhler O, Benros ME, Nordentoft M, *et al.* Effect of anti-inflammatory treatment on depression, depressive symptoms, and adverse effects: a systematic review and meta-analysis of randomized clinical trials. JAMA Psychiatry 2014; 71(12): 1381-91.
[http://dx.doi.org/10.1001/jamapsychiatry.2014.1611]

[171] Husain MI, Chaudhry IB, Husain N, *et al.* Minocycline as an adjunct for treatment-resistant depressive symptoms: A pilot randomised placebo-controlled trial. J Psychopharmacol 2017; 31(9): 1166-75.
[http://dx.doi.org/10.1177/0269881117724352]

[172] Levine J, Cholestoy A, Zimmerman J. Possible antidepressant effect of minocycline. Am J Psychiatry 1996; 153(4): 582b-.
[http://dx.doi.org/10.1176/ajp.153.4.582b]

[173] Miyaoka T, Wake R, Furuya M, *et al.* Minocycline as adjunctive therapy for patients with unipolar psychotic depression: An open-label study. Prog Neuropsychopharmacol Biol Psychiatry 2012; 37(2): 222-6.
[http://dx.doi.org/10.1016/j.pnpbp.2012.02.002]

[174] Emadi-Kouchak H, Mohammadinejad P, Asadollahi-Amin A, *et al.* Therapeutic effects of minocycline on mild-to-moderate depression in HIV patients. Int Clin Psychopharmacol 2016; 31(1): 20-6.
[http://dx.doi.org/10.1097/YIC.0000000000000098]

[175] Dean OM, Kanchanatawan B, Ashton M, *et al.* Adjunctive minocycline treatment for major depressive disorder: A proof of concept trial. Aust N Z J Psychiatry 2017; 51(8): 829-40.
[http://dx.doi.org/10.1177/0004867417709357]

[176] Soczynska JK, Kennedy SH, Alsuwaidan M, *et al.* A pilot, open-label, 8-week study evaluating the efficacy, safety and tolerability of adjunctive minocycline for the treatment of bipolar I/II depression. Bipolar Disord 2017; 19(3): 198-213.
[http://dx.doi.org/10.1111/bdi.12496]

[177] Singh T, Thapliyal S, Bhatia S, *et al.* Reconnoitering the transformative journey of minocycline from an antibiotic to an antiepileptic drug. Life Sci 2022; 293: 120346.
[http://dx.doi.org/10.1016/j.lfs.2022.120346]

[178] Beheshti Nasr SM, Moghimi A, Mohammad-Zadeh M, Shamsizadeh A, Noorbakhsh SM. The effect of minocycline on seizures induced by amygdala kindling in rats. Seizure 2013; 22(8): 670-4.
[http://dx.doi.org/10.1016/j.seizure.2013.05.005]

[179] Górska N, Słupski J, Cubała WJ, Wiglusz MS, Gałuszko-Węgielnik M. Antidepressants in epilepsy. Neurol Neurochir Pol 2018; 52(6): 657-61.
[http://dx.doi.org/10.1016/j.pjnns.2018.07.005]

[180] Peričić D, Lazić J, Švob Štrac D. Anticonvulsant effects of acute and repeated fluoxetine treatment in unstressed and stressed mice. Brain Res 2005; 1033(1): 90-5.
[http://dx.doi.org/10.1016/j.brainres.2004.11.025]

[181] Prendiville S, Gale K. Anticonvulsant effect of fluoxetine on focally evoked limbic motor seizures in rats. Epilepsia 1993; 34(2): 381-4.

[http://dx.doi.org/10.1111/j.1528-1157.1993.tb02425.x]

[182] Wada Y, Shiraishi J, Nakamura M, Hasegawa H. Prolonged but not acute fluoxetine administration produces its inhibitory effect on hippocampal seizures in rats. Psychopharmacology (Berl) 1995; 118(3): 305-9.
[http://dx.doi.org/10.1007/BF02245959]

[183] Ugale RR, Mittal N, Hirani K, Chopde CT. Essentiality of central GABAergic neuroactive steroid allopregnanolone for anticonvulsant action of fluoxetine against pentylenetetrazole-induced seizures in mice. Brain Res 2004; 1023(1): 102-11.
[http://dx.doi.org/10.1016/j.brainres.2004.07.018]

[184] Zienowicz M, Wisłowska A, Lehner M, *et al.* The effect of fluoxetine in a model of chemically induced seizures—behavioral and immunocytochemical study. Neurosci Lett 2005; 373(3): 226-31.
[http://dx.doi.org/10.1016/j.neulet.2004.10.009]

[185] Prasher VP. Seizures associated with fluoxetine therapy. Seizure 1993; 2(4): 315-7.
[http://dx.doi.org/10.1016/S1059-1311(05)80148-7]

CHAPTER 6

Progress on the Development of Oxime Derivatives as a Potential Antidote for Organophosphorus Poisoning

Manjunatha S. Katagi[1,*], **M.L Sujatha**[2], **Girish Bolakatti**[3], **B.P. Nandeshwarappa**[2], **S.N. Mamledesai**[4] and **Jennifer Fernandes**[5]

[1] *Department of Pharmaceutical Chemistry, Bapuji Pharmacy College, Davangere - 577 004, Karnataka, India*

[2] *Department of Studies in Chemistry, Davangere University, Shivagangothri, Tholhunase - 577 007, Karnataka, India*

[3] *Department of Pharmaceutical Chemistry, GM Institute of Pharmaceutical Sciences and Research, Davangere - 577 006, Karnataka, India*

[4] *Department of Pharmaceutical Chemistry, PES's Rajaram & Tarabai Bandekar College of Pharmacy, Farmagudi-Ponda - 403 401, Goa, India*

[5] *Department of Pharmaceutical Chemistry, NGSM Institute of Pharmaceutical Sciences, Mangalore-574 160, Karnataka, India*

Abstract: Nowadays, organophosphorus poisoning is the most common emergency throughout the world. Two functionally different types of drugs are used in common to treat such intoxication cases. The first type includes the reactivators of acetylcholinesterase (AChE)-oximes, which have the capability to restore the physiological function of inhibited AChE. The second type includes anticholinergic, such as atropine that antagonizes the effects of excessive ACh by blocking muscarinic receptors. Alternatively, anticholinergic and reactivators may be co-administered to get synergistic effects. At muscarinic and nicotinic synapses, organophosphorus compounds inhibit AChE release by phosphoryl group deposition at the enzyme's active site very quickly. AChE regenerative process can be accelerated by detaching the OP compound at -OH group of the enzyme. OP compound combines with the AChE enzyme forming a complex and making it inactive. After ageing of the inactive state of AChE, it is difficult to break the complex to regenerate the enzyme resulting in acetylcholine accumulation at synapses. To counter the effect of OP compound, oximes catalyse the reactivation of active AChE by exerting nucleophilic attack on the phosphoryl group. Oximes theoretically remove OP compound from the complex by acting on phosphoryl bond resulting in enzyme reactivation. Reactivation of AChE inhibited by OP compounds through the above mentioned approach poses certain limitations. There is no universal antidote capable of effectively restoring AChE inhibi-

* **Corresponding author Manjunatha S. Katagi:** Department of Pharmaceutical Chemistry, Bapuji Pharmacy College, Davangere-577 004, Karnataka, India; Tels: +91 8192 221459, +91 9886499160; E-mail: manju_mpharm@rediffmail.com

Zareen Amtul (Ed.)

ted by wide-ranging OP compounds. The oxime reactivators are efficient only when administered before the "ageing" of AChE-OP complex. Anticholinergic drugs, like atropine, are effective only on muscarinic receptors but not on nicotinic receptors (nAChRs).

Keywords: Acetylcholine, Acetylcholinesterase, Atropine, Obidoxime, Butyrylcholinesterase, HI-6, Organophoshosphorus, Poisoning: Oxime, Reactivation, 2-PAM.

INTRODUCTION

History

The history of organophosphate (OP) compounds began in 1800, when Moschnine [1, 2] synthesized a mono ester named tetraethyl pyrophosphate (TEPP). The process was first published in 1854 by de Clermont [3]. Nearly 80 years later in 1934, Dr. Gerhard Schrader, a German chemist, synthesized hundreds of OPs including parathion and tabun (dimethyl phosphoroamidocyanidate), sarin (isopropyl methylphosphonofluoridate), and soman (O-Pinacolylmethylphosphonofluoridate). Schrader later also synthesized a series of fluorine-containing esters including diisopropylfluorophosphate (DFP) and sarin, pyrophosphate esters including TEPP and octamethylpyrophosphortetramide (OMPA), and thio and thiono phosphorus esters including parathion and its oxygen analog paraxon [4, 5] led by observations of Lange and Kruger, who described the synthesis of two OP compounds and noted that their vapour inhalation produced certain health effects, engaged in the exploration of this type of compounds. Their work resulted in the synthesis of parathion; one of the most frequently used OP pesticide in recent decades. After II World War, thousand of OP compounds were synthesized worldwide for various purposes (pesticides, nerve agents in medicine and in chemical warfare, flame retardants and parasiticides in veterinary medicine). Due to the lack of persistence in the environment and in exposed individuals and due to lesser insect resistance development in comparison to organochlorine pesticides, the OP pesticides are today the most commonly used group of pesticides throughout the world. It should be emphasized, that from several points of view (public health and intensive agriculture), their use today is a must and not an option. Although their persistence in environment is relatively low, the extensive use of OP pesticides in modern agriculture has raised several problems regarding environmental and food safety issues [6 - 8].

The OP compounds have a wide variety of applications, hence is a serious threat for occupational hazard, self-poisoning, unintentional misuse, terrorist attack and threats of warfare use, not only for army rather civilian targets as well. Organophosphates are mainly used for civilian purposes as a pesticides or acaricides, *etc.* but their acute toxicity is comparable to the organophosphonates, developed for military purposes.

OP poisoning has been a frequent cause of admission of people to hospitals and Intensive Care Units (ICU) in developing countries [9]. OP poisoning causes about 3 million acute intoxications annually, 0.3 million of which lead to fatalities [10]. OPs that were developed as chemical warfare nerve agents (CWAs) are all highly toxic and dangerous [11 - 13]. OP nerve agents have been used and are most likely to be used in the future by terrorists and dictators around the world because of their relatively easy synthesis and availability of suitable delivery systems [14]. Warfare and terrorist use of CWAs include the Aum Shinrikyo terrorist attack in the Tokyo subway in 1995, the 1980–1988 Iraq–Iran war where Iraq reportedly used nerve agents against Iranian troops and later on Kurd civilians and the murder of a family member of the North Korea Leader, Kim Jong Nam on February 2017 in Kuala Lumpur [15 - 17]. Despite international efforts aimed at regulating and lessening the use of these environmentally toxic compounds, more than 100 different OP compounds are still being used intensively as pesticides, with only unreliable or sporadic monitoring of the environment and workers involved in their use.

Organophosphorus Compounds

Organophosphorus compounds are esters, amides or thiol derivatives of phosphoric, phosphonic, phosphinic acids, and phosphorothioic or phosphonothioic acids. The phosphonic acid derivatives are more toxic than the phosphoric acids (Inchem.org) whose oxygen atom can be substituted by sulphur or nitrogen atoms [18]. In other words, it is an organic compound that contains phosphorus as an integral part of the molecule and formed by the reaction of alcohol and phosphoric/phosphonic/phosphinic acids.

General Structure Of An OP

The basic structure of an OP compound consists of the following;

a. A central phosphorus atom (P).
b. P is double bonded to either oxygen or sulphur.
c. A leaving group which is specific to the individual organophosphorus. It is a labile acyl residue (halide, cyano, phenol, or thio group).
d. R_1 and R_2 groups which are ethyl or methyl, alkyl, alkoxy, alkylthio or amino group.

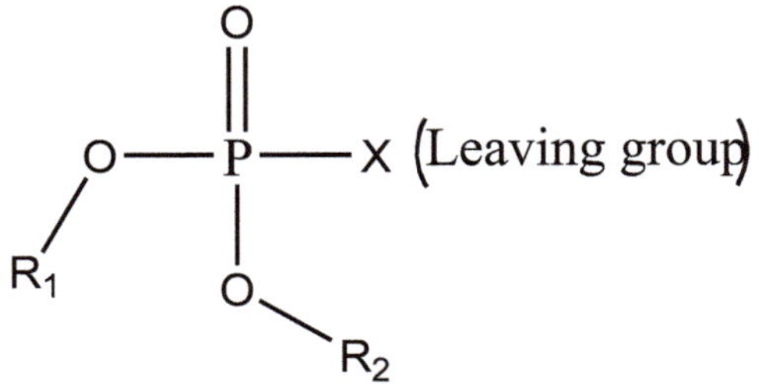

General structure of an OP compound

Classification of OP Compounds [19 - 23]

The organophosphorus compounds are hundreds in number, which are of common use. Depending upon toxicity and clinical uses, OPs are categorized into the following types:

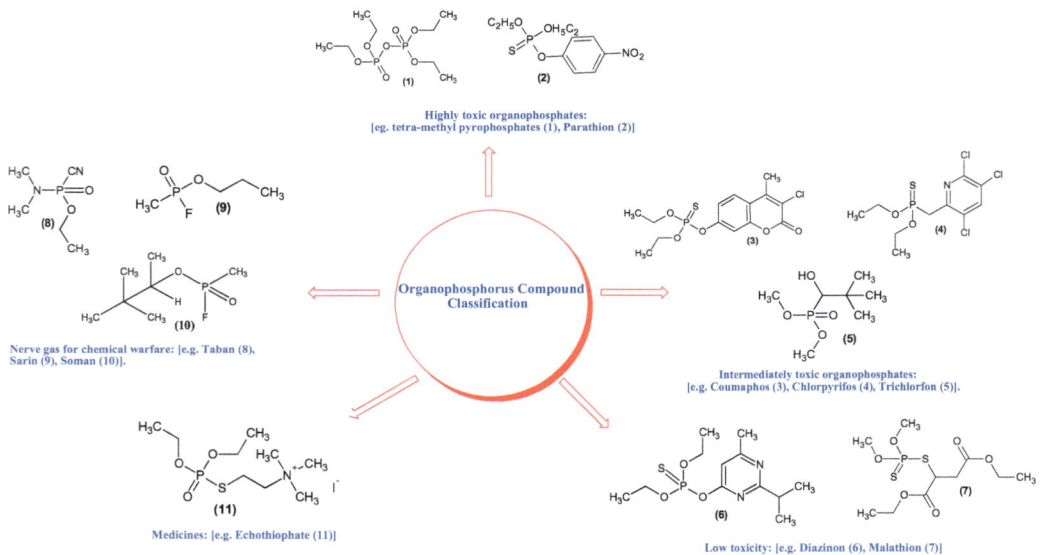

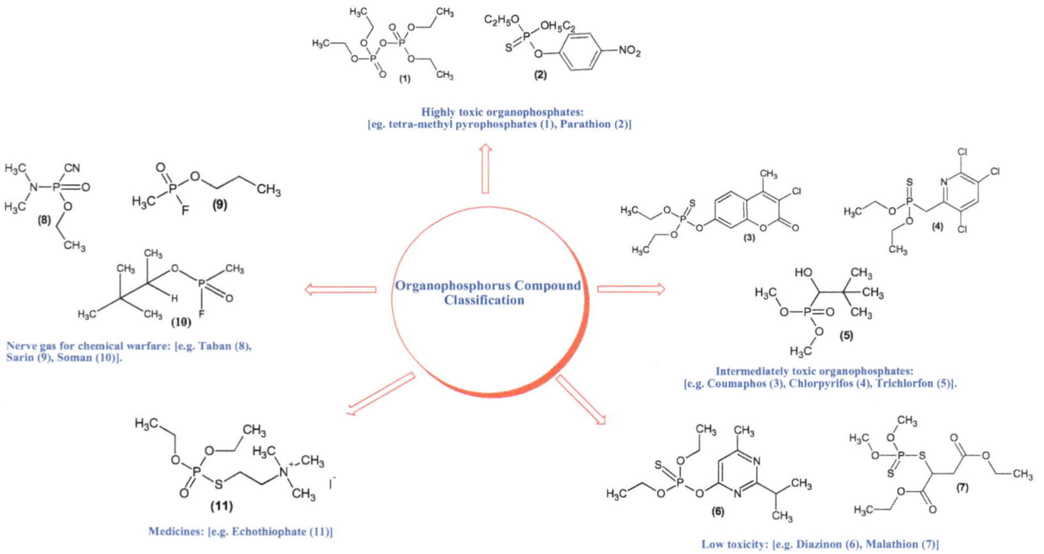

Highly toxic organophosphates: [eg. tetra-methyl pyrophosphates (1), Parathion (2)]

Organophosphorus Compound Classification

Nerve gas for chemical warfare: [e.g. Taban (8), Sarin (9), Soman (10)].

Intermediately toxic organophosphates: [e.g. Coumaphos (3), Chlorpyrifos (4), Trichlorfon (5)].

Medicines: [e.g. Echothiophate (11)]

Low toxicity: [e.g. Diazinon (6), Malathion (7)]

Acetylcholine

In the mammalian central nervous system (CNS), one of the oldest and best understood neurotransmitters is Acetylcholine (ACh). In the CNS, especially in cholinergic neurons, this ACh is synthesized, secreted in vessels and released at the nerve endings. All processes in the brain which involves gaining knowledge, thinking, remembering, judging and problem solving are carried out by cholinergic neurons which are projected from nuclei of the basal forebrain.

Whereas the body language such as gestures, motions and so on are all depended on the cholinergic neurons in the corpus-striatum.

The enzyme Choline O-acetyltransferase (ChAT) catalyses the formation ACh from choline and acetyl CoA in the cholinergic neurons of presynaptic cleft. (Scheme **1**).

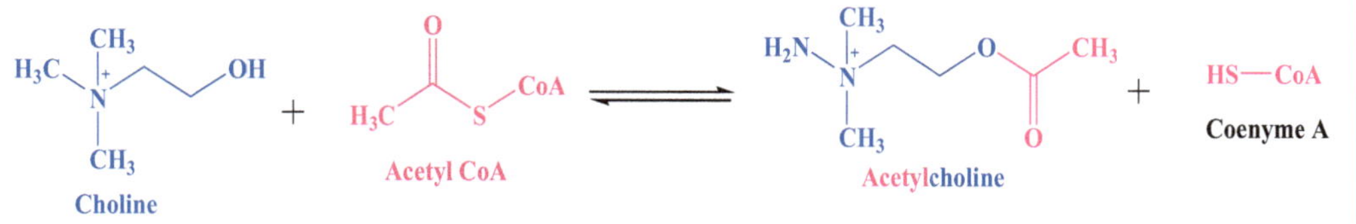

Scheme 1. Synthesis of Acetylcholine

Acetylcholinesterase

The enzyme acetylcholinesterase (acetylcholine acetylhydrolase, EC 3.1.1.7; abbreviated herein AChE), is a serine hydrolase that belongs to the esterases family within the higher eukaryotes as shown in Scheme **2**. This family acts on different types of carboxylic esters. In the cholinergic synapses of the CNS, there occurs the hydrolytic cleavage of acetylcholine molecule into choline and acetic acid, thereby terminating the impulse transmissions by the AChE [25].

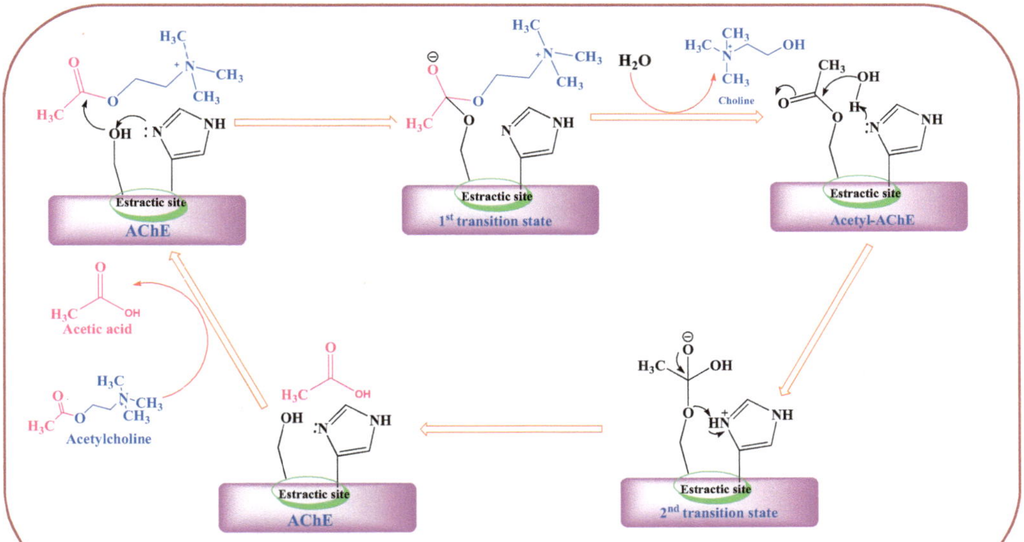

Scheme 2. Mechanism of ACh hydrolysis by AChE

The AChE which is a monomer having molecular weight approximately around 60,000 is of ellipsoidal in shape, with the size around 45 × 60 × 65 A° in which 14 alpha helices surrounds 12 stranded central mixed beta sheet [26].

The structure of AChE consists of one catalytic center with two compartments: one is esteratic subsite and another one is anionic subsite. The esteratic subsite having catalytic triad and anionic subsite consists of a positive quaternary compartment for acetylcholine.

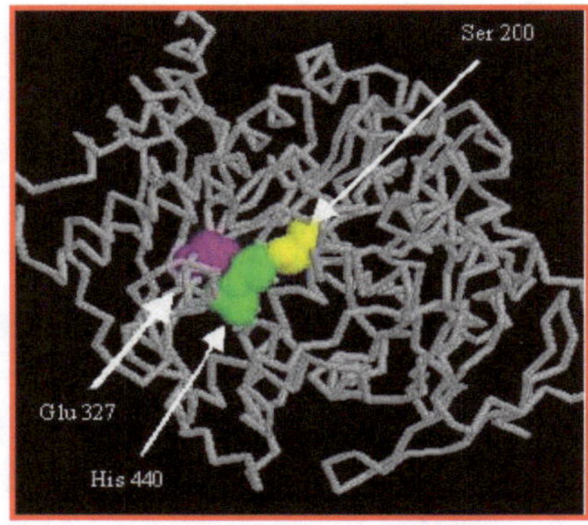

Fig. (1). Important residues in the catalytic anionic esteratic subsite [27, 28].

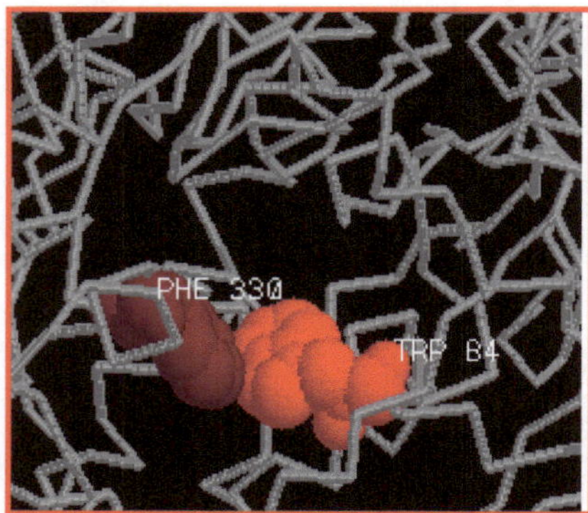

Fig. (2). The catalytic anionic subsite [27, 28].

The esteratic subsite contains the catalytic machinery of the enzyme: a catalytic triad of Ser 200, His 440, and Glu 327 as shown in Fig. (**1**). Fig. (**2**) reveals that there are Trp 84, Phe 330, and Phe 331in the anionic subsite. The vital role of the Trp residue is to direct the charged part of the substrate towards the active centre of the enzyme [26]. The x-ray crystallographic study of AChE reveals that the active catalytic site is placed deep within a gorge-like fold of the protein.

The anionic subsite is defined by Trp 84, Phe 330, and Phe 331 as shown in Fig. (**2**). Its role is to orient the charged part of the substrate that enters the active center. This is the main function of the Trp residue [26].

The recent rendition of the x-ray structure for AChE reveals the active catalytic site deep within a gorge-like fold of the protein.

Fig. (**3**) reveals that there is a 20 A° deep aromatic gorge in the protein which creeps into the middle of the enzyme. Here the active site is in turn labeled as a 'active gorge' because the active site is located at the base of this gorge only 4 A° above the base and 40% of this gorge mainly composed of 14 aromatic residue, which is highly conserved among different species of AChE [28]. Further studies also proposed the presence of hydrophobic and anionic binding sites independent of the active site. It also clearly explains the stabilization of the complex due to the presence of aromatic residues [29]. Unlike other enzymes, AChE with its electrostatic potential orients its charged substrate toward its gorge and into the active site.

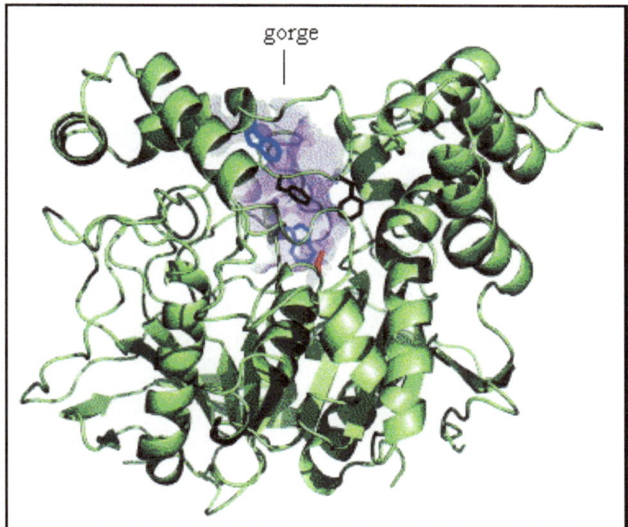

Fig. (3). In the structure of AChE, the purple colour indicates surface of the gorge in AChE [27].

The 'peripheral' anionic site is at the entry of active center gorge and consists of Asp 74 and Trp 286 residues. This comprises binding sites for various inhibitors as well as activators. The amino acid residues of 'peripheral' anionic site is the ligand binding site, in turn, act as key to the allosteric modulation of AChE catalytic activity [30].

OP Poisoning

The organophosphorus compounds are of different groups, in which they differ in their structural and toxicological features. Even though OP compounds differ in their structural and toxicological features, the inhibition of enzymes acetylcholinesterase (EC 3.1.1.7) is of irreversible mode [31 - 36]. The mechanism involved in toxification of OP compounds shown in (Scheme **3**).

In all the cases, OP compounds act as inhibitors for AChE, it should be in oxon (P=O) form rather than in P=S. If OP compounds possess P=S, then at first it will transform in O form (oxon) in the body and then exhibit the anticholinesterase action. There are mainly four reaction steps involved in theof interaction of OP oxon with AChE [37].

Mechanism of Toxicity

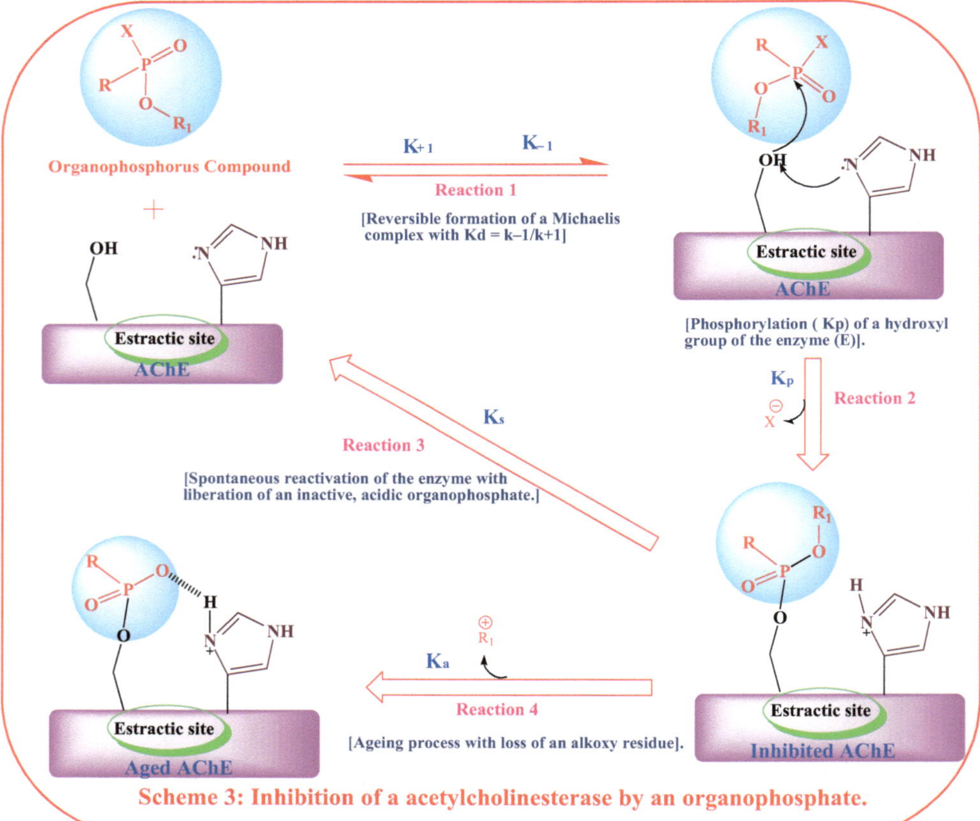

Scheme 3: Inhibition of a acetylcholinesterase by an organophosphate.

Reaction 1: In the first step, a Michaelis complex forms due to the reaction between enzyme and OP oxon. Amino acids, named serine in enzyme are phosphorylated with the loss of the leaving group.

Reaction 2: In this step, there occurs the inhibition of catalytic activity of enzyme by the formation of a strong covalent bond between OP compound and enzyme. Thereby organophosphorylation of the enzyme (esterase) takes place.

Reaction 3: Spontaneous reactivation of enzyme occurs slowly which is dependent on the nature of the attached moiety and on the enzyme protein. The rate reactivation of enzyme is depended upon the nucleophilic reagents added, such as oximes thereby acting as antidotes.

Note: Values of (k_s) differ greatly between substrate and inhibitors. For the hydrolysis of ACh by AChE, k_s is approximately 3 X 10^6 min^{-1} so that the acetyl-enzyme is rapidly deacetylated and catalytic activity is regenerated. For OP compounds and AChE, k_s is in the range of 10^{-1}-10^{-6} min^{-1}, and the spontaneous dephosphorylation of inhibited AChE to restore active enzyme is about 10^7–10^{12} fold slower than deacetylation.

Reaction 4: Ageing process occurs with the active loss of an alkoxy moiety. This can be clearly explained by the mechanism in which there occurs cleavage of one or other bond in the R-O-P chain with the loss of R group(alkoxy residues) resulting in the formation of a charged mono substituted phosphoric acid. This overall phenomenon is called as ageing. Here enzyme losses its ability to reactivate by any one of the active nucleophilic agents like oxime.

Since enzyme acetylcholinesterase (AChE) is responsible for terminating the action of neurotransmitter acetylcholine (ACh) at nerve synapse, which results in the over accumulation of acetylcholine at cholinergic synapses and stimulates both nicotinic receptors which are found at neuromuscular junctions and at sympathetic and parasympathetic nerve endings. In the central nervous system, both nicotinic as well as muscarinic receptors are overstimulated [38]). Over accumulation of acetylcholine results in death due to respiratory failure, cardiovascular collapse and/or generalized seizure [39 - 41].

Pharmacokinetics Of Organophosphorus Compounds [42–44]

Most of the organophosphates are lipid-soluble compounds, can easily get absorbed into the skin, gastrointestinal tract, respiratory tract, oral mucous membranes and in the liver and kidneys. Hence it quickly redistribute to all tissues of the body. Its effect is mainly on the CNS, as it can easily cross the blood/brain barrier because of its high lipid solubility nature. Once absorbed into the body, OP compounds undergoes oxidation in the liver with conjugation and esterase hydrolysis producing a half-life of minutes to hours. {(malaoxon [14] and paraoxon [15]} are the oxidative metabolites of malathion [12] and parathion [13] which are later hydrolyzed into inactive metabolites followed by elimination of its metabolites mainly *via* urine, bile and feaces.

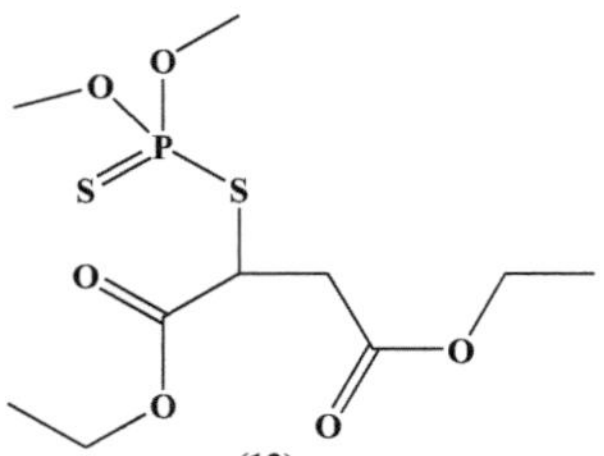

(12)
***S*-1,2-di(ethoxycarbonyl)ethyl *O,O*-dimethyl phosphorodithioate**

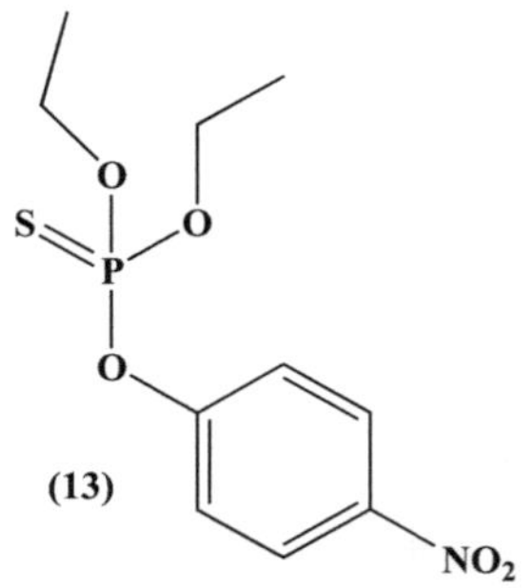

(13)
***O,O*-diethyl *O*-4-nitrophenyl phosphorothioate**

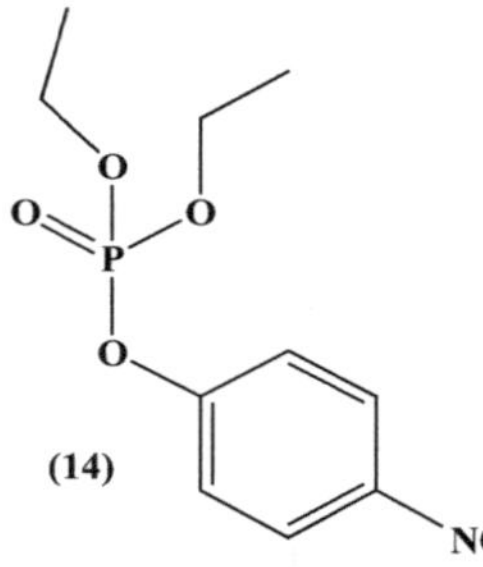

(14)
diethyl 4-nitrophenyl phosphate

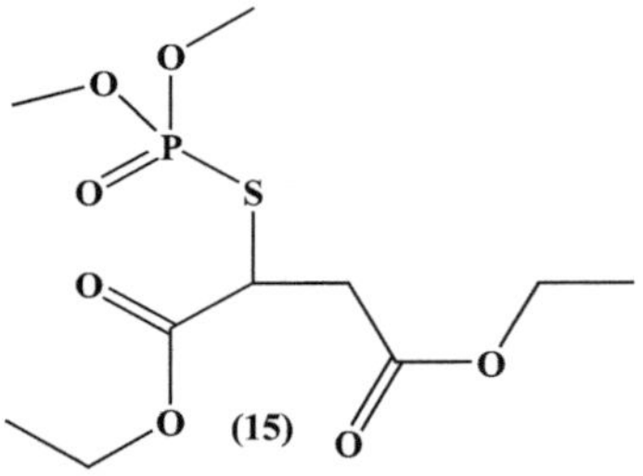

(15)
***S*-1,2-di(ethoxycarbonyl)ethyl *O,O*-dimethyl phosphorothioate**

Clinical Features of Organophosphorus Poisoning

Once exposed to organophosphorus compounds, its effects may start within 30 minutes and usually extents up to 3 hours. Sometimes in some cases, the toxic effects may be delayed depending on the extent of exposure and amount of systemic absorption. Most of the patient cases have a history of intentional ingestion of organophosphorus compounds but sometimes it might be accidental also. Once ingested, toxicity is produced by the rapid absorption of the compound through the gastrointestinal, respiratory tract and skin. The clinical symptoms and signs are non-specific and will depend on the specific agent, the quantity and the route of entry Table **1**.

Usually, patients will have the symptoms of vomiting, diarrhea and abdominal pain, whilst others may be unconscious while coming to the hospital. Primarily cases present a characteristic garlic smell with parasympathetic over-activity. But the end result may be a multi-system manifestation which hampers the gastrointestinal, respiratory, cardiovascular and nervous systems, as well as skeletal muscle. Most deaths occur within 24 hours and those who recover usually do so within 10 days [45].

Table 1. Signs and Symptoms [19, 46]

Muscarinic Receptor	Nicotinic Receptor	Central Receptor
Cardiovascular Bradycardia Hypotension **Respiratory** Rhinrrhoea Bronchorrhea Bronchospasm Cough **Gastrointestinal** Nausea/ Vomiting Increased salivation Abdominal cramps Diarrhea Fecal incontinence **Genitourinary** Urinary continence **Eyes** Blurred vision Increased lacrimation Miosis **Glands** Excessive salivation	**Cardiovascular** Tachycardia Hypertension **Musculoskeletal** Weakness Fasciculation Cramps Paralysis	**General Effects** Anxiety Restlessness Ataxia Convulsion Insomnia Dysarthria Tremors Coma Absent Reflexes Respiratory depression Circulatory collapse

Management and Treatment [19, 20, 46]

General supportive measure:

a. Termination of exposure, by removal of patient or application of gas mask; removal and destruction of contaminated clothings, copious washing of contaminated skin or mucous membrane with water, or gastric lavage.
b. Activated Charcoal 0.5-1mg/kg every 4 hours.
c. Anticholinesterase: Atropine/Glycopyrrolate.
d. Cholinesterase Reactivator: Pralidoxime.
e. Feeding –Enternal/Parental.
f. Tidal volume/vital capacity.

g. Maintenance of a patient airway, including endobronchial aspiration.
h. Artificial respiration if required.
i. Respiratory rate.
j. Administration of oxygen.
k. Allevation of persistent convulsions with diazepam.
l. Treatment of shock.
m. Arterial blood gas analysis.
n. Neck muscle weakness.
o. Occular muscle involvement *e.g.* Diplopia.

Muscarinic Antagonist Drug

Various studies are in progress to find out a potent muscarinic antagonists, but atropine (40) [16] remains the cornerstone of therapy worldwide. But the question is whether those novel drugs are capable of crossing the blood brain barrier. Because Glycopyrronium bromide [17] and hyoscine methobromide [18] do not enter the CNS, but hyoscine has excellent penetration. Atropine enters the CNS, but not to the same degree as hyoscine [47].

When atropine administered in high dose, anticholinergic delirium is observed in patients, which is the main adverse-effect of atropine. Therefore, physicians refer Glycopyrronium to treat the peripheral effects of organophosphorus with no confusion. But the drawback of its poor CNS penetration is rendering it ineffective at countering coma and reduced respiration is seen in patients with the cholinergic syndrome. A small randomized controlled trial comparing glycopyrronium with atropine noted no significant difference in mortality or ventilation rates, but it does not have sufficient control on detecting small differences between treatments [48].

Patients were successfully treated with Hyoscine but with severe extra-pyramidal features and few peripheral signs. Experiments were carried out to control seizures induced by inhaled organophosphorus nerve agents. Animal studies suggest that Hyoscine is more effective than atropine. However, extra-pyramidal effects and seizures are not common features of organophosphorus poisoning [40, 47, 49].

Until the discovery any other muscarinic antagonist which fulfill all the criteria's of Atropine such as widely available, affordable and moderately able to penetrate into the CNS, Atropine will be an antimuscarinic agent of choice for the treatment. Using different regimens of atropine, no known randomized controlled trials have compared either loading or continuation therapy. More than 30 dosing regimen have been observed, some of which would take many hours to give the full loading dose of atropine in as found in the review(2004) [47, 49].

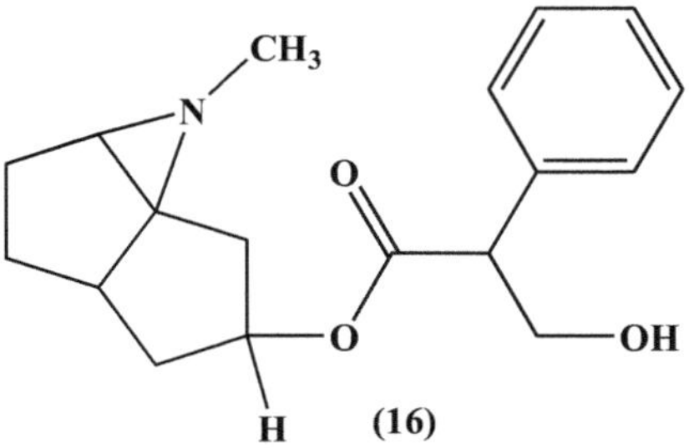

(16)

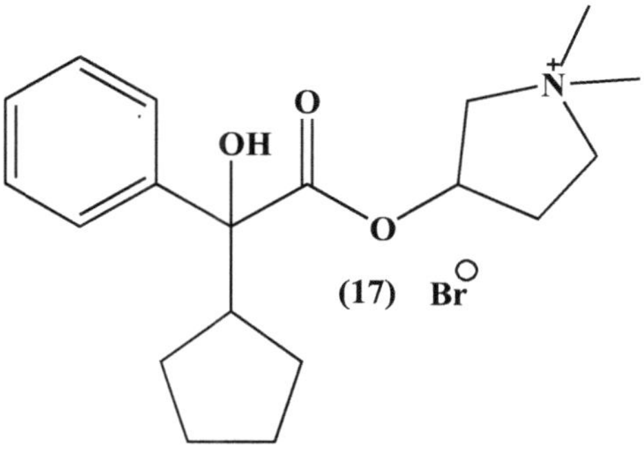

(17) Br

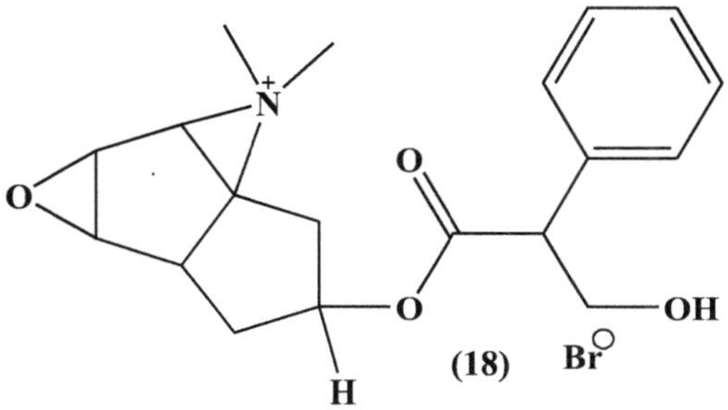

(18) Br

Oxime As Acetylcholinesterase Reactivator

Hydrolytic regeneration occurs at the phosphorylated esteratic site of AChE at a slower or negligible rate. The nucleophilic agents are hydroxylamine (NH_2OH), hydroximic acids (RCONH-OH), and oxime (RCH=NOH). Use of these nucleophilic agents reactivates the enzyme more rapidly than those spontaneous hydrolysis [19].

Mechanism of Action of Oxime

The inhibited enzyme can be restored by treating with reactivators. Reactivators are nucleophilic substances, capable of restoring the enzyme catalytic activity by cleaving the covalent bond formed during the inhibition process between OP and AChE. Still some challenges remained for future studies. The first effective reactivator discovered in 1955 was Pralidoxime, which used as a chemotherapeutic agent in treating OP intoxication and it is widely used till date [48, 50, 51]. It is to be understood that not all the OP compounds inhibit the enzyme and modifies the enzyme active site in same manner. Instead each OP residue formed after inhibition differently modifies the enzyme active site, thereby creating a selectivity of reactive molecules with the potential to bind to the enzymatic cavity [52, 53]. The reactivation of the OP inhibited enzyme occurs in two phases: In the first phase, the reactivator forms a pentacoordinate transition state with the phosphorus atom of the AChE-OP adduct. In the second phase, the enzyme function is restored, by release of the OP-reactivator conjugate. (Scheme **4**) [54, 55].

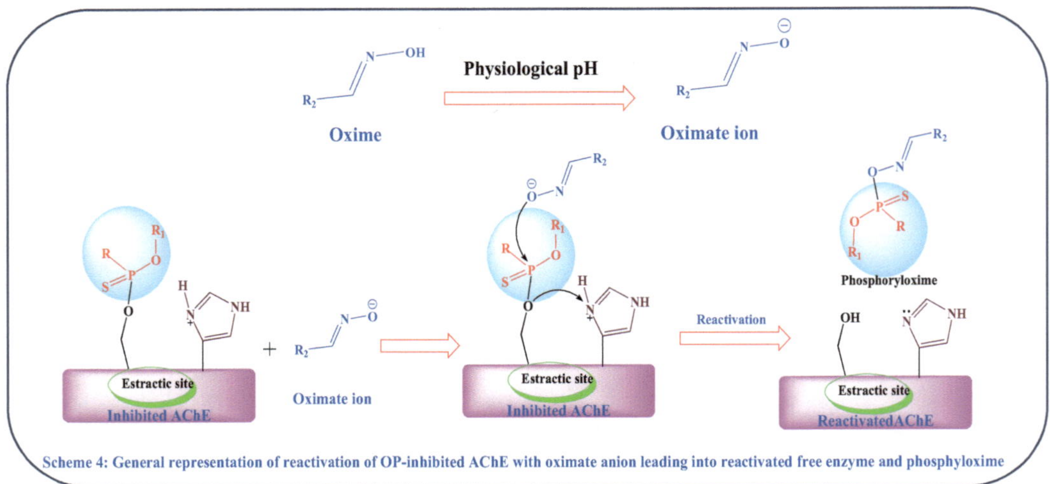

Scheme 4: General representation of reactivation of OP-inhibited AChE with oximate anion leading into reactivated free enzyme and phosphyloxime

In general, reactivators may have one, two, three, or no pyridine ring. The AChE peripheral anionic site interacts with quaternary pyridine rings, thereby stabilizing the reactivator in the cavity and thus favoring the nucleophilic attack.

The presence of quaternary nitrogen in the reactivator structure; the length and rigidity of the connection chain between them, in case of bis-pyridinium oxime; the presence of the oxime functional group; the position of the oxime group on the pyridinium ring; and the number of nucleophilic groups in the reactivator structure are some of the factors listed in the literature that may influence the reactivator efficiency [56].

Structure Activity Relationships

The affinity of the AChE reactivators toward inhibited AChE is influenced by four important structural factors.

a. A reactivator with quaternary nitrogen in the molecule is the prime known factor. Moreover, the presence of two quaternary nitrogens in a molecule has higher affinity toward the inhibited AChE as compared to monoquaternary compounds.
b. The second structural factor is the bisquaternary pyridinium reactivator that

plays an important role in connecting the chain between both pyridinium rings. There is dependence between the length of the linking chain and the nerve agent used. The length of 3 or 4 methylene groups is the ideal length of connecting chain for reactivation of VX, sarin or tabun inhibited AChE is an example for this case. On the other side, one methylene group containing methoxime seems to be the most potent reactivator of cyclosarin inhibited AChE. The other main substantial structural factor is the presence of the oxime group in the structure of the reactivator. The potential nucleophilic agents for breaking the bond between inhibitor and enzyme are hydroxyiminoacetone, hydroxamic acids, geminal dioles, and ketoximes.

c. Currently the most used group is the aldoxime group; hence it is involved in all newly synthesized AChE reactivators.
d. It is the position of the oxime group at the quaternary pyridinium ring. It is generally known that positions two and four are more suitable compared to position three. This fact is due to the difference in *pK*a between oximes in position two and four as against those in position three. The position of the oxime group is nerve agent dependent as well as the length of the connecting chain between the quaternary pyridinium rings. Reactivators with oxime in position is the best example for reactivation of cyclosarin inhibited AChE. At present, reactivators with the oxime group at the pyridinium ring in position four are the most potent for reactivation of tabun-inhibited AChE [57, 58].

Among the many classes of oxime, the monopyridinium and bispyridinium oximes are the two groups which can be divided under clinical application. Pralidoxime (PAM-2) is the currently used monopyridinium oxime [19]. On the other hand, the most significant bispyridinium oxime comprises trimedoxime (TMB-4) [20], obidoxime (LuH-6 Toxoginin) [21] and as asoxime (HI-6) [22] [59].

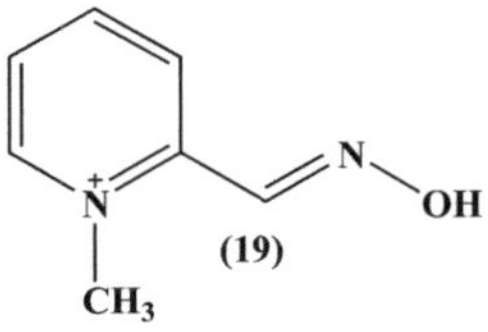

(19)

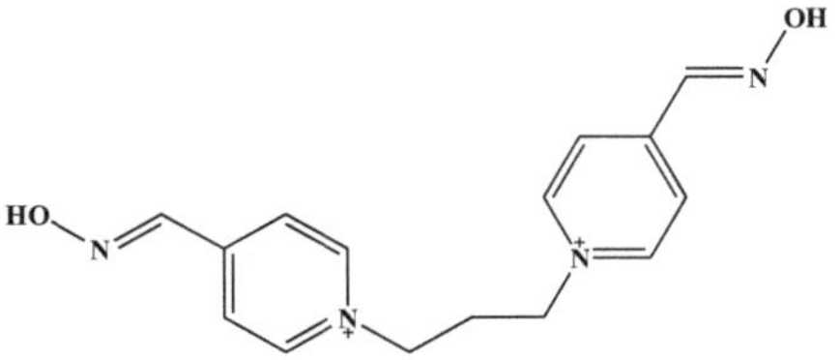

(20)

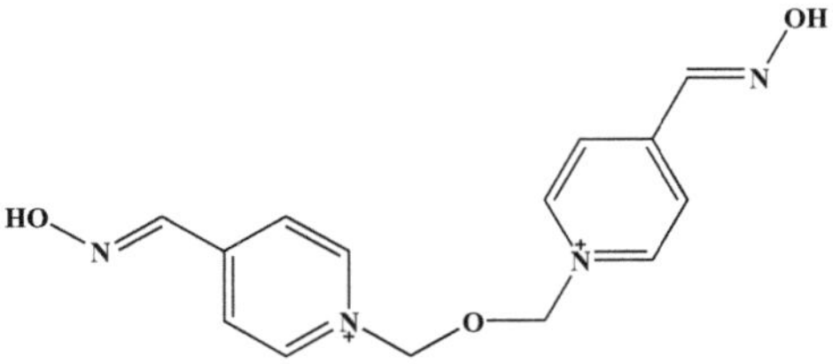

(21)

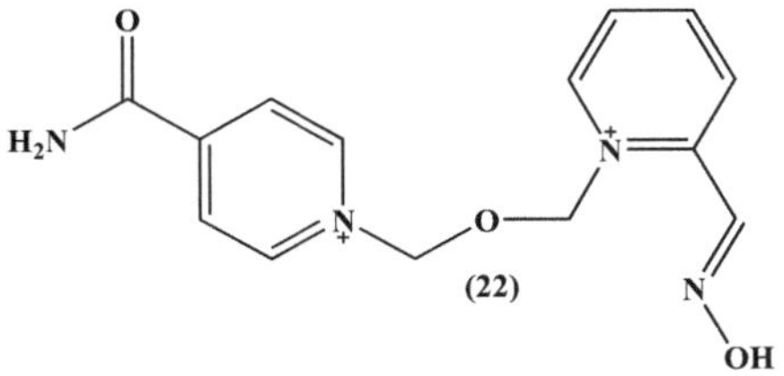

(22)

Other Therapies

Benzodiazepines [41]

Agitated delirium develops in patients poisoned with organophosphorus and the cause is complex may due to contributions from the pesticide itself, atropine toxicity, hypoxia, alcohol ingested with the poison, and medical complications. Prevention is the only mainstay of management for OP poisoning. If poisoned, pharmacotherapy is needed. Acutely agitated patients will benefit from treatment with diazepam.

For seizures, first-line therapy used is Diazepam [23, 59], but however in well oxygenated patients with pesticide poisoning, seizures are not common. Whereas seizures seem to be more common in patients poisoned with soman and tabun (organophosphorus nerve agents). It is suggested from animal studies, that diazepam reduces neural damage and prevents respiratory failure and death, but studies on humans are few.

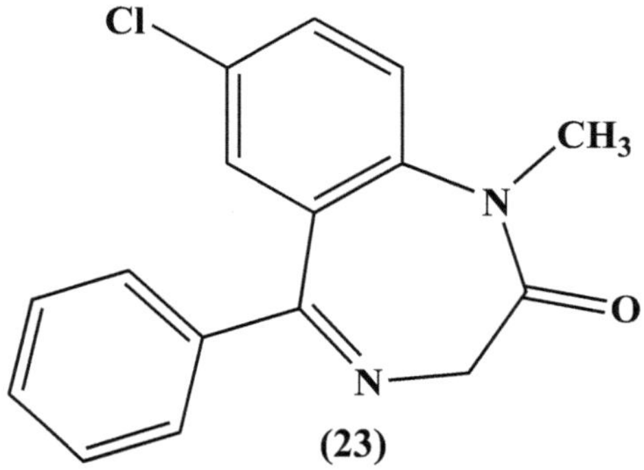

Cl
CH_3
N
O
N

(23)

Gastrointestinal Decontamination

Patients poisoned with organophosphorus when admitted to the hospital, Gastric lavage is often the first step carried out with the patient. Gastric lavage is the gastrointestinal decontamination technique for organophosphorus poisoning despite the absence of randomized controlled trials to confirm benefit. When the patient has been stabilized and treated with oxygen, atropine, and an oxime then only Gastric decontamination is done.

The extent of absorption of organophosphorus from the human bowel is not clearly known; while some pesticides show rapid absorption, occurring within minutes of ingestion followed by the rapid onset of poisoning in animals and humans. Therefore, the time window for effective lavage is probably short. There are certain guidelines for the treatment of drug for self-poisoning that lavage should be considered only if the patient arrives within 1 hour of ingesting poison. The relevance of these guidelines to organophosphorus poisoning is unclear but lavage should probably only be considered for patients who present soon after ingestion of a substantial amount of toxic pesticide who are conscious and willing to cooperate [41].

In organophosphorus pesticide poisoning, Ipecacuanha-induced emesis should not be used as patients can rapidly become unconscious, risking aspiration. The rate of absorption may increase probably, if mechanically induced emesis with large quantities of water risks pushing fluid through the pylorus and into the small bowel.

A randomized controlled trial of single and multiple doses of super activated charcoal failed to find a significant benefit of either regimen over placebo patients poisoned with pesticides. Because activated charcoal binds organophosphorus *in vitro*, the absence of effect in patients might be due to rapid absorption of pesticides into the blood. Alternatively, the ingested dose in fatal cases could be too large for the amount of charcoal given, the charcoal might be given too late, or the solvent might interfere with binding. No evidence suggests that patients with pesticide poisoning benefit from treatment with activated charcoal [41].

Others

Current therapy works through only a few mechanisms. Several new therapies have been studied but results were inconclusive. However, future research might reveal several affordable therapies working at separate sites that could complement present treatments.

Ligand-gated calcium channels is blocked by Magnesium sulphate, which results in low concentration of acetylcholine from pre-synaptic terminals, thus improves the function at neuromuscular junctions, and reduces CNS overstimulation mediated *via* NMDA receptor activation. A trial was recorded in people poisoned with organophosphorus pesticides treated with magnesium sulphate (0/11 [0%] *vs* 5/34 [14·7%]; $p<0{\cdot}01$) which resulted in reduced mortality. However, the study was small, allocation was not randomized (every fourth patient received the intervention), and the publication incompletely described the dose of magnesium sulphate used and other aspects of the methodology; therefore, these results should be interpreted with caution.

Acetylcholine synthesis and release from presynaptic terminals was also reduced the alpha2-adrenergic receptor agonist clonidine. Clonidine treatment, especially in combination with atropine showed benefit in case of animal studies, but effects on human beings are unknown.

In place of oximes, sometimes Sodium bicarbonate is used for the treatment of organophosphorus poisoning. Increases in blood pH (7.45–7.55) have been reported to improve outcome in dogs through an unknown mechanism.

In removing organophosphorus from the blood, the roles of haemodialysis and haemofiltration are not yet clear; however, there is a benefit of haemofiltration after poisoning with dichlorvos, which has poor solubility in fat, and therefore should have a relatively small volume of distribution. A systematic review of these therapies in organophosphorus poisoning is underway, but randomized controlled trials will be needed to establish good evidence-based treatment guidelines [51, 59, 60].

Organophosphorus in plasma is scavenged by butyrylcholinesterase, thereby reducing the amount of OP available to inhibit acetylcholinesterase in synapses.

Research on military now aims to inject soldiers with the enzyme before exposure to organophosphorus nerve gases. But such a prophylactic approach is not practical for self-poisoning with organophosphorus because we cannot predict when a person is going to ingest the pesticide.

The effect of scavenging organophosphorus by butyrylcholinesterase in frozen plasma is unclear. But butylcholinesterase seems to be an effective treatment for pesticide poisoning since it binds stoichiometrically to organophosphorus and will be overpowered by the amount of pesticides commonly ingested. This, if completely absorbed and transformed into the oxon, it would need an equivalent number of moles of butyrylcholinesterase for inactivation.

The use of recombinant bacterial phosphotriesterases or hydrolases in treating OP poisoning might be a better approach than the use of butylcholinesterase. Organophosphorus pesticides can be enzymatically broken down by using these proteins and thus protect animals from pesticide poisoning. Future clinical development of such enzymes could reduce blood concentrations of organophosphorus, allowing optimum activity of other treatments [41].

Design And Synthesis Of New AChE Reactivators

Since the mid-1950s, various attempts have been made to synthesize and enhance the reactivating potency and therapeutic efficacy of the oxime-based reactivators

[61 - 65]. The Food and Drug Administration (FDA) has approved pralidoxime (2-PAM) as the gold standard of oxime therapy of OP induced poisoning [66, 67]. Obidoxime, trimedoxime (TMB-4), methoxime (MMB-4) and HI-6 are the other important oximes. Generally, the oxime therapy has three major limitations:

1. **The lack of broad-spectrum oxime:** Even after 60 years of research [68, 69], not even a single oxime antidote as a ubiquitous reactivator against different types of organophosphorus compounds has been discovered [70 - 72]. For example, for toxicity of VX and sarin agents, pralidoxime is an effective antidote, but it can hardly reactivate the cyclosarin, soman or tabun inhibited AChE [73]. The determinative reason is due to the differences in the chemical structures of the mentioned OPNAs. During the reactivation process, there occurs a nucleophilic attack by the oxime. And steric hindrance is the main factor in nucleophilic attacks. From structural point of view, in case of pralidoxime, there is a pyridine ring that creates this preference to attack to the certain OP structures with lower steric hindrances. In the structures of cyclosarin, soman, and tabun, there is a cyclohexane ring, O pinacolyl, and dimethyl amine group respectively which create steric hindrance with the mentioned pyridine ring in pralidoxime. In the case of VX, sarin, tabun and diisopropyl fluorophosphate (DFP) intoxications, trimedoxime is considered an efficient reactivator, but it shows limited protection against soman [74].

In this case, it is guessed that the effectiveness of the oximes depends on their ability to make a proper orientation in the active site of the inhibited enzyme. On the other hand, the non-aged enzyme active site configuration varies depending on the type of OP. As a result, it is quite reasonable to verify that the effectiveness of a specific antidote depends on the type of poisoning agent [75]. However, some applicable solutions are suggested to address this issue (lack of broad-spectrum oxime). For instance, to prepare novel effective structures, it is required to merge two or more oximes that have complementary activities. Henceforth, a search for new other chemical compounds that have structural similarity with oximes and its SAR also is needed. This suggestion may also require some docking studies and precise predictions on the novel desirable structures.

2. **Inability to activate aged acetylcholinesterase:** As mentioned, based on the type of organophosphorus compound, secondary reaction occurs after a certain period of time with inhibited AChE which in turn leads to the formation of an anionic phosphylated serine residue termed as "aged" form of AChE. The aged conjugate of acetylcholinesterase is refractory to reactivation by traditional oximes owing to various reasons such as the repulsion between the negatively charged phosphylated serine residue, stabilizing salt-bridge formation with the catalytic histidine and oxime based antidotes [52, 76]. One of the major

challenges in the treatment process of OP-induced poisoning is aging. But various effects are in progress in the reactivation of the inhibited and aged enzyme by using uncharged oximes, other neutral nucleophiles, non-oxime reactivators such as Mannich bases and aromatic general bases [77].

3. Inefficiency in crossing the blood-brain barrier: The low permeability of antidotes into the BBB is the most important issue in the oxime therapy of OP induced poisoning, which is the main drawback of the drugs in reactivation of the inhibited enzyme in the central nervous system. The presence of permanent charge on oximes limits their ability to cross the BBB. In recent years, various efforts have been made to deal with the problems of BBB permeability of drugs [78]. Nonetheless, oximes are still the main treatment option.

Elaine da *et al.* (2015) [79] carried out the Ellman's test, nuclear magnetic resonance (NMR) and molecular docking studies to evaluate the AChE's inhibitory or reactivator potency of the synthesized compounds with a certain similarity to pralidoxime. 1-methylpyridine-2-carboxaldehyde hydrazone [24], 1-methylpyridine-2-carboxaldehyde guanylhydrazone [25], 1-pyridine-2-carboxaldehyde guanylhydrazone [26] and six other guanylhydrazones [27 - 32] obtained from different benzaldehydes were the analogs. The authors concluded that the compounds were weak AChE reactivators but relatively good AChE inhibitors, as reactivation of paraoxon-inhibited AChE with 1-methylpyridine-2-carboxaldehyde hydrazone indicates that, in general, hydrazones may be ineffective. This result suggests that for the reactivation of phosphorylated AChE, it is necessary to use better nucleophiles, especially compounds that could be easily deprotonated by the diverse basic groups of the amino acids in this enzyme.

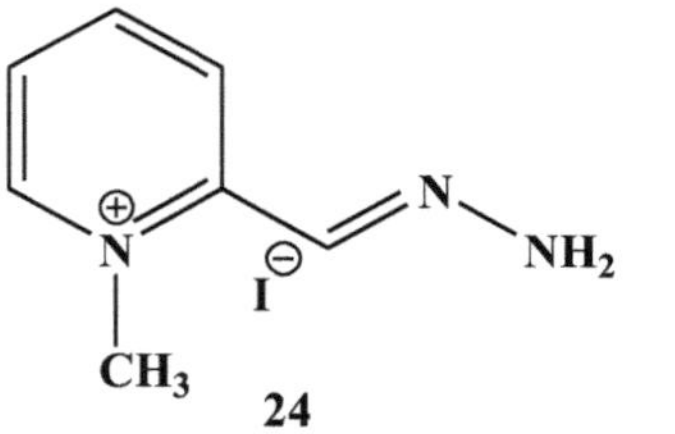

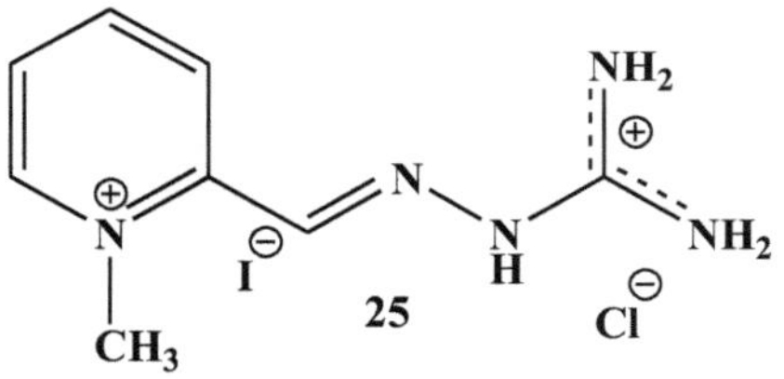

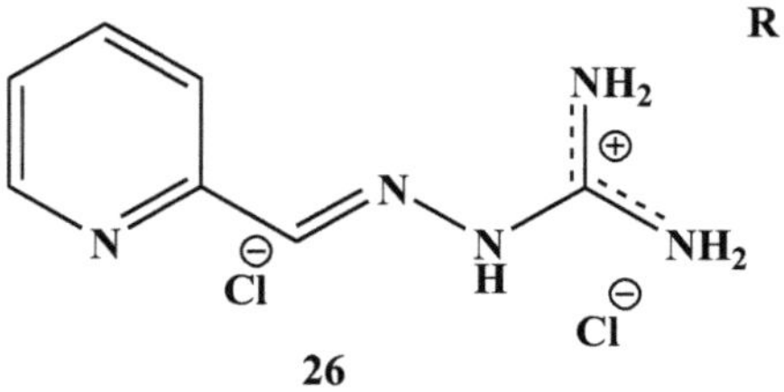

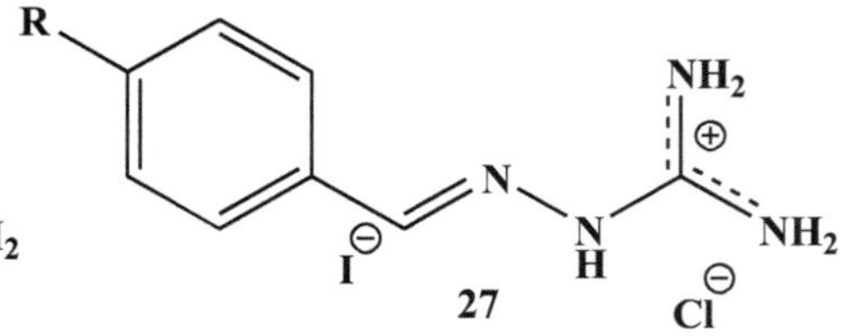

R = H (27), NO_2 (28),Cl (29), CH_3 (30), -N-$(CH_3)_2$ (31), OH (32).

The molecules and methods that can be used for applications in reactivation processes of the aged-AChE adduct were provided by the patent from Quinn and Topczewski, published in 2016 under the Pub. No.: US 2016/0151342 A1 [80]. It is suggested that the molecule is useful to counteract the intoxication caused by OP nerve agents. The oxime is capable of displacing the bound OP compound by releasing the serine residue, if oxime is administered soon after the exposure to the OP poisoned patients. However, as discussed previously, if the administration is delayed, there occurs solvolytic loss of an alkyl group from the AChE-OP adduct resulting in aging. To the aged AChE enzyme when treated with the oxime, reactivation becomes ineffective due to the stabilization of the aged adduct by diverse interactions with the enzyme active site [81]. Till date, scientists worldwide are striving to discover a new potent antidote against aged AChE-OP adducts [82]. So far, this patent brings about the discovery of a class of molecules that can reactivate an aged AChE-OP adduct, the process by which is denominated by resurrection. With the use of alkylating agents, adduct reactivation can be achieved (the method from Quinn and Topczewski). According to these new findings, along with the administration of ACh receptor antagonist and/or anti-seizure agent, an alkylating agent is also administered for treating a mammal suffering from OP intoxication. Formula I is the general formula for discovered compounds. In accordance to that formula, a series of molecules is possible to be synthesized. For the development of different oxime molecules, formula I were used, in which X be any suitable counter ion, whereas R_1 and R_2 stand for a range of chemical groups. As per pharmaceutical compositions, the Formula I based molecules can be formulated, with a posterior administration to a mammalian host, such as human beings. The compound can be administer orally or parenterally, by intramuscular, intravenous, topical, or subcutaneous routes. By making a comparison with respect to their *in vitro* and *in vivo* activity in animal models, proper dosages of compounds from Formula I can be set. For compound **33** and **34,** the patent outlines the synthesis route. In point of fact, these molecules have revealed good potential for applications in the resurrection process of the aged-AChE. The synthesis of N-methy--methoxypyridinium [**33** & **34**] can occur through the exposure of starting pyridines to trialkoxonium tetrafluoroborate (or another alkylating agent, such as methyl triflate (MeOTf)) in an appropriate solvent before or after the oxime formation. For instance, this patent brings about promising experimental results by employing Compound **33**, demonstrating its good performance for reactivation.

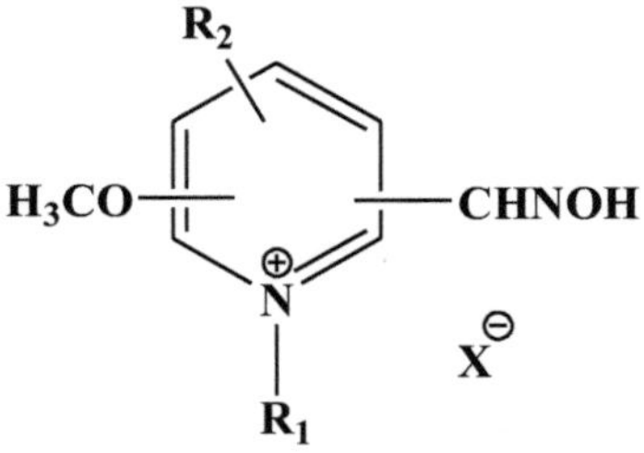

Formula I of the compound developed in this patent

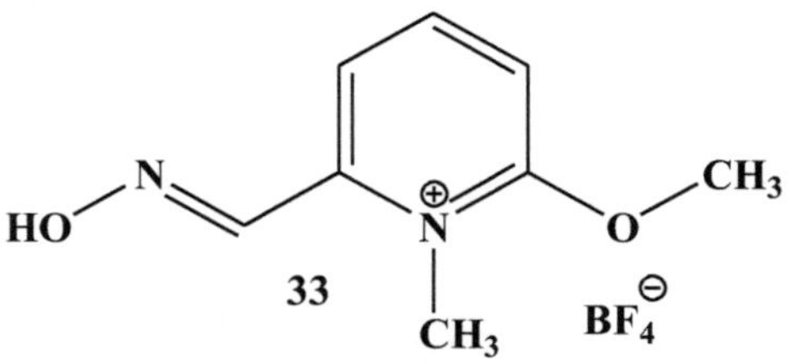

The best results were achieved in long periods of incubation, when the human AChE (HssAChE) was inhibited by exposure to an OP agent analog of Sarin, 7-(isopropyl methylphosphonyl)-4-methylumbelliferone, being incubated with Compound **33** and 2-PAM.

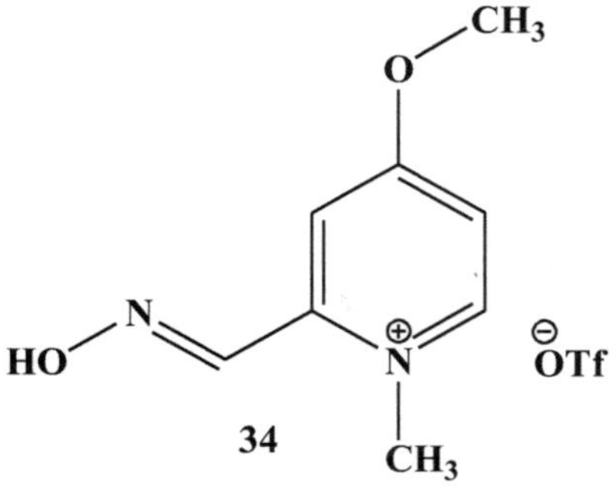

Six derivatives of chalcone-oximes were successfully synthesized and characterized on the basis of physicochemical and spectral studies by Katagi Manjunatha S *et al.* (2016) [82]. By using pralidoxime (2-PAM) as standard reference, the synthesized compound was tested for *in vitro* reactivation of chlorpyrifos or methyl parathion inhibited AChE enzyme. Compounds **35** and **36** showed promising reactivation; whereas **35** and **37** exhibited moderate reactivation against chlorpyrifos inhibited AChE. The compounds chloro substituted [**36** & **37**] have showed promising reactivation against methyl parathion as compared to standard at concentration 0.001 M.

According to Chambers *et al.* 2016 [83], the developed oximes were able to restore the inhibited AChE in the peripheral and central nervous systems by counteracting the harmful effects of OP poisoning. The inhibited BChE activity was also reactivated by some of these oximes. The serum BChE from human, guinea pig, and rat was used for the study.

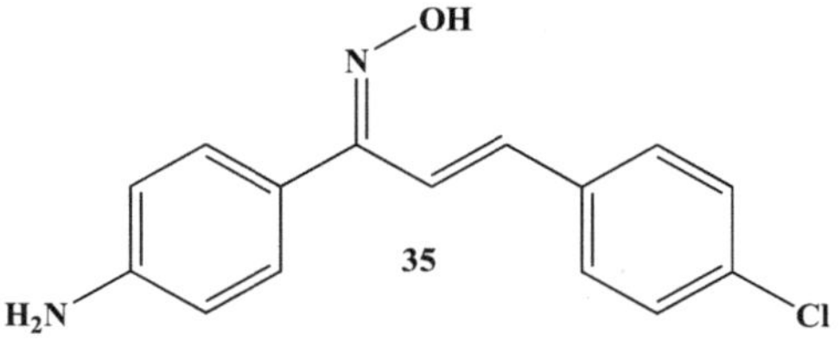

35

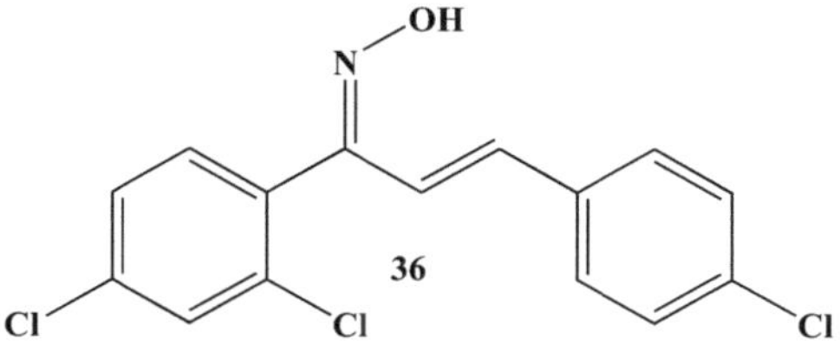

36

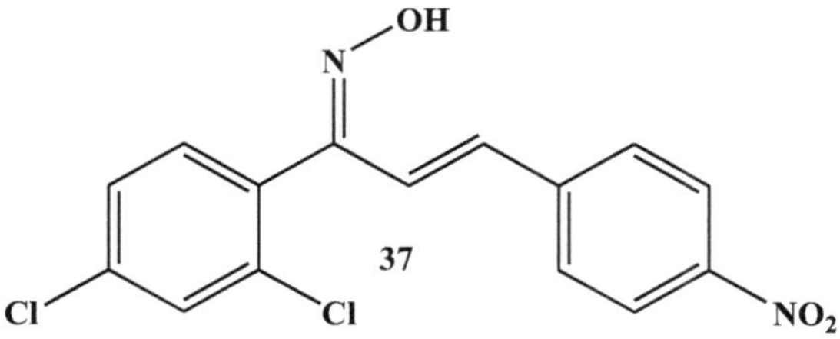

37

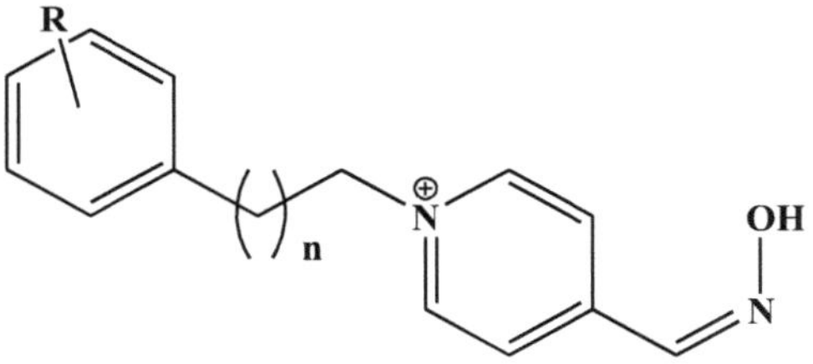

38. General structure of Phenoxyalkyl pyridinium molecules

In the structure of phenoxyalkyl pyridinium molecule [38], R = hydrogen, alkyl, alkenyl, aryl, acyl, nitro, or halo; n is an integer selected from 3, 4, or 5. The antidote for the intoxication caused by neurotoxic OP agents in military is able to come up with this investigation. OP pesticides-based poisoning and terrorist attacks against civilians can get direct benefit from this. The novel oximes exhibited significant broad-spectrum capability to reactivate both AChE and BChE after exposure to these nerve agents. The tested oxime molecules differ in the alkyl linker chain length (n) and/or the phenoxy ring substitution moiety (R).

Maja Katalini *et al.* 2017 [84], reported the synthesis of three derivatives of cinchona oxime compounds. It was proven that cinchona structure reversibly inhibits their activity, as cinchona structure fit with the cholinesterases active site. Among the three Cinchona oximes [**39**, **40**, and **41**], investigation was carried out for the synthesized derivatives of the 9-oxocinchonidine and its reactivation potency against various OP compounds such as VX, sarin, cyclosarin, tabun and paraoxon was looked into for both inhibited AChE and BChE. Compared to the known pyridinium oximes which used as antidotes in medical practice today, the tested oximes were more efficient in the reactivation of BChE and 70% enzyme activity was reactivated with reactivation rates similar to known pyridinium oximes. They showed somewhat promising results in case of BChE reactivation against various OPs but this is not the same in case of inhibited AChE. Adding to the cytotoxic effect of Cinchona oximes on two cell lines Hep G2 and SH-SY5Y were determined to know the possible limits for *in vivo* application.

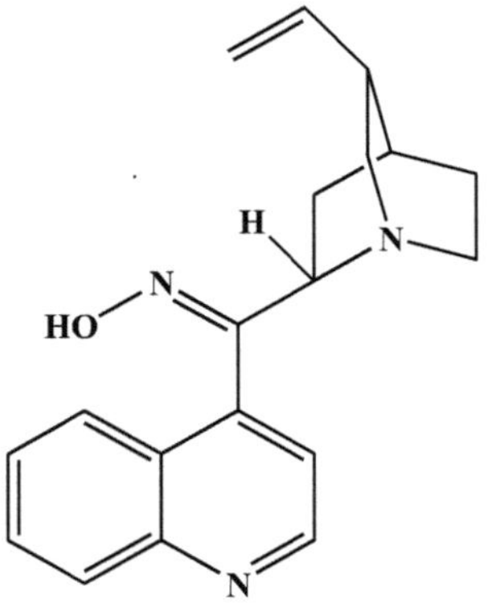

39

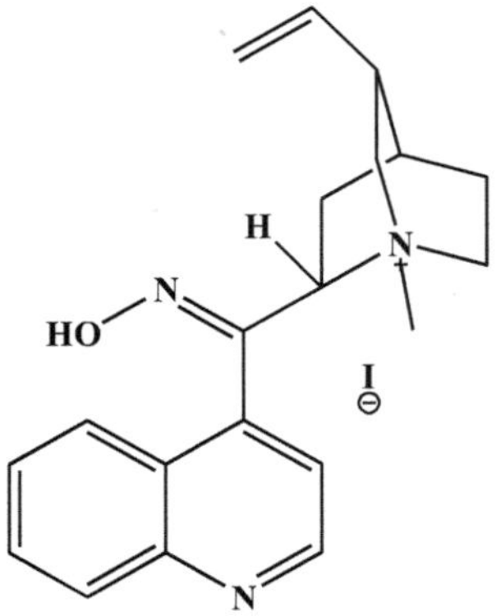

40

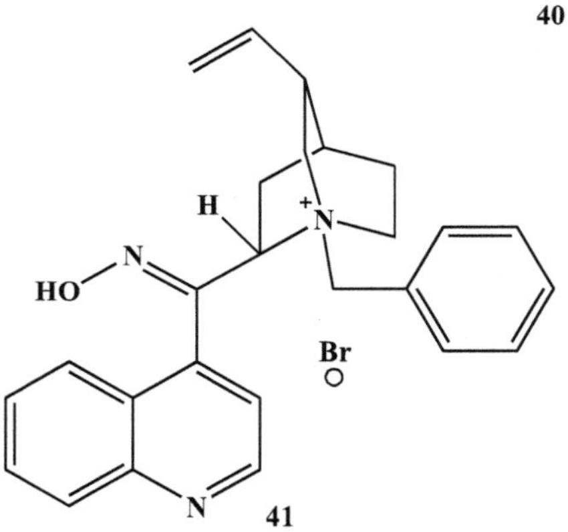

41

Chambers *et al.* (2017) [85] carried out diverse investigation through BChE with different OP agents and oximes. The paraoxon-inhibited BChE assay depicted that the novel oximes exhibited a reactivation efficacy of 33–72%, while 2-PAM exhibited a reactivation efficacy up to 32%. Except compound 43, all oximes presented better results compared to 2-PAM. The efficiency of **42** decreased significantly due to the alkyl linker chain length (from 4 to 5) in its structure. On the other hand, this difference in efficiency between **42** and **43** was smaller for the reactivation of the PIMP-inhibited BChE. Depending on the type of bound OP agent, the active site of enzyme can give rise to different conformations, thus interfering with the interaction modes of these oximes. For better performance on these oximes during the reactivation process, an enzyme should adopt an appropriate and/or correct conformation in its active site. For 2-PAM inhibited AChE, the new oximes presented an average of 46% reactivation and for PIMP-inhibited AChE, the reactivation range was 45–73%. When NEMP used as the inhibitor, the reactivation potency range was 18–62% (novel oximes), withan average of 8% (2-PAM). According to this experimental essay with NEMP as an inhibitor, **42** and **43** showed similar efficiencies. The new oximes did not confirm excellent result for phorate–oxon-inhibited BChE, with the reactivation percentage of 2-PAM being noticeably superior. Further studies regarding the interaction modes and reactivity of these oximes through BChE inhibited by different OP agents is necessary. All new oximes and 2-PAM did not show good reactivation potency when tested against BChE inhibited by NCMP and DFP. In comparison with the traditional 2-PAM, these examples exhibited a better performance for many of the developed oximes against OP inhibitors. However, the performance of each oxime may deeply shift dependent on the type of the bound OP agent.

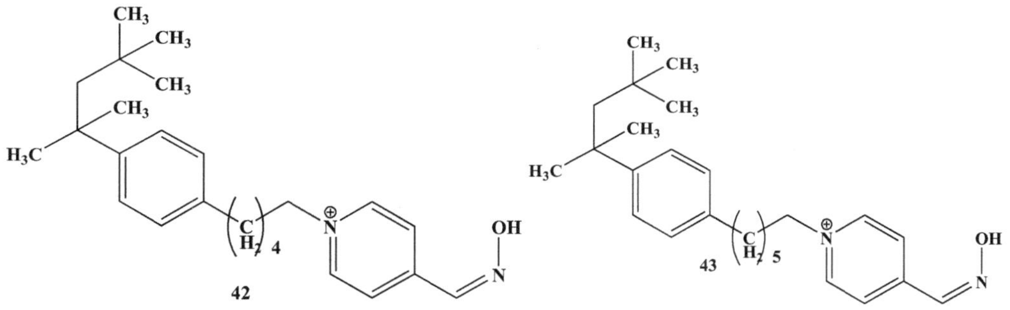

42 43

Mihail S. Tudosie *et al.* 2017 [86]. The work emphasized on the synthesis of imidazolquinuclidine-oxime and evaluating its reactivation efficacy against reactivating the phosphorylated acetylcholinesterase using the synthesized compound which in turn is able to decrease acute neurotoxic compounds toxicity.

An *in vivo* screening test was carried out to discover more potent chemical compounds, active in counteracting the nerve agent's acute toxicity. The antidotal efficiency of some compounds containing imidazolium, or quinuclidinium rings, which equimolar replace obidoxime chloride in the antidotal formula, also with atropine, was analyzed. The protective ratio was observed against an organophosphorus compound compared to obidoxime chloride, 1,3-dimethyl-2-hydroxyethyl-imidazolyliodide [44], 3-oxime-[3-(2-hidroxyimino-meth-l-1-imidazolyl-)-2oxapropyl] quinuclidin-dichloride [45], 1-methyl-quinuclidi-

-3-iodide [46]. The erythrocyte values of acetylcholinesterase are as follows: the unpoisoned and untreated study group:3,45±0,13mmol/dl; the poisoned and untreated study group: 0,89 ±0,09 mmol/dl; the poisoned and 3-oxime-[3-(--hidroxyimino-methyl-1-imidazolyl-)2oxapropyl]quinuclidin dichloride treated study group: 2,89 ±0.11 mmol/dl; the poisoned and obidoxime treated study group: 2,53±0,15 mmol/dl. With reference to obidoxime, compound **45** has shown a better protective ratio and a more prolonged continued existence time.

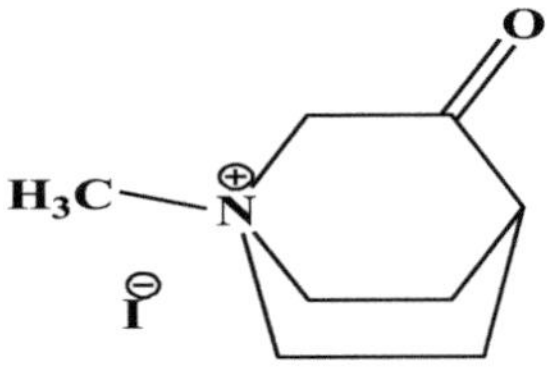

44 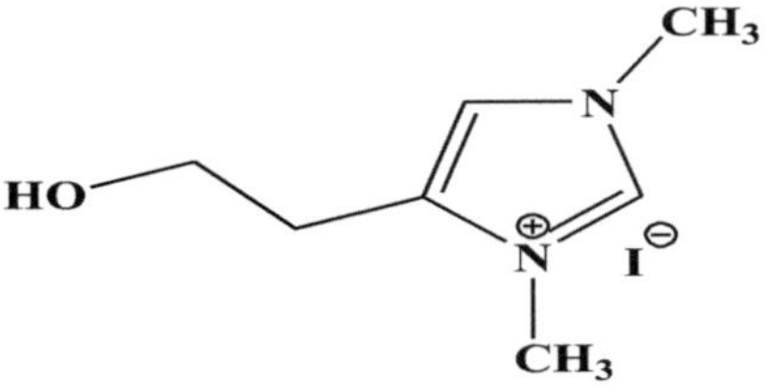 **45**

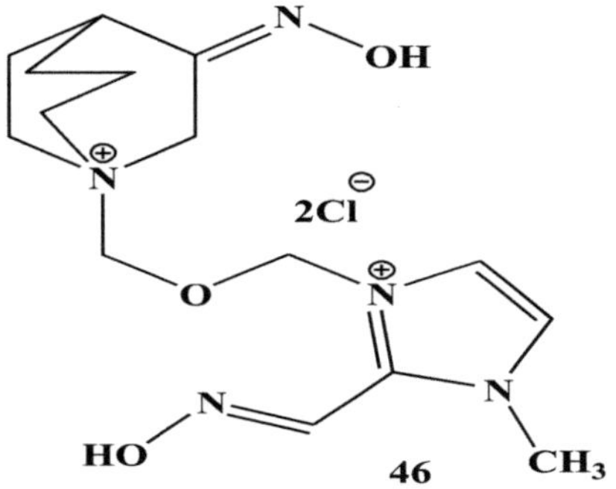

46

Tamara Zorbaz *et al.* (2018) [87]. A novel chlorinated double-charged mono-oxime reactivator was designed, synthesized and evaluated for its reactivating efficacy against human AChE inhibited by nerve agents such as cyclosarin, sarin, and VX. By comparing the standard reference compounds, novel reactivators were designed to keep the structural characteristics of their potent reference compounds but with increased lipophilicity that would possibly increase their penetration through BBB and result in higher brain concentrations. The analogous nonchlorinated oximes pKa was found to be higher than the compound under study. The human AChE inhibited by nerve agents such as cyclosarin, sarin, and VX, were efficiently reactivated by the novel compounds. The most potent was the dichlorinated analogue of oxime K027 [47] with significantly improved ability to reactivate the conjugated enzyme due to improved binding affinity and molecular recognition. Its overall reactivation of sarin-, VX-, and cyclosarin-inhibited AChE was 3-, 7-, and 8-fold higher than by K027, respectively. Its universality, PAMPA permeability, favorable acid dissociation constant coupled with its negligible cytotoxic effect, and successful ex vivo scavenge ability of nerve agents in whole human blood warrant further analysis of this compound as an antidote for organophosphorus poisoning.

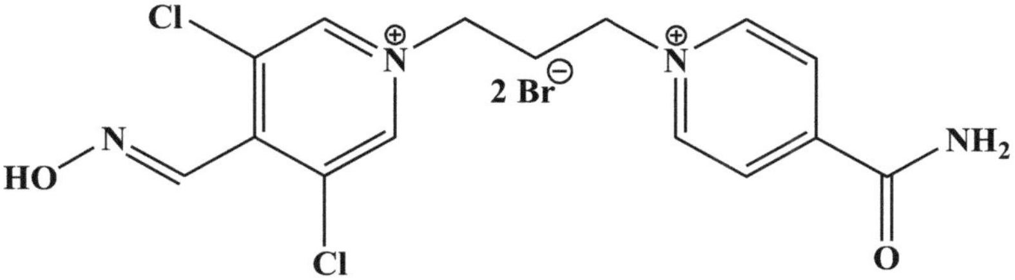

47. Dichlorinated analogues of K027

David Malinak *et al.* 2018 [88], prepared fifteen new compounds and studied them on the *in vitro* model of GA, POX, MePOX and DFP-inhibited human AChE (HssAChE). Inspired by former K-oxime compounds K048 [48] and K074 [49], they outlined their molecular design with remaining oxime part of the molecule and modified part with heteroarenium moiety. Corresponding to the results with the earlier prepared series with 3-membered linkers with analogous non-oxime moieties. The newly prepared compounds with 4-membered linkers exhibited the identical results. For proper reactivation effects in the case of various Ops, heteroarenium part of the molecule without oxime moiety plays a vital role. For interactions with aromatic amino acid residues in the PAS region, the subsequent heteroarenium ring was found to be a prime moiety. Modifications made by pyridazinium, quinolinium or isoquinolinium moieties [50 - 52] in no way led to enhanced reactivation of GA or OP pesticide inhibited HssAChE. Compared to the known and analogous bis-oxime, modification by various moieties bound in the 4-position of pyridinium scaffold showed better reactivation potency. However, some newly prepared compounds with hydrophobic (**53**-tertbutyl) or hydrophilic moieties (**54**-carbonitrile and **55**-amidoxime) were found to be promising reactivators of selected OP pesticide inhibited HssAChE. For this reason, the second heteroarenium part of the molecule that is the essential molecular fragment, and should be properly designed for enhanced reactivation of OP inhibited HssAChE. The reactivation results for tabun-inhibited HssAChE showed extensive differences, even in the group of commercial and known compounds. The reactivation potency of the novel compounds was compared with the reference standards and also with the previously known compounds which showed promising results. Some novel compounds presented enhanced ability to reactivate OP-inhibited HssAChE, while at least two of them (**53**, **54**) were able to exceed the reactivation potency of the already known compounds for POX-inhibited HssAChE. One compound (**55**) presented promising reactivation of GA-inhibited HssAChE.

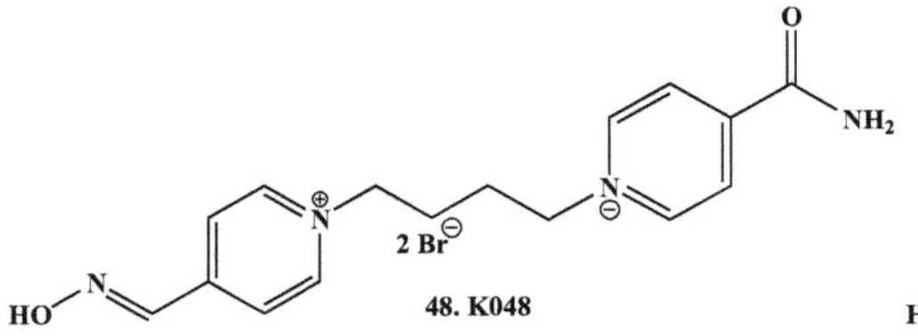

48. K048

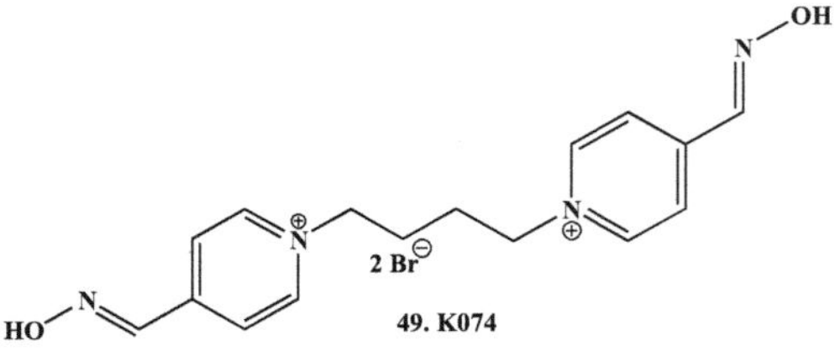

49. K074

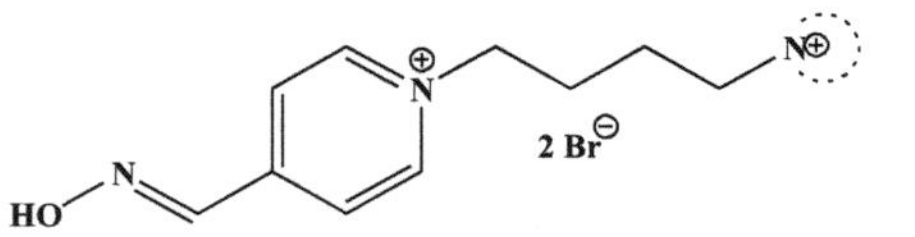

50. Pyridazinium, 51. quinolinium, 52. Isoquinolinium

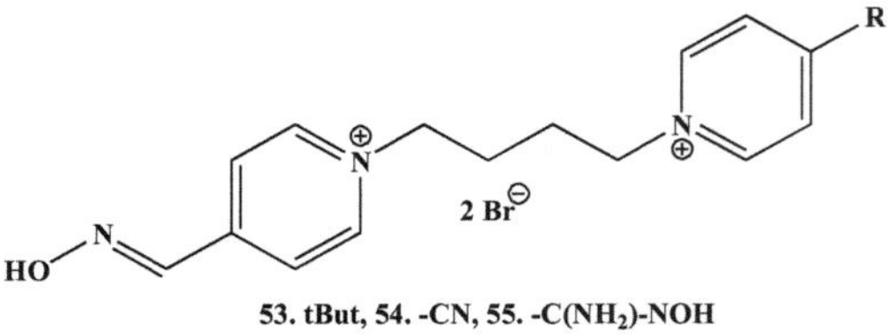

53. tBut, 54. -CN, 55. -C(NH$_2$)-NOH

Jarosław Kalisiak *et al.* 2018 [89] reported that using thioimidate intermediates, twenty two derivatives in two series of novel cyclic and acyclic amidine oximes were synthesized. The synthesized compounds were evaluated for the *in vitro* reactivation of inhibited hAChE and hBChE by nerve agent model compounds and the pesticide ETP. Compared to monoisonitrosoacetone (MINA), the amidine-oximes proved to have superior functional activity in reactivating OP-inhibited AChE and BChE. Among the novel compounds synthesized, **58** (*i.e.*, containing a cyclopropyl group) proved to be the most potent oxime reactivator of GB-AChE and exhibited slightly greater reactivation potency compared to **59** (*i.e.*, cyclobutyl group), **56** (*i.e.*, propyl group), or **57** (*i.e.*, butyl group). Usually, the reactivation potency of amidine-oxime can be enhanced by an increase in the lipophilic character of the substituent on the amidine group of the amidine-oxime, thereby the reactivation potency for Sp-GBC- or Sp-GF-SMe-inhibited BChE increased. However, along with lipophilicity, the size of the reactivator and nucleophilicity of the oxime undoubtedly prove to be an important contribution to the reactivation efficiency. As compared to 2-PAM, amidine-oximes in no case reveal greater reactivation rates for OP-inhibited AChE *in vitro* studies. However, two amidine oximes were effective in protecting mice *in vivo* from lethal doses of toxicity from a synthetic nerve agent model compound.

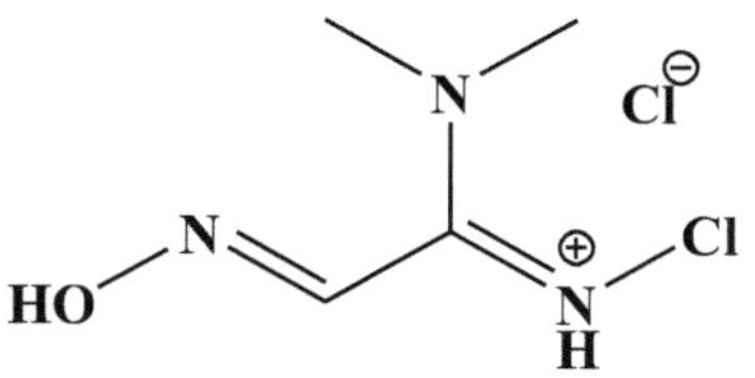

56-Propyl, 57-butyl, 58-Cyclopropyl, 59-Cyclobutyl.

In search of a new reactivator based on *in vitro* and *in vivo* rat experiments, Kamil Kuca *et al.* 2018 [90] came out with K203 (**60**) which appeared to be much more effective as a acetylcholinesterase (AChE) reactivator, in the treatment of tabun poisonings than currently fielded oximes. *In vitro* studies were carried out using human brain homogenate as the source of AChE to examine the potency of K203 as compared to commercially available oximes; pralidoxime, obidoxime and asoxime (HI-6). Reactivation studies revealed that compound **60** was the most effective reactivator with a second order kinetic constant (kr) of 2142 $min^{-1}. M^{-1}$, which was 51 times higher than that obtained for obidoxime (kr = 42 $min^{-1}. M^{-1}$). Both pralidoxime and asoxime (HI-6) failed to significantly reactivate tabun-inhibited human AChE. As reported by the author, oxime K203 appears to be the most effective reactivator of tabun-inhibited cholinesterase in several species including humans.

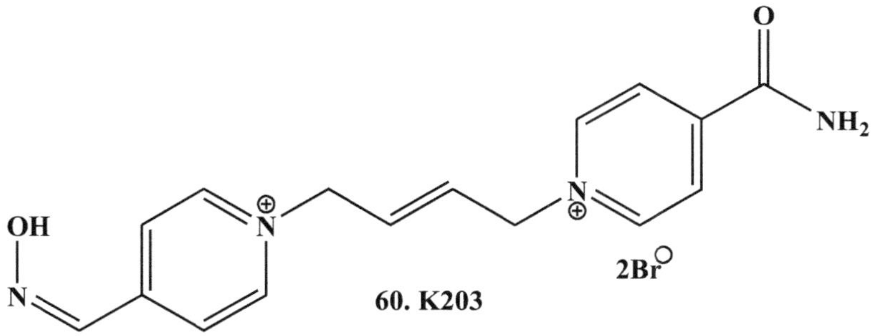

60. K203

Zhao Wei *et al.* 2018 [91], designed and synthesized 17 new tertiary salicylaldoxime and *in vitro* studies was performed to evaluated their reactivation potency against nerve agents and pesticides inhibited hAChE. In comparison to quaternary reactivators, they have advantage of greatly improved lipophilicity and increased *in vitro* reactivation efficacy of nerve OP-inhibited AChE. Conjugates with piperidine linked to the *ortho* position and methyl or halogen substituted in the *para* position of salicylaldoxime emerged as more efficient reactivators, some of them even surpassed the currently approved bis-pyridinium oximes HI-6 and obidoxime, such as L6M1R3 **(61),** L6M1R5 to L6M1R7 **(62-64)**, L4M1R3 **(65)** and L4M1R5 **(66)** and L4M1R7 **(67)**. While those conjugates containing isonicotinamide as PSL such as L3M2R2, L3M2R3, L3M2R5 and L3M2R6 were proved to be poor reactivators for most OP-inhibited hAChE. Some compounds were identified as equal or more efficient reactivators than 2-PAM, but they were less efficient than obidoxime, while HI-6 only exhibited slight reactivation ability. In most cases, the introduction of PSLs could increase oximes' binding affinity for inhibited hAChE, which in turn contribute to the reactivation efficiency. Due to their improved lipophilicity, these novel reactivators seem likely to be effective in crossing BBB and reactivating inhibited hAChE efficiently in the CNS. In future, the *in vivo* study would be conducted. With expectation by exploring an efficient centrally activating reactivators, structural transformation and modification of these non-quaternary oximes are in process.

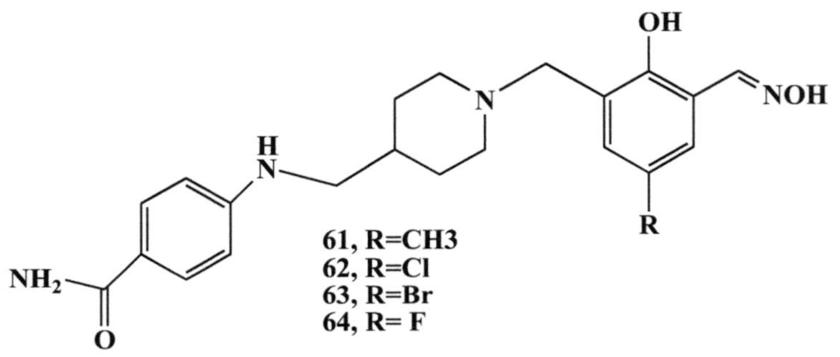

OH NOH H N N NH$_2$ O R

61, R=CH3
62, R=Cl
63, R=Br
64, R= F

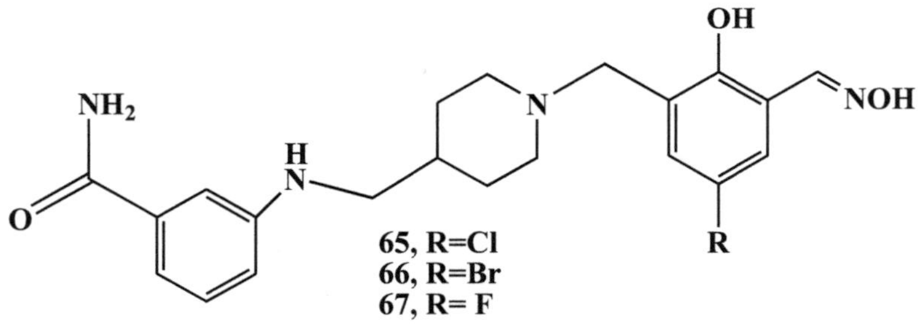

OH NOH NH$_2$ H N N O R

65, R=Cl
66, R=Br
67, R= F

Amna Iqbal *et al.* 2019 [92], outlined the synthesis of two newly developed oximes, K456 [68] and K733 [69] and their reactivation potency against paraoxon (POX)-inhibited human-RBC-AChE and human-plasma-BChE as compared with the reference molecule pralidoxime (2-PAM). *In Vitro* studies showed higher intrinsic toxicities of both oximes than 2-PAM for AChE. No substantial reactivation of hBChE was noted by tested concentration. *In silico* study predicted lower binding for both oximes contrary to 2-PAM. However, the detailed study revealed that, in contrast to 2-PAM, oximes' inability to interact with catalytic anionic site of AChE and hBChE. Both *in vitro* and *in silico* studies conclude that **68** and **69** are unlikely to be used as reactivators of paraoxon-inhibited AChE or BChE.

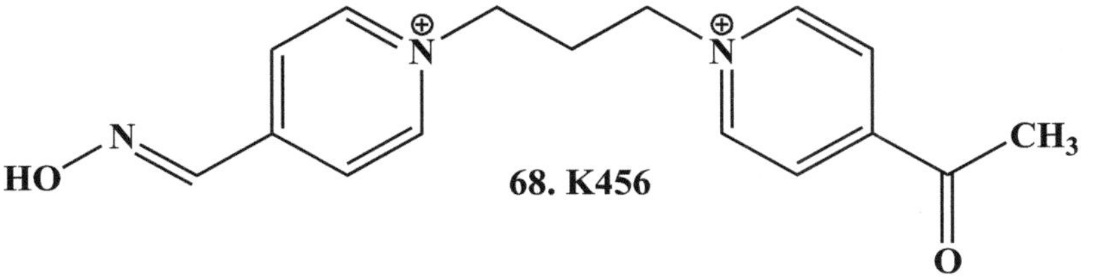

68. K456

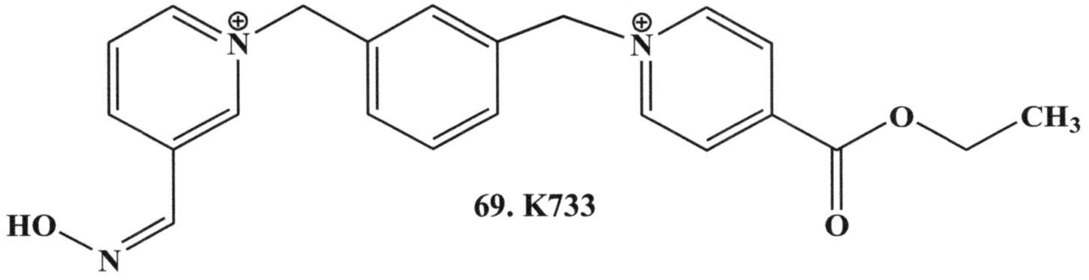

69. K733

Manjunatha S. Katagi *et al.* 2019 [93], designed and synthesized six novel schiff base oxime derivatives. The synthesized compounds were biologically evaluated to check the reactivation potency against chlorpyrifos-inhibited rat brain AChE by Ellman's method. The amino group of 4-amino acetophenone was exploited by treating with substituted benzaldehyde in the presence of glacial acetic acid to form Schiff base. The titled compounds were prepared by treating Schiff base with hydroxylamine hydrochloride in the presence of alcohol. The structure of compounds was confirmed by physical and spectral analysis. The results confirm worse reactivation observed for all the six derivatives as compared to pralidoxime (2-PAM) against chlorpyrifos-inhibited AChE at the concentration tested (0.001 M). However, it is worth notice that compounds having chloro [70] and nitro [71] substitution on the 4th position showed good activity against chlorpyrifos-inhibited AChE. Moreover, these Schiff base oximes appear to be very promising due to their sufficient reactivation strength at a lower concentration (10^{-3} M).

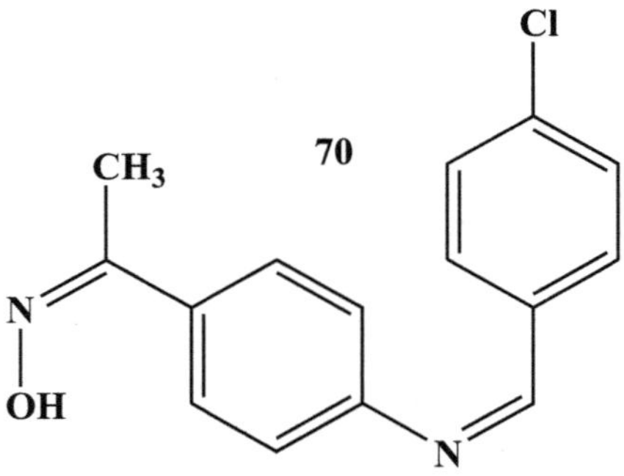

70

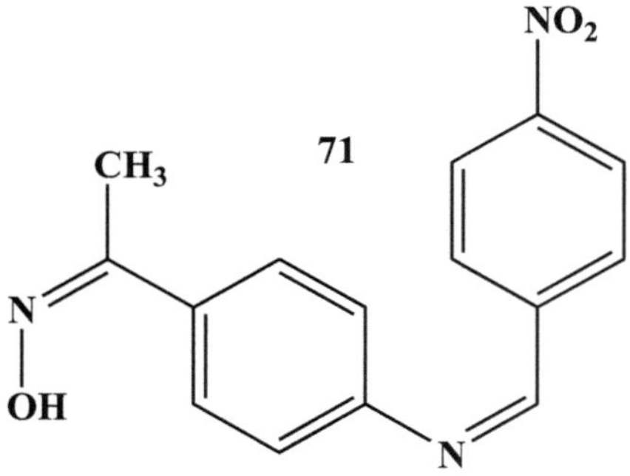

71

In the year 2019, Tamara Zorbaz *et al.* [94] apart from an earlier study, focused on the novel mono- and dichlorinated pyridinium oximes and their reactivating efficacy against OP-inhibited BChE(OP such as sarin, cyclosarin, VX and tabun). The investigation discloses that *ortho*-dichlorinated oxime K868 [72] is a very potent reactivator of human AChE. This study on reactivation of human BChE inhibited by different nerve agents validated that compound **72** is a promising novel antidote for nerve agent poisoning.

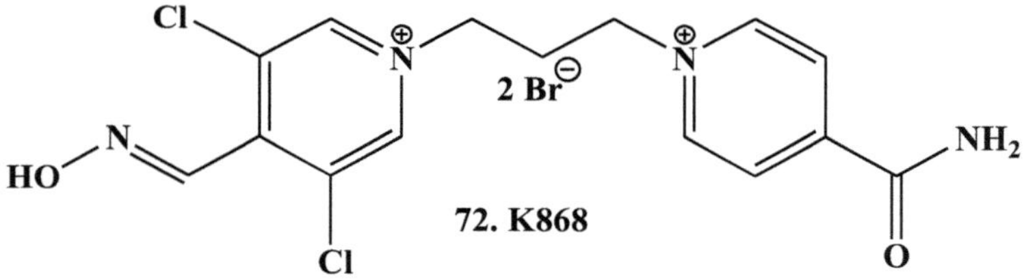

72. K868

Tamara Zorbaz *et al.* 2018 [17], put forwarded a scheme and synthesized a new series of uncharged 3-hydroxy-2-pyridine aldoxime reactivators connected through an aliphatic linker (with four to five methylene groups) to simple protonatable tertiary amines (morpholine, piperidine) that can be fused with dimethoxybenzene-(dimethoxytetrahydroisoquinoline) and substituted with N,N-dimethylaniline(1-[4-(N,N-dimethylamino)phenyl]-1,2,3,4tetrahydroisoquinoline, and studied *in vitro*, *in silico*, and *ex vivo* as reactivators of human acetylcholinesterase (hAChE) and butyrylcholinesterase (hBChE) inhibited by organophosphates (OPs such as VX, sarin, cyclosarin, tabun, and paraoxon). The reactivation rates of three oximes were determined to be appreciable than that of 2-PAM and comparable to that of HI-6, two pyridinium aldoximes. Among the evaluated compounds, the promising reactivators of OP-inhibited cholinesterases were two [**73** and **74**] morpholine-3-hydroxypyridine aldoxime conjugates.

Based on *in silico* predictions and on the *in vitro* brain membrane permeability test, oximes **73** and **74** are anticipated to be efficient in crossing the BBB. They could, therefore, potentially achieve significant concentrations both at the neuromuscular junction and in the brain. This, in turn, could result in an overall improved therapeutic outcome after OP poisoning.

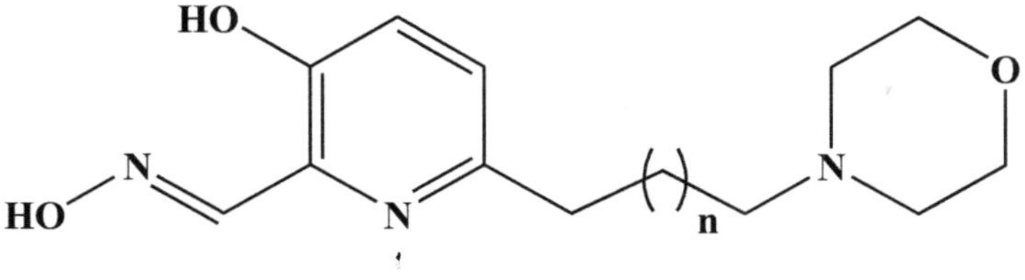

n=2, (73) n=3, (74)

Rahul Sharma *et al.* 2020 [95], designed and synthesized a series of non-quaternary and quaternary glycosylated imidazolium oximes with different alkane linkers because some natural compounds like glucose and certain amino acids such as glutamate, the anion of glutamic acid can easily cross the blood brain barrier although they are highly polar. Later, the synthesized compounds were evaluated for their *in-vitro* reactivation ability against pesticide (paraoxon-ethyl and paraoxon-methyl) inhibited-AChE and compared with standards antidote pralidoxime and obidoxime. Despite the fact that synthesized compound are not able to excel the activities of the standards, but the molecules have been found to have good lipophilicity in comparison with the standards and may be further modified in search of effective reactivator for organophosphate poisoning. Besides all the tested compounds, compound **75** with propane linker quaternary sugar oximes, showed good affinity towards OP-AChE adducts which is reflected in their lower *K*D values as compared to butane and pentane linked oximes.

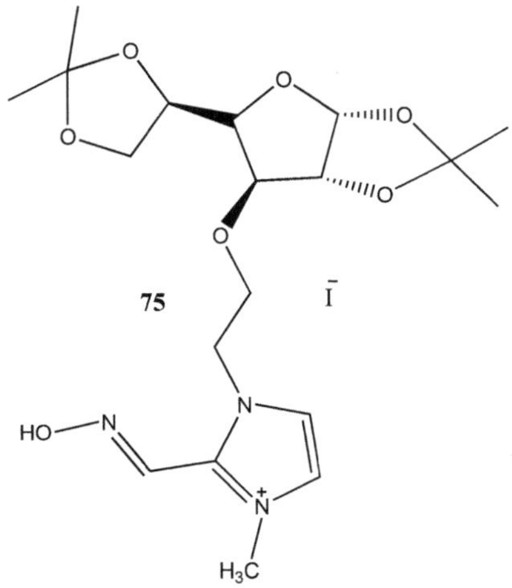

75

Antonio Zandona *et al.* 2020 [96], designed, synthesized and characterized 14 mono-oxime quinuclidinium-based compounds with alkyl or benzyl substituent and evaluated their reactivation potency against OP (VX, cyclosarin, sarin and paraoxon)inhibited human acetylcholinesterase (AChE) and butyrylcholinesterase (BChE) and assessed their reversible inhibition and also molecular docking studies were carried out to known their interactions. Among all the synthesized compounds, oxime **76** [4-bromobenzyl-3-(hydroxyimino) quinuclidinium bromide] was marked as having the highest determined the overall reactivation rate of approximately 20,000 M^{-1} min^{-1} for cyclosarin-inhibited BChE.

Furthermore, this oxime in combination with BChE exhibited a capability to act as a bioscavenger of cyclosarin, degrading within 2 h up to 100-fold excess of cyclosarin concentration over the enzyme. Molecular modeling revealed that the position of the cyclohexyl moiety conjugated with the active site serine of BChE directs the favorable positioning of the quinuclidinium ring and the bromophenyl moiety of **76**, which makes phosphorylated-serine easily accessible for the nucleophilic displacement by the oxime group of **76.**

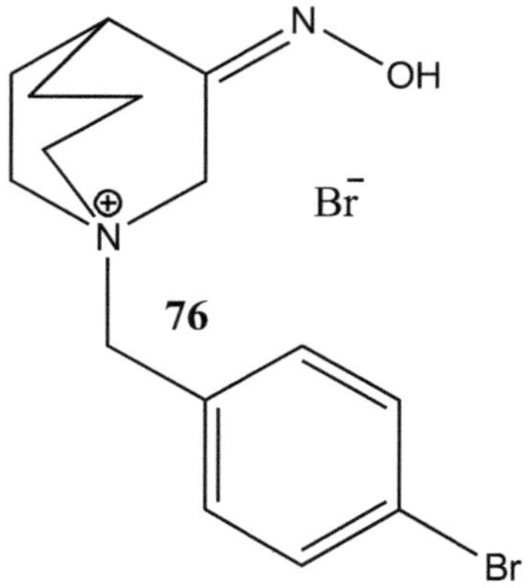

N OH Br^- N **76** Br

Manjunatha S. Katagi *et al.* 2020 [97], highlighted the synthesis of novel 2-quinolon-3-oxime by 3-acetyl-4-methoxy-1-phenyl/methyl-quinolin-2(1H)-one with different benzaldehyde, followed by oximation with hydroxylamine hydrochloride. The synthesized compounds were evaluated for their reactivation efficacy against OP inhibited AChE and the procured results were compared with standard pralidoxime. Among the tested compounds, the compound having a nitro substitution at 3rd (**77**) and 4th (**78**) positions gave potent activity against chlorpyrifos and for methyl parathion inhibited AChE, the 2-quinolon-3-oxime derivatives **77** and **78** showed promising reactivation as compared to standard inhibited AChE. Moreover, these oximes at a lower concentration (10^{-3} M) seems to be very promising reactivator.

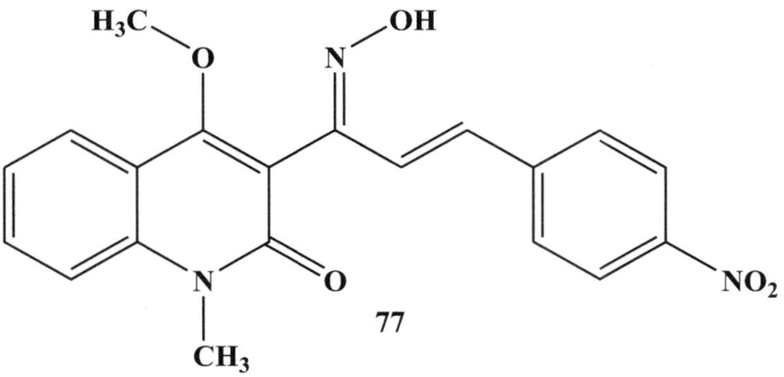

77

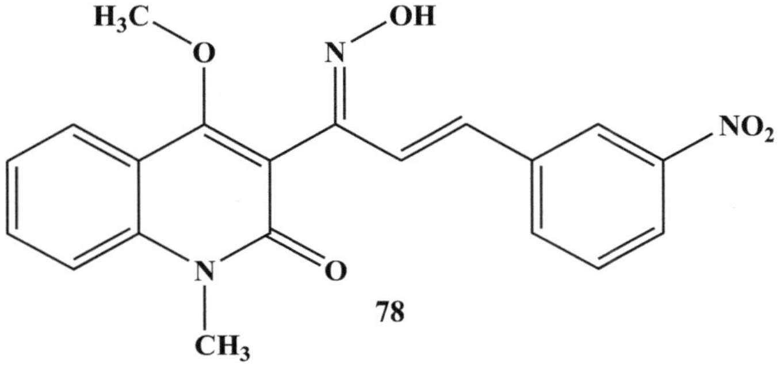

78

Jagadeesh Yerri *et al.* 2020 [98], emphasized on the successful synthesis of a chemoselective hydrogenation and semi-hydrogenation of 6-alkynyl-3-fluo-o-2-pyridinaldoximes that affords first-in-class new 6-alkyl- and 6-alkenyl-3-fluoro-2-pyridinaldoximes. Compound **79** with low molecular weight revealed higher affinity for sarin inhibited acetylcholinesterase (*h*-AChE), and greater reactivation potency or resurrection for sarin-inhibited *h*-AChE, compared to 2-pyridinaldoxime (2-PAM) and HI-6. As compared to 2-PAM, HI-6 and obidoxime, the uncharged 3-fluorinated bifunctional hybrid **79** exhibited higher *in vitro* brain blood-barrier permeability and surpassed the efficiency of both 2-PAM and HI-6. This unique synthetic strategy allows the expedient and efficient synthesis of new fluorinated hybrid reactivators for biological profiling and improved therapeutic window compared to medically approved pyridinium aldoximes.

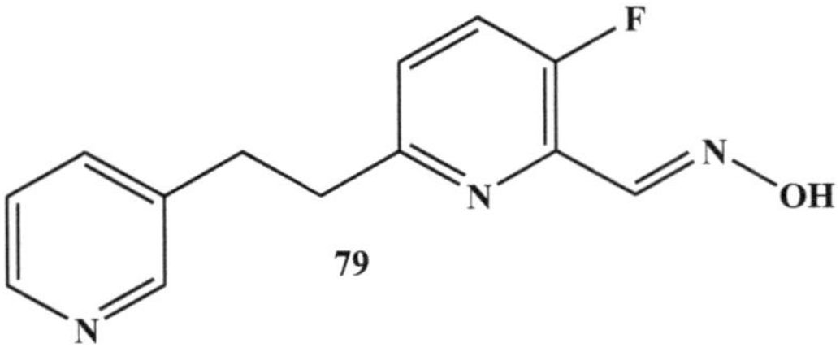

79

Hyun Myung Lee *et al.* 2020 [99], emphasized on the design, synthesis and detailed *in vitro*/*in silico* evaluation of charged pyridinium-2-carbaldoximes with quinolinium carboxamide moiety on human recombinant AChE (hrAChE) and human recombinant BChE (hrBChE) inhibited by nerve agent surrogates 4-nitrophenyl isopropyl methylphosphonate (NIMP, sarin surrogate; >95%), 4-nitrophenyl ethyl methylphosphonate (NEMP, VX surrogate, >95%), and 4-nitrophenyl ethyl dimethyl phosphoramidate (NEDPA, tabun surrogate, >95%) and pesticide metabolite paraoxon. The novel oximes were designed from formerly effective and simultaneous AChE/BChE reactivators K117 or K127 with changed structural motif for binding in cholinesterase active site. The compound **80** (KR-26352) was able to reactivate inhibited hrAChE by above mentioned organophosphates and two novel compounds **80** and **81** (KR-26354) were able to reactivate NIMP/NEMP-hrBChE. The reactivation kinetics revealed that compound **80** proved to be an outstanding reactivator of paraoxon-hrAChE compared to obidoxime and revealed increased reactivation potency against NIMP/NEMP-hrBChE.

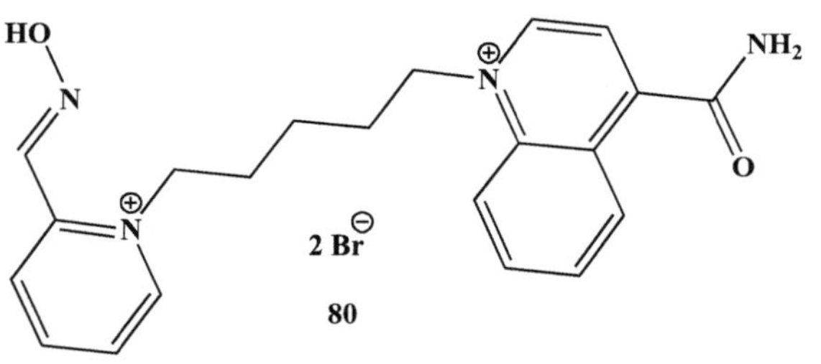

HO N N NH_2 O N 2 Br

80

HO N N N 2 Br NH_2 O

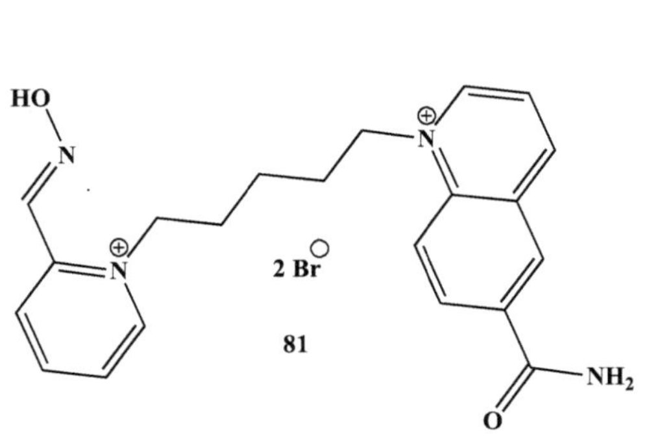

81

Lukas Gorecki *et al.* 2020 [100], highlighted the design, synthesis, and *in vitro* characterization of seven uncharged heterocyclic bisoxime reactivators with a conceptually novel scaffold as promising prototypes for the creation of a next generation of adaptable, accelerated CNS-active antidotes against OP intoxication. Seven novel compounds such as two piperidine, three piperazine, and two homopiperazine doubly substituted alkyl acetamido oximes were synthesized by computational docking analysis and the compounds synthesized in 20–150 mg quantities. In reactivation of sarin-, cyclosarin-, VX-, and paraoxon-inhibited hAChE, the designed derivatives exceeded the efficacy of RS194B. To develop a structural template for the design of accelerated uncharged antidotes of OP-conjugated hAChE, the author solved X-ray crystal structures of RS194B, one of the currently most promising uncharged oxime reactivators, in complex with apo-hAChE and with VX-hAChE conjugate. For one of the representative bis-oximes, LG-703 [82], the structure was confirmed by solving the X-ray structure, that one of its oxime groups bound 5 Å close and pointing toward Ser-203, in an orientation productive for reactivation. By solving an X-ray structure of the bis-oxime **82**, in complex with hAChE, we confirmed that improved reactivation efficiency was consistent with the productive orientation of one of the two nucleophilically reactive aldoxime groups. Reactive aldoxime-bearing "arms" of **82** assumed an asymmetric conformation in the complex, despite two-dimensional symmetry of the LG-703 molecule.

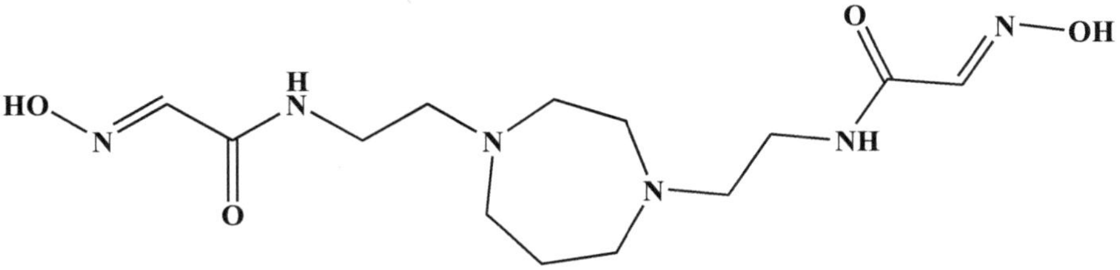

82

Daniel A. S. Kitagawa *et al.* 2021 [101], designed and synthesized five novel isatin derivatives, linked to a pyridinium 4-oxime moiety by an alkyl chain with improved calculated properties, and assessed their reactivation efficacy against paraoxon (PXN) and (O-(4-nitrophenyl) O-ethyl methyl phosphonate), a VX surrogate (NEMP)-inhibited acetylcholinesterase in comparison to the standard antidote pralidoxime. Theoretical and experimental investigations revealed that the isatin derivatives **83** and **85** proved to be promising candidates to address AChE inhibition caused by PXN and NEMP, respectively. Promising results were shown by the derivative **84** in *in vitro* studies for both OP at both reactivator concentrations used, being better than 2-PAM for PXN and comparable for VX.

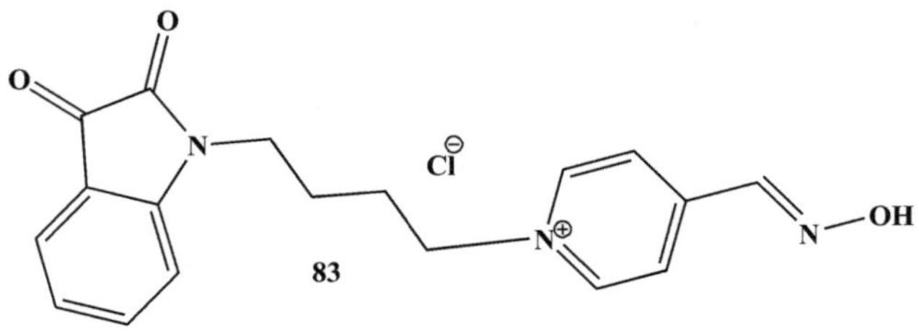

83

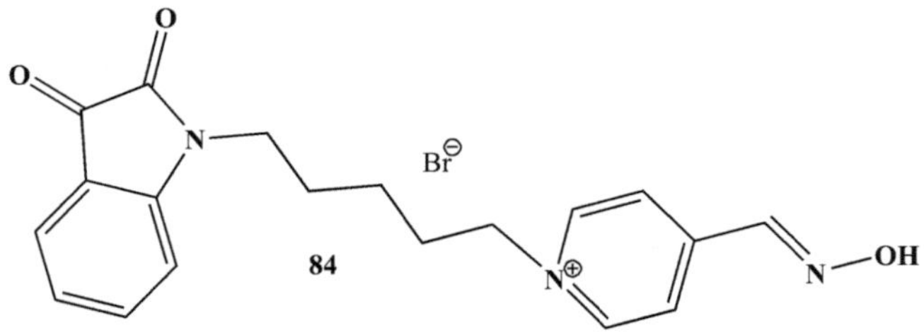

84

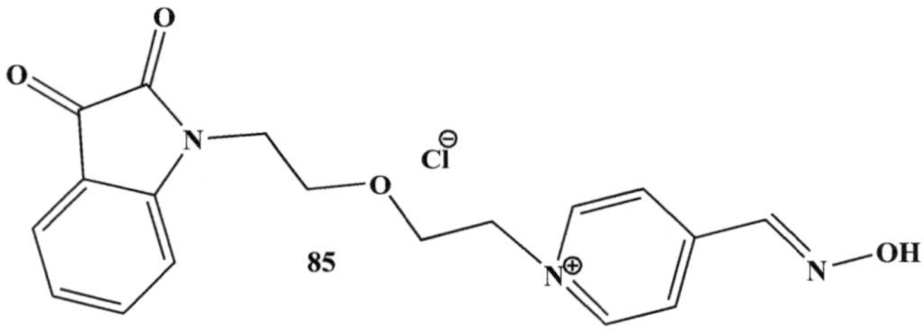

85

Gabriel Amitai *et al.* 2021 [102], reported two high-throughput screening (HTS) campaigns with >150,000 small molecules, each performed using fluorogenic O-ethyl methylphosphonyl O-4-methyl-3-cyano-coumarin (EMPMeCyC) as a surrogate of OP nerve agents (OPNAs) for the discovery of new OPNA detoxifiers and OPNABChE reactivators. Four new oximes were synthesized for the detoxification of VX and Sarin, and the reactivation of VX and Sarin-inhibited BChE. The synthesized oximes having distinctive chemical structures in which benzaldoxime (PCM-0211955) (**86**) and benzamidoxime (PCM-0211088) (**87**) were devoid of any nitrogen atom in the oxime-carrying aromatic ring, and also included a pyridine amidoxime (PCM-0212399) (**88**) and a dialkyl ketoxime (PCM-0211338) (**89**). Compared to the standard quaternary 2-PAM, nonquaternary oximes which were synthesized exhibited faster reactivation efficacy against VX- and Sarin-inhibited BChE. The reactivation reaction rates of these novel molecules were indistinguishable to those observed with known bis-quaternary reactivators for VX-inhibited BChE and faster than mono-quaternary pyridinium oximes.

Preliminary toxicological studies demonstrated that the newly discovered non-quaternary oximes were relatively non-toxic in mice. The authors inferred that non-quaternary oximes open the door to the design of novel therapeutics and decontamination agents following OPNA exposure. In particular, authors discovered the first ketoxime reactivator of OP-ChEs and detoxifier of OPNAs.

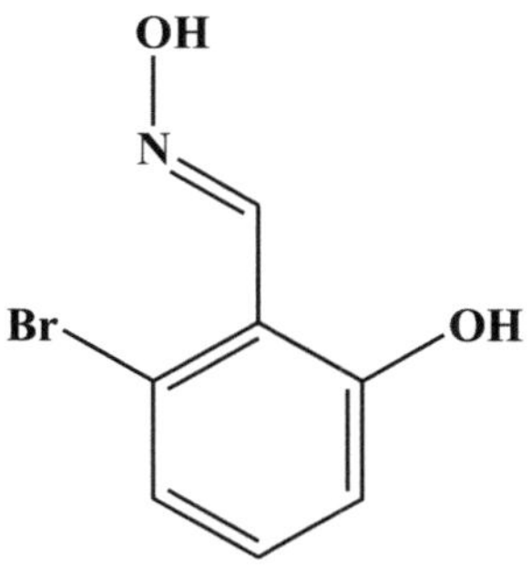

86 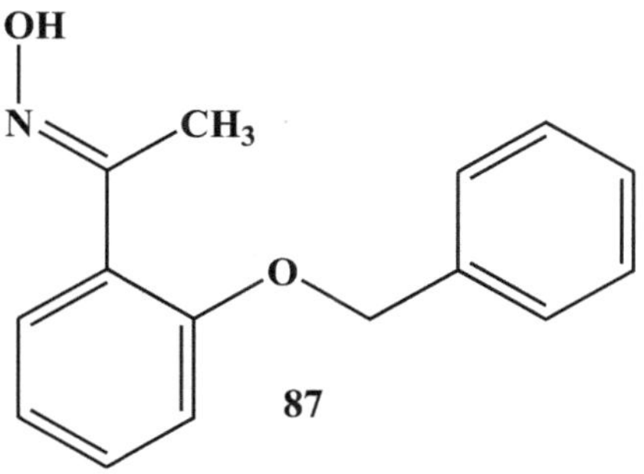 **87**

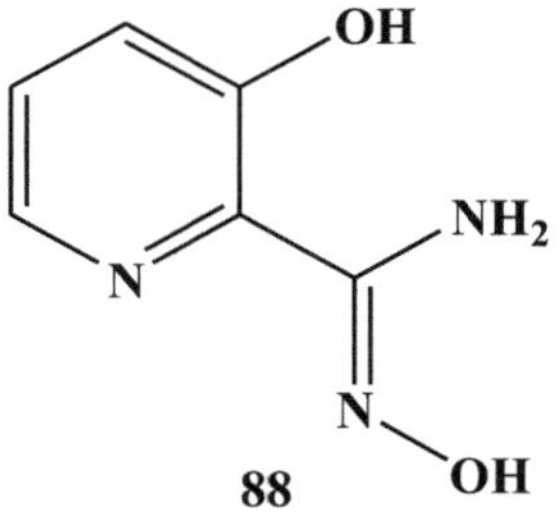

88 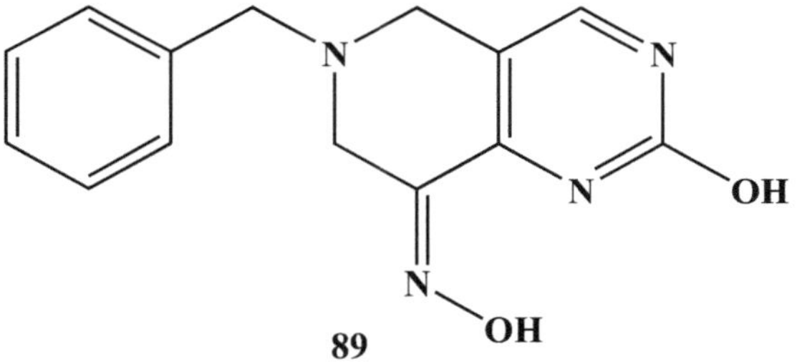 **89**

R A Mohamed *et al.* 2021 [103], described a novel approach that is In silico approach in discovering a new AChE reactivator against malathion- inhibited AChE. With reference to it, fourteen potential compounds were selected to assess their binding energies and nucleophilic attack distances with Malathion inhibited AChE complexes to govern their antidote potentiality. 2-PAM were used as a reference standard. A bioinformatics tool, YASARA was used to dock malathion-inhibited human AChE and reactivator-malathion inhibited AChE complexations. The Insilco study presented that the potential reactivators can be selected based on their binding energy, bound position and the distance of the nucleophilic attack of the reactivators towards the inhibited AChE. 4-hydroxybenzohydrazide which has a good binding energy value with malathion inhibited-AChE as well as shorter nucleophilic attacks distance could be an alternative antidote for malathion AChE poisonings. This *In silico* screening method is helpful in narrowing down the selection of the potential reactivator. The promising results of 4-

hydroxybenzohydrazide (**90**) shown in this preliminary study, can be tested experimentally in future.

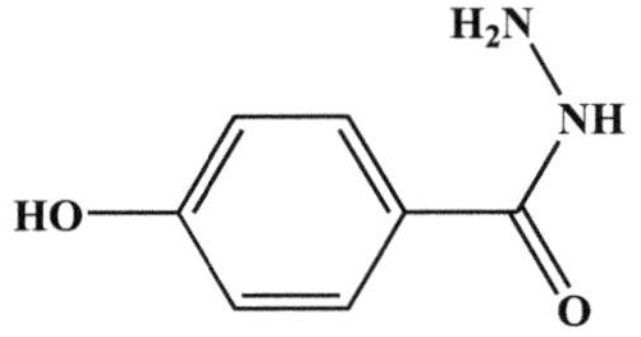

90

Jennifer Fernandes *et al.* 2021 [104], emphasized on the synthesis, characterization and *in vitro* reactivation of chlorpyrifos inhibited acetylcholinesterase by pyrazole-oxime derivatives. By adopting the standard procedure, five derivatives of (*E*)-1-(phenyl)-3-(4-(dimethylamino) phenyl prop-2-en-1-ones were synthesized from substituted acetophenone and aldehydes. By treating with hydrazine hydrate in glacial acetic acid under the reflux, the intermediates 1-(5-(Phenyl)-3-(4-(dimethylamino)phenyl)-4,5-dihydropyrazol-1-yl)ethanones were synthesized.

Along with the condensation of hydroxylamine hydrochloride in ethanol and pyridine as a base, the title compounds were synthesized. By adopting Ellman's method, all the synthesized compounds were tested for their intoxification against chlorpyrifos inhibited rat brain AChE. Among all the synthesized compounds under test, only two compounds of pyrazole oxime, **91** and **92** were found to reactivate the chlorpyrifos inhibited rat brain AChE as compared to standard.

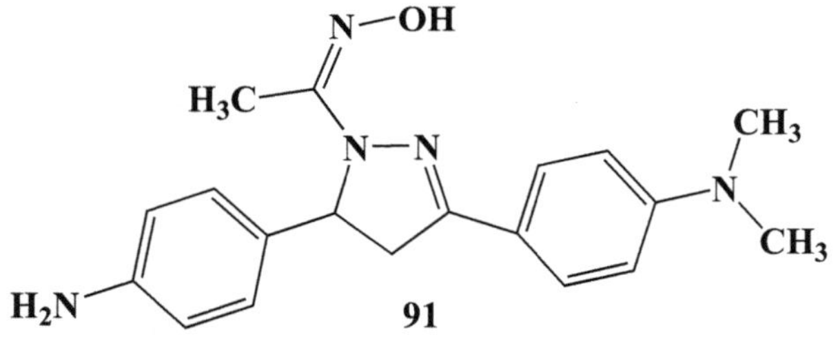

91

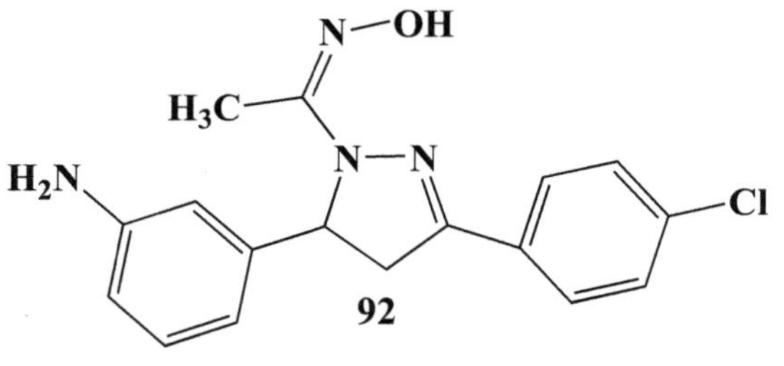

92

DISCUSSION

The elementary comprehension of NAs and pesticides or insecticides is necessary; in order to ensure efficient detection, identification, and urgent therapy especially when two groups of NAs are known and their properties differ from one another. The major threat of non-volatile V-agents is their easy penetrations into the skin. Whereas, G-agents represent volatile compounds imposing major threat by inhalation. Both categories possess high lipophilicity which in turn constitute additional toxicity in the BBB penetration and subsequent assemblage in the lipid tissues of the organism. This extensive lethal effect may occur within minutes in the case of OP inhalation or penetration through the eye to about an hour for the skin penetration [105, 106]. Based on this study, it is presumed that instant administration of antidotes is the only way to save life.

Developments of these oximes are a turning point in the research after many years of searching for novel and more efficient bis-pyridinium reactivators. HL7, K027, and K203 are some of the compounds which have brought some favorable outcomes in this field. Due to the strain associated with clinical testing and ethical issues, their potential use in clinical practice still remains only at the experimental level [107 - 112]. Recent inventions and publications deflect from the dogma of bis-pyridinium oximes and propose a novel concept of uncharged reactivators. In point of fact, a fascinating gadget that might surpass the limitations of common oxime therapy is the uncharged reactivators. They can own sufficient BBB penetration with the ability to recover AChE in the CNS [72]. Such a corresponding strategy acquired success in some cases, especially AChE reactivators reported by the groups of Taylor *et al.* [113] and Cashman and Kalisiak [114]. French group of Baati et al., proposed dual binding site strategy applied to the development of novel reactivators of some promising compounds [115]. Few drawbacks limited the beneficial lipophilic properties of the uncharged reactivators that enhance BBB penetration. Intravenous or intramuscular administration of such compounds may be hampered by low solubility in hydrophilic solutions given by increased lipophilicity [116]. Moreover, it should be borne in mind that escalated lipophilicity is often bounded to increased off-target toxicity especially in the CNS [117]. The solution of the drawbacks coupled with the uncharged reactivators may offer mono-charged compounds. Since BBB penetration of monocharged 2-PAM is appointed as 10% of plasma level, such an amount might be sufficient to deal with central symptoms of NAs intoxication [118]. Chambers *et al.* [119] synthesized phenoxyalkyl pyridinium oximes, which brought attenuated neurological symptoms of NAs as comparable to 2-PAM.

In the literature and patent sources, often 2-PAM accounted as standard for reactivation potency with novel reactivators against NAs-inhibited AChE. Such results might be regarded as 'purpose-built' and are often criticized by the scientific community for the lack of more relevant data (i.e. comparison with more potent AChE reactivators such as HI-6, HLö-7, and others is missing).

In order to initiate novel reactivators to clinical practice, complete assessment and comparison against the extensive NA and comparison with commonly used oximes are necessary to achieve.

In addition, it is hard to compare reactivation outcome of each reactivator for a specific target (NA) from each research group. This inconsistency lies in the individual AChE sources used in the *in vitro* testing together with variances in experimental protocols. That leads to significant distinctions of *in vitro* reactivation potencies and a pharmacokinetic property of the reactivators and data becomes hard to transfer to *in vivo* studies. Such a heterogeneity in data is well described elsewhere [120].

Alkylating agents are likely to reactivate aged-AChE [80]. This alkylation there in enhances reactivation; leading to a recovery of the AChE function. Cancerogenicity, the significant off-target toxicity is led by the administration of such an alkylating agent in *in vivo* studies. The reliability of aged-AChE reactivators further requires a direct proof of concept before assessing these compounds in *in vivo* models.

To conclude, the main concern for all potential compounds and methods should be mentioned. It is hard to anticipate the effective dose of the reactivator requisite to save life as OPs are highly toxic compounds; nonetheless, human *in vivo* data are only hardly obtainable. If the statistics on the pharmacokinetics, tolerability, and efficacy of reactivators *in vivo* are mislaid, then the development of novel oximes may not reach its therapeutic goal. Such ethical trouble may only be overcome with the accomplishment of *in vitro* and *in vivo* therapeutic results [51].

Besides, the prerequisite of instant administration of an antidote after the exposure to OP, its on the spot detection is essential. Along with the monitoring basic symptoms of intoxication, diagnosis of OP exposure can be accomplished. When such symptoms occur, it can be too late for causal therapy.

Ultimately, the complication associated with OP intoxication can be overcome by discovering broad-spectrum reactivators that may partly solve this drawback. On the other hand, even faultless and fast detection of OP does not ensure efficient treatment due to the 'aging' process or bioavailability of reactivators.

CONCLUSION AND OUTLOOK

As discussed in this chapter, the threat of exploitation of OP nerve agents persists. Pesticides' requirement is also extending day by day for agricultural production, consequently death rates is also increasing by the uses of these compounds in agricultural field. There is no ubiquitous oxime that might perhaps be used in every case where a nerve agent or OP pesticide intoxication is needed. Henceforth, the development of novel, successful and safe reactivators is a requisite.

Ongoing therapy in OP poisoning is inadequate and probable abuse of NA may lead to a great tragedy for the mankind.

Therein, we unveiled current innovations dealing with the strains related to OP poisoning. In the course of three decades, after the discovery of monopyridinium and bispyridinium oximes as reactivators for OP inhibited AChE, tons of variations have been synthesized and assessed.

All reactivators have three major disadvantages.

1. Their everlasting positive charge prevents them from crossing the BBB to reactivate brain inhibited AChE.
2. With different sorts of OP, reactivators exhibit different rates of reactivation efficacy against inhibited AChE.
3. They are ineffective in reactivating "aged" AChE. The latest research has evolved with new and efficient uncharged reactivators that are able to cross the BBB.

Hence, the development of a broad-spectrum reactivator which is satisfactory for the wide range of Ops is necessary. In no way, the existing pyridinium oximes is a true broad-spectrum reactivator [121]. The solution to explore a broad-spectrum reactivator would be to merge two or more oximes that have complementary activities. In this regard, combining obidoxime with HI-6 is a promising approach [122]. With respect to "aged" AChE, further research is necessary, as no existing reactivator is able to reactivate it. Additional studies are required in this area of research which will lead to the development of broad-spectrum reactivators, which will, in turn, benefit the public from OPs used in both pest control and warfare.

ACKNOWLEDGEMENTS

This work was supported by financial assistance received from Rajiv Gandhi University of Health Sciences, Bangalore [Project code: 21PHA314].

REFERENCES

[1] Antonijevic B, Stojiljkovic MP. Unequal efficacy of pyridinium oximes in acute organophosphate poisoning. Clin Med Res 2007; 5(1): 71-82. http://www.clinmedres.org/cgi/doi/10.3121/cmr.2007.701 [http://dx.doi.org/10.3121/cmr.2007.701] [PMID: 17456837]

[2] Petroianu GA, Hasan MY, Nurulain SM, Arafat K, Sheen R, Nagelkerke N. Comparison of two pre-exposure treatment regimens in acute organophosphate (paraoxon) poisoning in rats: Tiapride *vs.* pyridostigmine. Toxicol Appl Pharmacol 2007; 219(2-3): 235-40. Available from: https://linkinghub.elsevier.com/retrieve/pii/S0041008X06003048 [http://dx.doi.org/10.1016/j.taap.2006.09.002] [PMID: 17056080]

[3] Katz KDD. Organophosphate toxicity. Medscape 2008. Available from: https://emedicine.medscape.com/article/167726-overview

[4] Chaudhry R, Lall SB, Mishra B, Dhawan B. Lesson of the week: A foodborne outbreak of organophosphate poisoning. BMJ 1998; 317(7153): 268-9. https://www.bmj.com/lookup/doi/10.1136/bmj.317.7153.268 [http://dx.doi.org/10.1136/bmj.317.7153.268] [PMID: 9677223]

[5] Ohayo-Mitoko GJA, Heederik DJJ, Kromhout H, Omondi BEO, Boleij JSM. Acetylcholinesterase inhibition as an indicator of organophosphate and carbamate poisoning in kenyan agricultural workers. Int J Occup Environ Health 1997; 3(3): 210-20. Available from: http://www.tandfonline.com/doi/full/10.1179/oeh.1997.3.3.210 [http://dx.doi.org/10.1179/oeh.1997.3.3.210] [PMID: 9891121]

[6] Chambers , Janice PL. Organophosphates Chemistry, Fate, and Effects. 1st ed.. 1992; pp. 1-440. Available from: https://www.elsevier.com/books/organophosphates-chemistry-fate--nd-effects/chambers/978-0-08-091726-9

[7] Abdel-Halim KY, Salama AK, El-khateeb EN, Bakry NM. Organophosphorus pollutants (OPP) in aquatic environment at Damietta Governorate, Egypt: Implications for monitoring and biomarker responses. Chemosphere 2006; 63(9): 1491-8. https://linkinghub.elsevier.com/retrieve/pii/S0045653505010969 [http://dx.doi.org/10.1016/j.chemosphere.2005.09.019] [PMID: 16289700]

[8] Konstantinou IK, Hela DG, Albanis TA. The status of pesticide pollution in surface waters (rivers and lakes) of Greece. Part I. Review on occurrence and levels. Environ Pollut 2006; 141(3): 555-70. https://linkinghub.elsevier.com/retrieve/pii/S0269749105004598 [http://dx.doi.org/10.1016/j.envpol.2005.07.024] [PMID: 16226830]

[9] Peter JV, Sudarsan T, Moran J. Clinical features of organophosphate poisoning: A review of different classification systems and approaches. Indian J Crit Care Med 2014; 18(11): 735-45. https://www.ijccm.org/doi/10.4103/0972-5229.144017 [http://dx.doi.org/10.4103/0972-5229.144017] [PMID: 25425841]

[10] Eddleston M, Phillips MR. Self poisoning with pesticides. BMJ 2004; 328(7430): 42-4. https://www.bmj.com/lookup/doi/10.1136/bmj.328.7430.42 [http://dx.doi.org/10.1136/bmj.328.7430.42] [PMID: 14703547]

[11] Antonijevic E, Musilek K, Kuca K, Djukic-Cosic D, Vucinic S, Antonijevic B. Therapeutic and reactivating efficacy of oximes K027 and K203 against a direct acetylcholinesterase inhibitor. Neurotoxicology 2016; 55: 33-9. https://linkinghub.elsevier.com/retrieve/pii/S0161813X16300845 [http://dx.doi.org/10.1016/j.neuro.2016.05.006] [PMID: 27177985]

[12] Kassa J. Review of oximes in the antidotal treatment of poisoning by organophosphorus nerve agents. J Toxicol Clin Toxicol 2002; 40(6): 803-16. http://www.tandfonline.com/doi/full/ 10.1081/CLT-120015840 [http://dx.doi.org/10.1081/CLT-120015840] [PMID: 12475193]

[13] Jokanović M. Biotransformation of Warfare Nerve Agents.Handbook of Toxicology of Chemical Warfare Agents. Elsevier 2015; pp. 883-94. Internet Available from:

https://linkinghub.elsevier.com/retrieve/pii/B9780128001592000592
[http://dx.doi.org/10.1016/B978-0-12-800159-2.00059-2]

[14] Gupta RC, Ed. Handbook of Toxicology of Chemical Warfare Agents. Elsevier 2015. Internet Available from: https://linkinghub.elsevier.com/retrieve/pii/C20130154025

[15] Kliachyna M, Santoni G, Nussbaum V, *et al.* Design, synthesis and biological evaluation of novel tetrahydroacridine pyridine- aldoxime and -amidoxime hybrids as efficient uncharged reactivators of nerve agent-inhibited human acetylcholinesterase. Eur J Med Chem 2014; 78: 455-67. https://linkinghub.elsevier.com/retrieve/pii/S0223523414002578
[http://dx.doi.org/10.1016/j.ejmech.2014.03.044] [PMID: 24704618]

[16] Kuca K, Korabecny J, Dolezal R, *et al.* Tetroxime: reactivation potency – *in vitro* and *in silico* study. RSC Advances 2017; 7(12): 7041-5. [Internet].
[http://dx.doi.org/10.1039/C6RA16499D]

[17] Zorbaz T, Braïki A, Maraković N, *et al.* Potent 3-hydroxy-2-pyridine aldoxime reactivators of organophosphate-inhibited cholinesterases with predicted blood-brain barrier penetration. Chemistry 2018; 24(38): 9675-91.
[http://dx.doi.org/10.1002/chem.201801394] [PMID: 29672968]

[18] Bosak A. Organophosphorus compounds: classification and enzyme reactions. Arh Hig Rada Toksikol 2006; 57(4): 445-57. https://hrcak.srce.hr/6047 [Organophosphorus compounds: classification and enzyme reactions]. [Internet].
[PMID: 17265684]

[19] Laurence L. Goodman & Gilman's: The pharmacological basis of therapeutics. New York: McGraw-Hill 2006.

[20] Tripati KD. Essential of medical pharmacology. Opioid Analgesics and Antagonists. 6., India 2008.
[http://dx.doi.org/10.5005/jp/books/10282]

[21] John M. Beale JJHB Organic Medicinal and Pharmaceutical Chemistry. Lippincott Williams & Wilkins 1998.

[22] Richardson RJ, Makhaeva GF. Organophosphorus Compounds, Encyclopedia of Toxicology. 3rd ed. Academic Press 2014; pp. 714-9.
[http://dx.doi.org/10.1016/B978-0-12-386454-3.00173-1]

[23] Parsons SM, Prior C, Marshall IG. Acetylcholine transport, storage, and release. International Review of Neurobiology. 1993; 35: pp. 279-390. Available from: https://linkinghub.elsevier.com/retrieve/pii/S0074774208605723

[24] Schumacher M, Camp S, Maulet Y, *et al.* Primary structure of Torpedo californica acetylcholinesterase deduced from its cDNA sequence. Nature 1986; 319(6052): 407-9. Available from: http://www.nature.com/articles/319407a0
[http://dx.doi.org/10.1038/319407a0] [PMID: 3753747]

[25] Sussman JL, Harel M, Frolow F, Oefner C, Goldman A, Toker L, *et al.* Atomic structure of acetylcholinesterase from torpedo californica : A prototypic acetylcholine-binding protein. Science (80-) 1991; 253(5022): 872-9. Available from: https://www.science.org/doi/10.1126/science.1678899
[http://dx.doi.org/10.1126/science.1678899]

[26] Silman I, Sussman JL. Acetylcholinesterase: 'classical' and 'non-classical' functions and pharmacology. Curr Opin Pharmacol 2005; 5(3): 293-302. https://linkinghub.elsevier.com/retrieve/pii/S1471489205000445
[http://dx.doi.org/10.1016/j.coph.2005.01.014] [PMID: 15907917]

[27] Harel M, Schalk I, Ehret-Sabatier L, *et al.* Quaternary ligand binding to aromatic residues in the active-site gorge of acetylcholinesterase. Proc Natl Acad Sci USA 1993; 90(19): 9031-5. http://www.pnas.org/cgi/doi/10.1073/pnas.90.19.9031
[http://dx.doi.org/10.1073/pnas.90.19.9031] [PMID: 8415649]

[28] Berman HA, Leonard K. Ligand exclusion on acetylcholinesterase. Biochemistry 1990; 29(47): 10640-9. https://pubs.acs.org/doi/abs/10.1021/bi00499a010 [http://dx.doi.org/10.1021/bi00499a010] [PMID: 2271673]

[29] Bourne Y, Taylor P, Radić Z, Marchot P. Structural insights into ligand interactions at the acetylcholinesterase peripheral anionic site. EMBO J 2003; 22(1): 1-12. http://emboj.embopress.org/cgi/doi/10.1093/emboj/cdg005 [http://dx.doi.org/10.1093/emboj/cdg005] [PMID: 12505979]

[30] Marrs TC. Organophosphate poisoning. Pharmacol Ther 1993; 58(1): 51-66. https://linkinghub.elsevier.com/retrieve/pii/016372589390066M [http://dx.doi.org/10.1016/0163-7258(93)90066-M] [PMID: 8415873]

[31] Fukuto TR. Mechanism of action of organophosphorus and carbamate insecticides. Environ Health Perspect 1990; 87: 245-54. https://ehp.niehs.nih.gov/doi/10.1289/ehp.9087245 [http://dx.doi.org/10.1289/ehp.9087245] [PMID: 2176588]

[32] Mileson BE, Chambers JE, Chen WL, *et al.* Common mechanism of toxicity: a case study of organophosphorus pesticides. Toxicol Sci 1998; 41(1): 8-20. Available from: https://linkinghub.elsevier.com/retrieve/pii/S1096608097924318 [Internet]. [PMID: 9520337]

[33] Pope CN. Organophosphorus pesticides: do they all have the same mechanism of toxicity? J Toxicol Environ Health B Crit Rev 1999; 2(2): 161-81. http://www.tandfonline.com/doi/abs/10.1080/109374099281205 [http://dx.doi.org/10.1080/109374099281205] [PMID: 10230392]

[34] Reiner E, Radić Z, Simeon-Rudolf V. Mechanisms of organophosphate toxicity and detoxication with emphasis on studies in croatia. Arch Ind Hyg Toxicol 2007; 58(3): 329-38. https://content.sciendo.com/doi/10.2478/v10004-007-0026-2 [http://dx.doi.org/10.2478/v10004-007-0026-2]

[35] Delfino RT, Ribeiro TS, Figueroa-Villar JD. Organophosphorus compounds as chemical warfare agents: a review. J Braz Chem Soc 2009; 20: 3. http://www.scielo.br/scielo.php?script=sci_arttext&pid=S0103-50532009000300003&lng=en&nrm=iso&tlng=en [Internet]. [http://dx.doi.org/10.1590/S0103-50532009000300003]

[36] Johnson MK, Jacobsen D, Meredith TJ, Eyer P, Heath AJ, Ligtenstein DA, *et al.* Evaluation of antidotes for poisoning by organophosphorus pesticides. Emerg Med Australas 2000; 12(1): 22-37. http://doi.wiley.com/10.1046/j.1442-2026.2000.00087.x [Internet].

[37] Gupta RC. Brain regional heterogeneity and toxicological mechanisms of organophosphates and carbamates. Toxicol Mech Methods 2004; 14(3): 103-43. http://www.tandfonline.com/doi/full/10.1080/15376520490429175 [http://dx.doi.org/10.1080/15376520490429175] [PMID: 20021140]

[38] Bajgar J. Organophosphates/nerve agent poisoning: Mechanism of action, diagnosis, prophylaxis, and treatment. Advances in Clinical Chemistry. 2004; 38: pp. 151-216. Available from: https://linkinghub.elsevier.com/retrieve/pii/S0065242304380066

[39] Eddleston M, Mohamed F, Davies JOJ, Eyer P, Worek F, Sheriff MHR, *et al.* Respiratory failure in acute organophosphorus pesticide self-poisoning. QJM An Int J Med 2006; 99(8): 513-22. Available from: https://academic.oup.com/qjmed/article-lookup/doi/10.1093/qjmed/hcl065 [http://dx.doi.org/10.1093/qjmed/hcl065]

[40] Eddleston M, Buckley NA, Eyer P, Dawson AH. Management of acute organophosphorus pesticide poisoning. Lancet 2008; 371(9612): 597-607. https://linkinghub.elsevier.com/retrieve/pii/S0140673607612021 [http://dx.doi.org/10.1016/S0140-6736(07)61202-1] [PMID: 17706760]

[41] Vale JA. Toxicokinetic and toxicodynamic aspects of organophosphorus (OP) insecticide poisoning.

Toxicol Lett 1998; 102-103: 649-52. https://linkinghub.elsevier.com/retrieve/pii/S037842749800277X [http://dx.doi.org/10.1016/S0378-4274(98)00277-X] [PMID: 10022329]

[42] Durham WF, Wolfe HR, Elliott JW. Absorption and excretion of parathion by spraymen. Arch Environ Heal An Int J 1972; 24(6): 381-7. Available from: http://www.tandfonline.com/doi/abs/10.1080/00039896.1972.10666113 [http://dx.doi.org/10.1080/00039896.1972.10666113]

[43] Bull DL. Metabolism of organophosphorus insecticides in animals and plants.Residue Reviews. New York, NY: Springer New York 1972; pp. 1-22. Internet http://link.springer.com/10.1007/978-1-4615-8485-8_1 [http://dx.doi.org/10.1007/978-1-4615-8485-8_1]

[44] Bardin PG, van Eeden SF, Moolman JA, Foden AP, Joubert JR. Organophosphate and carbamate poisoning. Arch Intern Med 1994; 154(13): 1433-41. http://www.ncbi.nlm.nih.gov/pubmed/8017998 [http://dx.doi.org/10.1001/archinte.1994.00420130020005] [PMID: 8017998]

[45] Tafuri J, Roberts J. Organophosphate poisoning. Ann Emerg Med 1987; 16(2): 193-202. https://linkinghub.elsevier.com/retrieve/pii/S019606448780015X [http://dx.doi.org/10.1016/S0196-0644(87)80015-X] [PMID: 3541700]

[46] Vance MV, Selden BS, Clark RF. Optimal patient position for transport and initial management of toxic ingestions. Ann Emerg Med 1992; 21(3): 243-6. https://linkinghub.elsevier.com/retrieve/pii/S0196064405808820 [http://dx.doi.org/10.1016/S0196-0644(05)80882-0] [PMID: 1536482]

[47] Jokanović M. Medical treatment of acute poisoning with organophosphorus and carbamate pesticides. Toxicol Lett 2009; 190(2): 107-15. https://linkinghub.elsevier.com/retrieve/pii/S0378427409013678 [http://dx.doi.org/10.1016/j.toxlet.2009.07.025] [PMID: 19651196]

[48] Petroianu GA. The history of pyridinium oximes as nerve gas antidotes: the British contribution. Pharmazie 2013; 68(11): 916-8. [PMID: 24380243]

[49] Wilson IB. Acetylcholinesterase. XI. Reversibility of tetraethyl pyrophosphate. J Biol Chem 1951; 190(1): 111-7. http://www.ncbi.nlm.nih.gov/pubmed/14841157 [http://dx.doi.org/10.1016/S0021-9258(18)56051-8] [PMID: 14841157]

[50] Wilson IB, Ginsburg S. A powerful reactivator of alkylphosphate-inhibited acetylcholinesterase. Biochim Biophys Acta 1955; 18(1): 168-70. https://linkinghub.elsevier.com/retrieve/pii/0006300255900408 [http://dx.doi.org/10.1016/0006-3002(55)90040-8] [PMID: 13260275]

[51] Worek F, Thiermann H. The value of novel oximes for treatment of poisoning by organophosphorus compounds. Pharmacol Ther 2013; 139(2): 249-59. https://linkinghub.elsevier.com/retrieve/pii/S0163725813000880 [http://dx.doi.org/10.1016/j.pharmthera.2013.04.009] [PMID: 23603539]

[52] Wolthuis OL, Kepner LA. Successful oxime therapy one hour after soman intoxication in the rat. Eur J Pharmacol 1978; 49(4): 415-25. [http://dx.doi.org/10.1016/0014-2999(78)90316-3] [PMID: 668812]

[53] Wang J, Gu J, Leszczynski J, Feliks M, Sokalski WA. Oxime-induced reactivation of sarin-inhibited AChE: a theoretical mechanisms study. J Phys Chem B 2007; 111(9): 2404-8. https://pubs.acs.org/doi/10.1021/jp067741s [http://dx.doi.org/10.1021/jp067741s] [PMID: 17298091]

[54] Artursson E, Akfur C, Hörnberg A, Worek F, Ekström F. Reactivation of tabun-hAChE investigated by structurally analogous oximes and mutagenesis. Toxicology 2009; 265(3): 108-14. https://linkinghub.elsevier.com/retrieve/pii/S0300483X09004648 [http://dx.doi.org/10.1016/j.tox.2009.09.002] [PMID: 19761810]

[55] Kuca K, Jun D, Musilek K. Structural requirements of acetylcholinesterase reactivators. Mini-Reviews Med Chem 2006; 6(3): 269-77. Available from: http://www.eurekaselect.com/openurl/content.php?genre=article&issn=1389-5575&volume=6&issue=3&spage=269
[http://dx.doi.org/10.2174/138955706776073510]

[56] Patočka J, Cabal J, Kuča K, Jun D. Oxime reactivation of acetylcholinesterase inhibited by toxic phosphorus esters: *in vitro* kinetics and thermodynamics. J Appl Biomed 2005; 3(2): 91-9. http://jab.zsf.jcu.cz/doi/10.32725/jab.2005.011.html [Internet].
[http://dx.doi.org/10.32725/jab.2005.011]

[57] Musilek K, Kuca K, Jun D, Dohnal V, Dolezal M. Synthesis of the novel series of bispyridinium compounds bearing (E)-but-2-ene linker and evaluation of their reactivation activity against chlorpyrifos-inhibited acetylcholinesterase. Bioorg Med Chem Lett 2006; 16(3): 622-7. https://linkinghub.elsevier.com/retrieve/pii/S0960894X05013223
[http://dx.doi.org/10.1016/j.bmcl.2005.10.059] [PMID: 16288867]

[58] Dickson EW, Bird SB, Gaspari RJ, Boyer EW, Ferris CF. Diazepam inhibits organophosphate-induced central respiratory depression. Acad Emerg Med 2003; 10(12): 1303-6. http://doi.wiley.com/10.1111/j.1553-2712.2003.tb00001.x
[http://dx.doi.org/10.1197/S1069-6563(03)00533-5] [PMID: 14644779]

[59] Eyer P, Buckley N. Pralidoxime for organophosphate poisoning. Lancet 2006; 368(9553): 2110-1. https://linkinghub.elsevier.com/retrieve/pii/S0140673606698437
[http://dx.doi.org/10.1016/S0140-6736(06)69843-7] [PMID: 17174692]

[60] Balali-Mood M, Saber H. Recent advances in the treatment of organophosphorous poisonings. Iran J Med Sci 2012; 37(2): 74-91. http://www.ncbi.nlm.nih.gov/pubmed/23115436 [Internet].
[PMID: 23115436]

[61] Kovarik Z, Maček N, Sit RK, *et al.* Centrally acting oximes in reactivation of tabun-phosphoramidated AChE. Chem Biol Interact 2013; 203(1): 77-80. https://linkinghub.elsevier.com/retrieve/pii/S0009279712001573
[http://dx.doi.org/10.1016/j.cbi.2012.08.019] [PMID: 22960624]

[62] Eyer PA, Worek F. Oximes.Chemical Warfare Agents. 305-29. Internet https://onlinelibrary.wiley.com/doi/10.1002/9780470060032.ch15

[63] Musilek K, Dolezal M, Gunn-Moore F, Kuca K. Design, evaluation and structure-Activity relationship studies of the AChE reactivators against organophosphorus pesticides. Med Res Rev 2011; 31(4): 548-75. https://onlinelibrary.wiley.com/doi/10.1002/med.20192
[http://dx.doi.org/10.1002/med.20192] [PMID: 20027669]

[64] Luettringhaus A, hagedorn I. Quaternary hydroxyiminomethylpyridinium salts. The dischloride of bis-(4-hydroxyiminomethyl-1-pyridinium-methyl)-ether (lueh6), a new reactivator of acetylcholinesterase inhibited by organic phosphoric acid esters. Arzneimittelforschung 1964; 14: 1-5. http://www.ncbi.nlm.nih.gov/pubmed/14223684 [Internet].
[PMID: 14223684]

[65] Kuca K, Hrabinova M, Soukup O, Tobin G, Karasova J, Pohanka M. Pralidoxime--the gold standard of acetylcholinesterase reactivators--reactivation *in vitro* efficacy. Bratisl Lek Listy 2010; 111(9): 502-4. http://www.ncbi.nlm.nih.gov/pubmed/21180265 [Internet].
[PMID: 21180265]

[66] Newmark J. Therapy for nerve agent poisoning. Arch Neurol 2004; 61(5): 649-52. http://archneur.jamanetwork.com/article.aspx?doi=10.1001/archneur.61.5.649
[http://dx.doi.org/10.1001/archneur.61.5.649] [PMID: 15148139]

[67] Worek F, Thiermann H, Wille T. Oximes in organophosphate poisoning: 60 years of hope and despair. Chem Biol Interact 2016; 259(Pt B): 93-8. https://linkinghub.elsevier.com/retrieve/pii/S0009279716301557

[http://dx.doi.org/10.1016/j.cbi.2016.04.032] [PMID: 27125761]

[68] Lundy PM, Raveh L, Amitai G. Development of the bisquaternary oxime HI-6 toward clinical use in the treatment of organophosphate nerve agent poisoning. Toxicol Rev 2006; 25(4): 231-43. [http://dx.doi.org/10.2165/00139709-200625040-00004] [PMID: 17288495]

[69] Worek F, Eyer P, Aurbek N, Szinicz L, Thiermann H. Recent advances in evaluation of oxime efficacy in nerve agent poisoning by *in vitro* analysis. Toxicol Appl Pharmacol 2007; 219(2-3): 226-34. https://linkinghub.elsevier.com/retrieve/pii/S0041008X06003498 [http://dx.doi.org/10.1016/j.taap.2006.10.001] [PMID: 17112559]

[70] Romano JA Jr, McDonough JH, Sheridan RSF. Health effects of low-level exposure to nerve agents Chemical warfare agents: Toxicity at low levels. CRC Press Taylor & Francis Group 2001; pp. 1-24.

[71] Kuca K, Jun D, Bajgar J. Currently used cholinesterase reactivators against nerve agent intoxication: comparison of their effectivity *in vitro*. Drug Chem Toxicol 2007; 30(1): 31-40. [http://dx.doi.org/10.1080/01480540601017637] [PMID: 17364862]

[72] Kuča K, Kassa J. A comparison of the ability of a new bispyridinium oxime--1-(4-hydroxyiminomethylpyridinium)-4-(4-carbamoylpyridinium)butane dibromide and currently used oximes to reactivate nerve agent-inhibited rat brain acetylcholinesterase by *in vitro* methods. J Enzyme Inhib Med Chem 2003; 18(6): 529-35. [http://dx.doi.org/10.1080/14756360310001605552] [PMID: 15008517]

[73] Korabecny J, Soukup O, Dolezal R, Spilovska K, Nepovimova E, Andrs M, *et al.* From pyridinium-based to centrally active acetylcholinesterase reactivators. Mini-Reviews Med Chem 2014; 14(3): 215-21. Available from: http://www.eurekaselect.com/openurl/content.php?genre=article& issn=1389-5575&volume=14&issue=3&spage=215 [http://dx.doi.org/10.2174/1389557514666140219103138]

[74] Mercey G, Verdelet T, Renou J, *et al.* Reactivators of acetylcholinesterase inhibited by organophosphorus nerve agents. Acc Chem Res 2012; 45(5): 756-66. https://pubs.acs.org/doi/10.1021/ar2002864 [http://dx.doi.org/10.1021/ar2002864] [PMID: 22360473]

[75] Kovacevic V, Maksimovic M, Pantelic D, *et al.* Protective and reactivating effects of HI-6 PAM-2 mixture in rats with nerve chemical warfare agents (nerve CWA). Acta Pharm 1989; 39: 161-5.

[76] Millard CB, Kryger G, Ordentlich A, *et al.* Crystal structures of aged phosphonylated acetylcholinesterase: nerve agent reaction products at the atomic level. Biochemistry 1999; 38(22): 7032-9. [http://dx.doi.org/10.1021/bi982678l] [PMID: 10353814]

[77] Kalász H, Nurulain SM, Veress G, *et al.* Mini review on blood-brain barrier penetration of pyridinium aldoximes. J Appl Toxicol 2015; 35(2): 116-23. https://onlinelibrary.wiley.com/doi/10.1002/jat.3048 [http://dx.doi.org/10.1002/jat.3048] [PMID: 25291712]

[78] Petronilho EC, Rennó MN, Castro NG, da Silva FMR, Pinto AC, Figueroa-Villar JD. Design, synthesis, and evaluation of guanylhydrazones as potential inhibitors or reactivators of acetylcholinesterase. J Enzyme Inhib Med Chem 2016; 31(6): 1069-78. https://www.tandfonline.com/doi/full/10.3109/14756366.2015.1094468 [http://dx.doi.org/10.3109/14756366.2015.1094468] [PMID: 26558640]

[79] Topczewski DMQJ. Compounds and methods to treat organophosphorus poisoning. US 9,884,052B2, 2018.

[80] Carletti E, Colletier JP, Dupeux F, Trovaslet M, Masson P, Nachon F. Structural evidence that human acetylcholinesterase inhibited by tabun ages through O-dealkylation. J Med Chem 2010; 53(10): 4002-8. https://pubs.acs.org/doi/10.1021/jm901853b [http://dx.doi.org/10.1021/jm901853b] [PMID: 20408548]

[81] Kalisiak J, Ralph EC, Zhang J, Cashman JR. Amidine-oximes: reactivators for organophosphate

exposure. J Med Chem 2011; 54(9): 3319-30. https://pubs.acs.org/doi/10.1021/jm200054r [http://dx.doi.org/10.1021/jm200054r] [PMID: 21438612]

[82] Katagi Manjunatha S, Jennifer F, Satyanarayana D, Girish B, Mamledesai SN. Synthesis of Chalcone-Oxime Derivatives And Evaluation Of Their *in vitro* Reactivation Efficacy Against Op Inhibited AChE. Univers J Pharm 2016; 05(03): 32-7.

[83] Chambers JE, Meek EC, Bennett JP, *et al.* Novel substituted phenoxyalkyl pyridinium oximes enhance survival and attenuate seizure-like behavior of rats receiving lethal levels of nerve agent surrogates. Toxicology 2016; 339: 51-7. https://linkinghub.elsevier.com/retrieve/pii/S0300483X15300585 [http://dx.doi.org/10.1016/j.tox.2015.12.001] [PMID: 26705700]

[84] Katalinić M, Zandona A, Ramić A, Zorbaz T, Primožič I, Kovarik Z. New cinchona oximes evaluated as reactivators of acetylcholinesterase and butyrylcholinesterase inhibited by organophosphorus compounds. Molecules 2017; 22(7): 1234. http://www.mdpi.com/1420-3049/22/7/1234 [http://dx.doi.org/10.3390/molecules22071234] [PMID: 28737687]

[85] Meek JECWCC. Novel oximes for reactivating butyrylcholinesterase. US20170258774A1, 2017. Available from: https://patents.google.com/patent/US20170258774A1/en

[86] Tudosie MS, Patrinich B, Negrea AR, Secară CA. New synthesized oximes active in nerve agents' hazards. Rom J Mil Med 2017; 120(2): 47-53. [http://dx.doi.org/10.55453/rjmm.2017.120.2.7]

[87] Zorbaz T, Malinak D, Maraković N, *et al.* Pyridinium oximes with *ortho* -positioned chlorine moiety exhibit improved physicochemical properties and efficient reactivation of human acetylcholinesterase inhibited by several nerve agents. J Med Chem 2018; 61(23): 10753-66. https://pubs.acs.org/doi/10.1021/acs.jmedchem.8b01398 [http://dx.doi.org/10.1021/acs.jmedchem.8b01398] [PMID: 30383374]

[88] Malinak D, Nepovimova E, Jun D, Musilek K, Kuca K. Novel group of ache reactivators—synthesis, *in vitro* reactivation and molecular docking study. Molecules 2018; 23(9): 2291. http://www.mdpi.com/1420-3049/23/9/2291 [http://dx.doi.org/10.3390/molecules23092291] [PMID: 30205495]

[89] Kalisiak J, Ralph EC, Cashman JR. Nonquaternary reactivators for organophosphate-inhibited cholinesterases. J Med Chem 2012; 55(1): 465-74. https://pubs.acs.org/doi/10.1021/jm201364d [http://dx.doi.org/10.1021/jm201364d] [PMID: 22206546]

[90] Kuca K, Musilek K, Jun D, *et al.* A newly developed oxime K203 is the most effective reactivator of tabun-inhibited acetylcholinesterase. BMC Pharmacol Toxicol 2018; 19(1): 8. https://bmcpharmacoltoxicol.biomedcentral.com/articles/10.1186/s40360-018-0196-3 [http://dx.doi.org/10.1186/s40360-018-0196-3] [PMID: 29467029]

[91] Wei Z, Bi H, Liu Y, *et al.* Design, synthesis and evaluation of new classes of nonquaternary reactivators for acetylcholinesterase inhibited by organophosphates. Bioorg Chem 2018; 81: 681-8. https://linkinghub.elsevier.com/retrieve/pii/S0045206818308241 [http://dx.doi.org/10.1016/j.bioorg.2018.09.025] [PMID: 30265992]

[92] Iqbal A, Malik S, Nurulain SM, *et al.* Reactivation potency of two novel oximes (K456 and K733) against paraoxon-inhibited acetyl and butyrylcholinesterase: *in silico* and *in vitro* models. Chem Biol Interact 2019; 310: 108735. https://linkinghub.elsevier.com/retrieve/pii/S0009279719300328 [http://dx.doi.org/10.1016/j.cbi.2019.108735] [PMID: 31276662]

[93] Manjunatha S. Schiff base oxime derivatives reactivate chlorpyrifos-induced acetylcholinesterase inhibition. INNOSC Theranostics Pharmacol Sci 2019; 2(1): 14-8. [http://dx.doi.org/10.26689/itps.v2i1.499]

[94] Zorbaz T, Malinak D, Kuca K, Musilek K, Kovarik Z. Butyrylcholinesterase inhibited by nerve agents is efficiently reactivated with chlorinated pyridinium oximes. Chem Biol Interact 2019; 307: 16-20. https://linkinghub.elsevier.com/retrieve/pii/S0009279719303941 [http://dx.doi.org/10.1016/j.cbi.2019.04.020] [PMID: 31004594]

[95] Sharma R, Upadhyaya K, Gupta B, *et al.* Glycosylated-imidazole aldoximes as reactivators of pesticides inhibited AChE: Synthesis and *in-vitro* reactivation study. Environ Toxicol Pharmacol 2020; 80: 103454. https://linkinghub.elsevier.com/retrieve/pii/S1382668920301307 [http://dx.doi.org/10.1016/j.etap.2020.103454] [PMID: 32645360]

[96] Zandona A, Katalinić M, Šinko G, Radman Kastelic A, Primožič I, Kovarik Z. Targeting organophosphorus compounds poisoning by novel quinuclidine-3 oximes: development of butyrylcholinesterase-based bioscavengers. Arch Toxicol 2020; 94(9): 3157-71. https://link.springer.com/10.1007/s00204-020-02811-5 [http://dx.doi.org/10.1007/s00204-020-02811-5] [PMID: 32583098]

[97] Katagi MS, Mamledesai S, Bolakatti G, Fernandes J, Ml S, Tari P. Design, synthesis, and characterization of novel class of 2-quinolon-3-oxime reactivators for acetylcholinesterase inhibited by organophosphorus compounds. Chemical Data Collections 2020; 30: 100560. [http://dx.doi.org/10.1016/j.cdc.2020.100560]

[98] Yerri J, Dias J, Nimmakayala MR, *et al.* Chemoselective hydrogenation of 6-alkynyl-3-fluoo-2-pyridinaldoximes: Access to first-in-class 6-alkyl-3-fluoro-2-pyridinaldoxime scaffolds as new reactivators of sarin-inhibited human acetylcholinesterase with increased blood–brain barrier permeability. Chemistry 2020; 26(65): 15035-44. https://onlinelibrary.wiley.com/doi/10.1002/chem.202002012 [http://dx.doi.org/10.1002/chem.202002012] [PMID: 32633095]

[99] Lee HM, Andrys R, Jonczyk J, *et al.* Pyridinium-2-carbaldoximes with quinolinium carboxamide moiety are simultaneous reactivators of acetylcholinesterase and butyrylcholinesterase inhibited by nerve agent surrogates. J Enzyme Inhib Med Chem 2021; 36(1): 437-49. https://www.tandfonline.com/doi/full/10.1080/14756366.2020.1869954 [http://dx.doi.org/10.1080/14756366.2020.1869954] [PMID: 33467931]

[100] Gorecki L, Gerlits O, Kong X, *et al.* Rational design, synthesis, and evaluation of uncharged, "smart" bis-oxime antidotes of organophosphate-inhibited human acetylcholinesterase. J Biol Chem 2020; 295(13): 4079-92. Available from: https://linkinghub.elsevier.com/retrieve/pii/S0021925817487394 [http://dx.doi.org/10.1074/jbc.RA119.012400] [PMID: 32019865]

[101] Kitagawa DAS, Rodrigues RB, Silva TN, dos Santos WV, da Rocha VCV. Design, synthesis, *in silico* studies and *in vitro* evaluation of isatin-pyridine oximes hybrids as novel acetylcholinesterase reactivators. J Enzyme Inhib Med Chem 2021; 36(1): 1370-7. [http://dx.doi.org/10.1080/14756366.2021.1916009]

[102] Amitai G, Plotnikov A, Chapman S, *et al.* Non-quaternary oximes detoxify nerve agents and reactivate nerve agent-inhibited human butyrylcholinesterase. Commun Biol 2021; 4(1): 573. http://www.nature.com/articles/s42003-021-02061-w [http://dx.doi.org/10.1038/s42003-021-02061-w] [PMID: 33990679]

[103] Mohamed RA, Ong KK, Halim NA. 4-Hydroxybenzohydrazide: A potential reactivator for malathion-inhibited human acetylcholinesterase. IOP Conf Ser Mater Sci Eng 2021; 1051(1): 012021. [http://dx.doi.org/10.1088/1757-899X/1051/1/012021]

[104] Fernandes J. Pyrazole-oxime as reactivator for Chlorpyrifos inhibited Acetylcholinesterase: Synthesis and *in vitro* reactivation study. Thai J Pharm Sci 2021.

[105] Watson A, Opresko DYR. Organophosphate nerve agents.Handbook of Toxicology of Chemical Warfare Agents. Elsevier 2015; pp. 1141-84. Internet Available from: https://linkinghub.elsevier.com/retrieve/pii/B9780128001592000907 [http://dx.doi.org/10.1016/B978-0-12-800159-2.00009-9]

[106] Kuča K, Pohanka M. Chemical warfare agents. 2010; pp. 543-58. Available from: http://link.springer.com/10.1007/978-3-7643-8338-1_16

[107] de Jong LPA, Verhagen MAA, Langenberg JP, Hagedorn I, Löffler M. The bispyridinium-dioxime HLö-7. Biochem Pharmacol 1989; 38(4): 633-40. https://linkinghub.elsevier.com/retrieve/pii/

0006295289902098
[http://dx.doi.org/10.1016/0006-2952(89)90209-8] [PMID: 2917018]

[108] Walt , David R, Bencic-Nagale S. Chemical switches for detecting reactive chemical agents. Wo/2008/ 048698, 2008.

[109] Device for detecting a cholinesterase-inhibiting substance comprising a hydrophilic photo-crosslinkable resin. US8329454B2, 2012. Available from: https://www.patentguru.com/US8329454B2

[110] Redinbo Matthew R, Hemmert Andrew C, Edwards Jonathan S. Methods and compositions for detection and identification of organophosphorus nerve agents, pesticides and other toxin. Wo/2011/072007, 2007.

[111] Musilek K, Jun D, Cabal J, Kassa J, Gunn-Moore F, Kuca K. Design of a potent reactivator of tabun-inhibited acetylcholinesterase--synthesis and evaluation of (E)-1-(4-carbamoylpyridinium)-4-(4-hydroxyiminomethylpyridinium)-but-2-ene dibromide (K203). J Med Chem 2007; 50(22): 5514-8. https://pubs.acs.org/doi/10.1021/jm070653r
[http://dx.doi.org/10.1021/jm070653r] [PMID: 17924614]

[112] Kuča K, Bielavský J, Cabal J, Kassa J. Synthesis of a new reactivator of tabun-inhibited acetylcholinesterase. Bioorg Med Chem Lett 2003; 13(20): 3545-7. https://linkinghub.elsevier.com/retrieve/pii/S0960894X03007510
[http://dx.doi.org/10.1016/S0960-894X(03)00751-0] [PMID: 14505667]

[113] Taylor P, Radic Z, Sharpless K B, Fokin V, Sit R. Centrally active and orally bioavailable antidotes for organophosphate exposure and methods for making and using them. Wo/2014/127315, 2014.

[114] Cashman , John R, Kalisiak J. Blood brain barrier-penetrating oximes for cholistenerases reactivation. Wo/2012/083261, 2012.

[115] Baati R, kliachyna M, Nussbaum V, Renard P-Y, Jean L, Weik M, *et al.* Ovel uncharged reactivators against op-inhibition of human acetylcholinesterase. WO/2015/075082, 2015.

[116] Calas AG, Dias J, Rousseau C, *et al.* An easy method for the determination of active concentrations of cholinesterase reactivators in blood samples: Application to the efficacy assessment of non quaternary reactivators compared to HI-6 and pralidoxime in VX-poisoned mice. Chem Biol Interact 2017; 267: 11-6.
[http://dx.doi.org/10.1016/j.cbi.2016.03.009] [PMID: 26972668]

[117] Soukup O, Jun D, Tobin G, Kuca K. The summary on non-reactivation cholinergic properties of oxime reactivators: the interaction with muscarinic and nicotinic receptors. Arch Toxicol 2013; 87(4): 711-9. http://link.springer.com/10.1007/s00204-012-0977-1
[http://dx.doi.org/10.1007/s00204-012-0977-1] [PMID: 23179755]

[118] Sakurada K, Matsubara K, Shimizu K, *et al.* Pralidoxime iodide (2-pAM) penetrates across the blood-brain barrier. Neurochem Res 2003; 28(9): 1401-7. http://www.ncbi.nlm.nih.gov/pubmed/12938863
[http://dx.doi.org/10.1023/A:1024960819430] [PMID: 12938863]

[119] Chambers Janice E, Chambers Howard W, Meek Edward C. Phenoxyalkyl pyridinium oxime therapeutics for treatment of organophosphate poisoning. WO/2011/142826, 2011.

[120] Zemek F, Drtinova L, Nepovimova E, *et al.* Outcomes of Alzheimer's disease therapy with acetylcholinesterase inhibitors and memantine. Expert Opin Drug Saf 2014; 13(6): 759-74. http://www.ncbi.nlm.nih.gov/pubmed/24845946 [Internet].
[PMID: 24845946]

[121] Saint-André G, Kliachyna M, Kodepelly S, *et al.* Design, synthesis and evaluation of new α-nucleophiles for the hydrolysis of organophosphorus nerve agents: application to the reactivation of phosphorylated acetylcholinesterase. Tetrahedron 2011; 67(34): 6352-61. https://linkinghub.elsevier.com/retrieve/pii/S0040402011008416 [Internet].
[http://dx.doi.org/10.1016/j.tet.2011.05.130]

[122] Louise-Leriche L, Păunescu E, Saint-André G, *et al.* A HTS assay for the detection of

organophosphorus nerve agent scavengers. Chemistry 2010; 16(11): 3510-23. https://onlinelibrary.wiley.com/doi/10.1002/chem.200902986 [http://dx.doi.org/10.1002/chem.200902986] [PMID: 20143367]

SUBJECT INDEX

A

B

C

Zareen Amtul (Ed.)

D

E

F

G

H

I

J

M

N

O

P

R

S

T

U

www.ingramcontent.com/pod-product-compliance
Lightning Source LLC
Chambersburg PA
CBHW050819220326
41598CB00006B/261
* 9 7 8 9 8 1 5 1 7 9 8 6 6 *